Particles of the Standard Model

Edited by Paul F. Kisak

Contents

5 Higgs boson 115

6 Composite particles 141

Chapter 1

Introduction

1.1 Standard Model

This article is about the Standard Model of particle physics. For other uses, see Standard model (disambiguation).

This article is a non-mathematical general overview of the Standard Model. For a mathematical description, see the article Standard Model (mathematical formulation).

For the Standard Model of Big Bang cosmology, Lambda-CDM model.

The **Standard Model** of particle physics is a theory con-

The Standard Model of elementary particles (more schematic depiction), with the three generations of matter, gauge bosons in the fourth column, and the Higgs boson in the fifth.

cerning the electromagnetic, weak, and strong nuclear interactions, as well as classifying all the subatomic particles known. It was developed throughout the latter half of the 20th century, as a collaborative effort of scientists around the world.[1] The current formulation was finalized in the mid-1970s upon experimental confirmation of the existence of quarks. Since then, discoveries of the top quark (1995), the tau neutrino (2000), and more recently the Higgs boson (2013), have given further credence to the Standard Model. Because of its success in explaining a wide variety of experimental results, the Standard Model is sometimes regarded as a "theory of almost everything".

Although the Standard Model is believed to be theoretically self-consistent[2] and has demonstrated huge and continued successes in providing experimental predictions, it does leave some phenomena unexplained and it falls short of being a complete theory of fundamental interactions. It does not incorporate the full theory of gravitation[3] as described by general relativity, or account for the accelerating expansion of the universe (as possibly described by dark energy). The model does not contain any viable dark matter particle that possesses all of the required properties deduced from observational cosmology. It also does not incorporate neutrino oscillations (and their non-zero masses).

The development of the Standard Model was driven by theoretical and experimental particle physicists alike. For theorists, the Standard Model is a paradigm of a quantum field theory, which exhibits a wide range of physics including spontaneous symmetry breaking, anomalies, non-perturbative behavior, etc. It is used as a basis for building more exotic models that incorporate hypothetical particles, extra dimensions, and elaborate symmetries (such as supersymmetry) in an attempt to explain experimental results at variance with the Standard Model, such as the existence of dark matter and neutrino oscillations.

1.1.1 Historical background

The first step towards the Standard Model was Sheldon Glashow's discovery in 1961 of a way to combine the electromagnetic and weak interactions.[4] In 1967 Steven Weinberg[5] and Abdus Salam[6] incorporated the Higgs mechanism[7][8][9] into Glashow's electroweak theory, giving it its modern form.

The Higgs mechanism is believed to give rise to the masses of all the elementary particles in the Standard Model. This includes the masses of the W and Z bosons, and the masses of the fermions, i.e. the quarks and leptons.

After the neutral weak currents caused by Z boson exchange were discovered at CERN in 1973,[10][11][12][13] the electroweak theory became widely accepted and Glashow,

Salam, and Weinberg shared the 1979 Nobel Prize in Physics for discovering it. The W and Z bosons were discovered experimentally in 1981, and their masses were found to be as the Standard Model predicted.

The theory of the strong interaction, to which many contributed, acquired its modern form around 1973–74, when experiments confirmed that the hadrons were composed of fractionally charged quarks.

1.1.2 Overview

At present, matter and energy are best understood in terms of the kinematics and interactions of elementary particles. To date, physics has reduced the laws governing the behavior and interaction of all known forms of matter and energy to a small set of fundamental laws and theories. A major goal of physics is to find the "common ground" that would unite all of these theories into one integrated theory of everything, of which all the other known laws would be special cases, and from which the behavior of all matter and energy could be derived (at least in principle).[14]

1.1.3 Particle content

The Standard Model includes members of several classes of elementary particles (fermions, gauge bosons, and the Higgs boson), which in turn can be distinguished by other characteristics, such as color charge.

Fermions

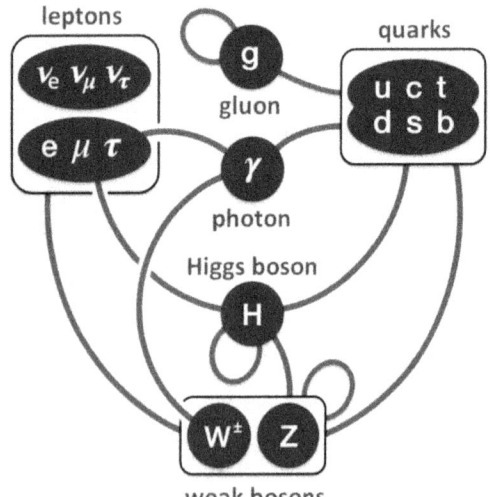

Summary of interactions between particles described by the Standard Model.

The Standard Model includes 12 elementary particles of spin-½ known as fermions. According to the spin-statistics theorem, fermions respect the Pauli exclusion principle. Each fermion has a corresponding antiparticle.

The fermions of the Standard Model are classified according to how they interact (or equivalently, by what charges they carry). There are six quarks (up, down, charm, strange, top, bottom), and six leptons (electron, electron neutrino, muon, muon neutrino, tau, tau neutrino). Pairs from each classification are grouped together to form a generation, with corresponding particles exhibiting similar physical behavior (see table).

The defining property of the quarks is that they carry color charge, and hence, interact via the strong interaction. A phenomenon called color confinement results in quarks being very strongly bound to one another, forming color-neutral composite particles (hadrons) containing either a quark and an antiquark (mesons) or three quarks (baryons). The familiar proton and the neutron are the two baryons having the smallest mass. Quarks also carry electric charge and weak isospin. Hence they interact with other fermions both electromagnetically and via the weak interaction.

The remaining six fermions do not carry colour charge and are called leptons. The three neutrinos do not carry electric charge either, so their motion is directly influenced only by the weak nuclear force, which makes them notoriously difficult to detect. However, by virtue of carrying an electric charge, the electron, muon, and tau all interact electromagnetically.

Each member of a generation has greater mass than the corresponding particles of lower generations. The first generation charged particles do not decay; hence all ordinary (baryonic) matter is made of such particles. Specifically, all atoms consist of electrons orbiting around atomic nuclei, ultimately constituted of up and down quarks. Second and third generations charged particles, on the other hand, decay with very short half lives, and are observed only in very high-energy environments. Neutrinos of all generations also do not decay, and pervade the universe, but rarely interact with baryonic matter.

Gauge bosons

In the Standard Model, gauge bosons are defined as force carriers that mediate the strong, weak, and electromagnetic fundamental interactions.

Interactions in physics are the ways that particles influence other particles. At a macroscopic level, electromagnetism allows particles to interact with one another via electric and magnetic fields, and gravitation allows particles with mass to attract one another in accordance with Einstein's theory

Standard Model Interactions
(Forces Mediated by Gauge Bosons)

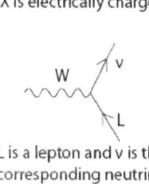

X is any fermion in the Standard Model.

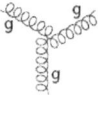

X is electrically charged.

X is any quark.

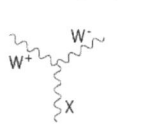

U is a up-type quark; D is a down-type quark.

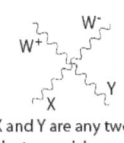

L is a lepton and ν is the corresponding neutrino.

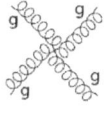

X is a photon or Z-boson.

X and Y are any two electroweak bosons such that charge is conserved.

The above interactions form the basis of the standard model. Feynman diagrams in the standard model are built from these vertices. Modifications involving Higgs boson interactions and neutrino oscillations are omitted. The charge of the W bosons is dictated by the fermions they interact with; the conjugate of each listed vertex (i.e. reversing the direction of arrows) is also allowed.

of general relativity. The Standard Model explains such forces as resulting from matter particles exchanging other particles, generally referred to as *force mediating particles*. When a force-mediating particle is exchanged, at a macroscopic level the effect is equivalent to a force influencing both of them, and the particle is therefore said to have *mediated* (i.e., been the agent of) that force. The Feynman diagram calculations, which are a graphical representation of the perturbation theory approximation, invoke "force mediating particles", and when applied to analyze high-energy scattering experiments are in reasonable agreement with the data. However, perturbation theory (and with it the concept of a "force-mediating particle") fails in other situations. These include low-energy quantum chromodynamics, bound states, and solitons.

The gauge bosons of the Standard Model all have spin (as do matter particles). The value of the spin is 1, making them bosons. As a result, they do not follow the Pauli exclusion principle that constrains fermions: thus bosons (e.g. photons) do not have a theoretical limit on their spatial density (number per volume). The different types of gauge bosons are described below.

- Photons mediate the electromagnetic force between electrically charged particles. The photon is massless and is well-described by the theory of quantum electrodynamics.

- The W+, W−, and Z gauge bosons mediate the weak interactions between particles of different flavors (all quarks and leptons). They are massive, with the Z being more massive than the W±. The weak interactions involving the W± exclusively act on *left-handed* particles and *right-handed* antiparticles. Furthermore, the W± carries an electric charge of +1 and −1 and couples to the electromagnetic interaction. The electrically neutral Z boson interacts with both left-handed particles and antiparticles. These three gauge bosons along with the photons are grouped together, as collectively mediating the electroweak interaction.

- The eight gluons mediate the strong interactions between color charged particles (the quarks). Gluons are massless. The eightfold multiplicity of gluons is labeled by a combination of color and anticolor charge (e.g. red–antigreen).[nb 1] Because the gluons have an effective color charge, they can also interact among themselves. The gluons and their interactions are described by the theory of quantum chromodynamics.

The interactions between all the particles described by the Standard Model are summarized by the diagrams on the right of this section.

Higgs boson

Main article: Higgs boson

The Higgs particle is a massive scalar elementary particle theorized by Robert Brout, François Englert, Peter Higgs, Gerald Guralnik, C. R. Hagen, and Tom Kibble in 1964 (see 1964 PRL symmetry breaking papers) and is a key building block in the Standard Model.[7][8][9][15] It has no intrinsic spin, and for that reason is classified as a boson (like the gauge bosons, which have integer spin).

The Higgs boson plays a unique role in the Standard Model, by explaining why the other elementary particles, except the photon and gluon, are massive. In particular, the Higgs boson explains why the photon has no mass, while the W and Z bosons are very heavy. Elementary particle masses, and the differences between electromagnetism (mediated by the photon) and the weak force (mediated by the W and Z bosons), are critical to many aspects of the structure of microscopic (and hence macroscopic) matter. In electroweak theory, the Higgs boson generates the masses of the leptons (electron, muon, and tau) and quarks. As the Higgs boson is massive, it must interact with itself.

Because the Higgs boson is a very massive particle and also decays almost immediately when created, only a very high-energy particle accelerator can observe and record it. Experiments to confirm and determine the nature of the Higgs boson using the Large Hadron Collider (LHC) at CERN began in early 2010, and were performed at Fermilab's Tevatron until its closure in late 2011. Mathematical consistency of the Standard Model requires that any mechanism capable of generating the masses of elementary particles become visible at energies above 1.4 TeV;[16] therefore, the LHC (designed to collide two 7 to 8 TeV proton beams) was built to answer the question of whether the Higgs boson actually exists.[17]

On 4 July 2012, the two main experiments at the LHC (ATLAS and CMS) both reported independently that they found a new particle with a mass of about 125 GeV/c^2 (about 133 proton masses, on the order of 10^{-25} kg), which is "consistent with the Higgs boson." Although it has several properties similar to the predicted "simplest" Higgs,[18] they acknowledged that further work would be needed to conclude that it is indeed the Higgs boson, and exactly which version of the Standard Model Higgs is best supported if confirmed.[19][20][21][22][23]

On 14 March 2013 the Higgs Boson was tentatively confirmed to exist.[24]

Total particle count

Counting particles by a rule that distinguishes between particles and their corresponding antiparticles, and among the many color states of quarks and gluons, gives a total of 61 elementary particles.[25]

1.1.4 Theoretical aspects

Main article: Standard Model (mathematical formulation)

Construction of the Standard Model Lagrangian

Technically, quantum field theory provides the mathematical framework for the Standard Model, in which a Lagrangian controls the dynamics and kinematics of the theory. Each kind of particle is described in terms of a dynamical field that pervades space-time. The construction of the Standard Model proceeds following the modern method of constructing most field theories: by first postulating a set of symmetries of the system, and then by writing down the most general renormalizable Lagrangian from its particle (field) content that observes these symmetries.

The global Poincaré symmetry is postulated for all relativistic quantum field theories. It consists of the familiar translational symmetry, rotational symmetry and the inertial reference frame invariance central to the theory of special relativity. The local SU(3)×SU(2)×U(1) gauge symmetry is an internal symmetry that essentially defines the Standard Model. Roughly, the three factors of the gauge symmetry give rise to the three fundamental interactions. The fields fall into different representations of the various symmetry groups of the Standard Model (see table). Upon writing the most general Lagrangian, one finds that the dynamics depend on 19 parameters, whose numerical values are established by experiment. The parameters are summarized in the table above (note: with the Higgs mass is at 125 GeV, the Higgs self-coupling strength $\lambda \sim 1/8$).

Quantum chromodynamics sector Main article: Quantum chromodynamics

The quantum chromodynamics (QCD) sector defines the interactions between quarks and gluons, with SU(3) symmetry, generated by T^a. Since leptons do not interact with gluons, they are not affected by this sector. The Dirac Lagrangian of the quarks coupled to the gluon fields is given by

$$\mathcal{L}_{QCD} = i\overline{U}(\partial_\mu - ig_s G_\mu^a T^a)\gamma^\mu U + i\overline{D}(\partial_\mu - ig_s G_\mu^a T^a)\gamma^\mu D.$$

G_μ^a is the SU(3) gauge field containing the gluons, γ^μ are the Dirac matrices, D and U are the Dirac spinors associated with up- and down-type quarks, and g_s is the strong coupling constant.

Electroweak sector Main article: Electroweak interaction

The electroweak sector is a Yang–Mills gauge theory with the simple symmetry group U(1)×SU(2)L,

$$\mathcal{L}_{EW} = \sum_\psi \bar{\psi}\gamma^\mu \left(i\partial_\mu - g'\frac{1}{2}Y_W B_\mu - g\frac{1}{2}\vec{\tau}_L \vec{W}_\mu \right)\psi$$

where $B\mu$ is the U(1) gauge field; Y_W is the weak hypercharge—the generator of the U(1) group; $\vec{W}_\mu$ is the three-component SU(2) gauge field; $\vec{\tau}_L$ are the Pauli matrices—infinitesimal generators of the SU(2) group. The subscript L indicates that they only act on left fermions; g' and g are coupling constants.

Higgs sector Main article: Higgs mechanism

In the Standard Model, the Higgs field is a complex scalar of the group SU(2)L:

$$\varphi = \frac{1}{\sqrt{2}} \begin{pmatrix} \varphi^+ \\ \varphi^0 \end{pmatrix} ,$$

where the indices + and 0 indicate the electric charge (Q) of the components. The weak isospin (Y_W) of both components is 1.

Before symmetry breaking, the Higgs Lagrangian is:

$$\mathcal{L}_H = \varphi^\dagger \left(\partial^\mu - \frac{i}{2} \left(g' Y_W B^\mu + g \vec{\tau} \vec{W}^\mu \right) \right) \left(\partial_\mu + \frac{i}{2} \left(g' Y_W B_\mu + g \vec{\tau} \vec{W}_\mu \right) \right) \varphi - \frac{\lambda^2}{4} \left(\varphi^\dagger \varphi - v^2 \right)^2 ,$$

which can also be written as:

$$\mathcal{L}_H = \left| \left(\partial_\mu + \frac{i}{2} \left(g' Y_W B_\mu + g \vec{\tau} \vec{W}_\mu \right) \right) \varphi \right|^2 - \frac{\lambda^2}{4} \left(\varphi^\dagger \varphi - v^2 \right)^2 .$$

1.1.5 Fundamental forces

Main article: Fundamental interaction

The Standard Model classified all four fundamental forces in nature. In the Standard Model, a force is described as an exchange of bosons between the objects affected, such as a photon for the electromagnetic force and a gluon for the strong interaction. Those particles are called force carriers.[26]

1.1.6 Tests and predictions

The Standard Model (SM) predicted the existence of the W and Z bosons, gluon, and the top and charm quarks before these particles were observed. Their predicted properties were experimentally confirmed with good precision. To give an idea of the success of the SM, the following table compares the measured masses of the W and Z bosons with the masses predicted by the SM:

The SM also makes several predictions about the decay of Z bosons, which have been experimentally confirmed by the Large Electron-Positron Collider at CERN.

In May 2012 BaBar Collaboration reported that their recently analyzed data may suggest possible flaws in the Standard Model of particle physics.[28][29] These data show that a particular type of particle decay called "B to D-star-tau-nu" happens more often than the Standard Model says it should. In this type of decay, a particle called the B-bar meson decays into a D meson, an antineutrino and a tau-lepton.

While the level of certainty of the excess (3.4 sigma) is not enough to claim a break from the Standard Model, the results are a potential sign of something amiss and are likely to impact existing theories, including those attempting to deduce the properties of Higgs bosons.[30]

On December 13, 2012, physicists reported the constancy, over space and time, of a basic physical constant of nature that supports the *standard model of physics*. The scientists, studying methanol molecules in a distant galaxy, found the change ($\Delta\mu/\mu$) in the proton-to-electron mass ratio μ to be equal to "$(0.0 \pm 1.0) \times 10^{-7}$ at redshift z = 0.89" and consistent with "a null result".[31][32]

1.1.7 Challenges

See also: Physics beyond the Standard Model

Self-consistency of the Standard Model (currently formulated as a non-abelian gauge theory quantized through path-integrals) has not been mathematically proven. While regularized versions useful for approximate computations (for example lattice gauge theory) exist, it is not known whether they converge (in the sense of S-matrix elements) in the limit that the regulator is removed. A key question related to the consistency is the Yang–Mills existence and mass gap problem.

Experiments indicate that neutrinos have mass, which the classic Standard Model did not allow.[33] To accommodate this finding, the classic Standard Model can be modified to include neutrino mass.

If one insists on using only Standard Model particles, this can be achieved by adding a non-renormalizable interaction of leptons with the Higgs boson.[34] On a fundamental level, such an interaction emerges in the seesaw mechanism where heavy right-handed neutrinos are added to the theory. This is natural in the left-right symmetric extension of the Standard Model[35][36] and in certain grand unified theories.[37] As long as new physics appears below or around 10^{14} GeV, the neutrino masses can be of the right order of magnitude.

Theoretical and experimental research has attempted to extend the Standard Model into a Unified field theory or a Theory of everything, a complete theory explaining all physical phenomena including constants. Inadequacies of the Standard Model that motivate such research include:

- It does not attempt to explain gravitation, although a theoretical particle known as a graviton would help explain it, and unlike for the strong and electroweak interactions of the Standard Model, there is no known way of describing general relativity, the canonical theory of gravitation, consistently in terms of quantum field theory. The reason for this is, among other things, that quantum field theories of gravity generally break down before reaching the Planck scale. As a consequence, we have no reliable theory for the very early universe;

- Some consider it to be *ad hoc* and inelegant, requiring 19 numerical constants whose values are unrelated and arbitrary. Although the Standard Model, as it now stands, can explain why neutrinos have masses, the specifics of neutrino mass are still unclear. It is believed that explaining neutrino mass will require an additional 7 or 8 constants, which are also arbitrary parameters;

- The Higgs mechanism gives rise to the hierarchy problem if some new physics (coupled to the Higgs) is present at high energy scales. In these cases in order for the weak scale to be much smaller than the Planck scale, severe fine tuning of the parameters is required; there are, however, other scenarios that include quantum gravity in which such fine tuning can be avoided.[38] There are also issues of Quantum triviality, which suggests that it may not be possible to create a consistent quantum field theory involving elementary scalar particles.

- It should be modified so as to be consistent with the emerging "Standard Model of cosmology." In particular, the Standard Model cannot explain the observed amount of cold dark matter (CDM) and gives contributions to dark energy which are many orders of magnitude too large. It is also difficult to accommodate the observed predominance of matter over antimatter (matter/antimatter asymmetry). The isotropy and homogeneity of the visible universe over large distances seems to require a mechanism like cosmic inflation, which would also constitute an extension of the Standard Model.

- The existence of ultra-high-energy cosmic rays are difficult to explain under the Standard Model.

Currently, no proposed Theory of Everything has been widely accepted or verified.

1.1.8   See also

- Fundamental interaction:

  - Quantum electrodynamics
  - Strong interaction: Color charge, Quantum chromodynamics, Quark model

- Weak interaction: Electroweak theory, Fermi theory of beta decay, Weak hypercharge, Weak isospin

- Gauge theory: Nontechnical introduction to gauge theory

- Generation

- Higgs mechanism: Higgs boson, Higgsless model

- J. C. Ward

- J. J. Sakurai Prize for Theoretical Particle Physics

- Lagrangian

- Open questions: BTeV experiment, CP violation, Neutrino masses, Quark matter, Quantum triviality

- Penguin diagram

- Quantum field theory

- Standard Model: Mathematical formulation of, Physics beyond the Standard Model

1.1.9   Notes and references

[1] Technically, there are nine such color–anticolor combinations. However, there is one color-symmetric combination that can be constructed out of a linear superposition of the nine combinations, reducing the count to eight.

1.1.10   References

[1] R. Oerter (2006). *The Theory of Almost Everything: The Standard Model, the Unsung Triumph of Modern Physics* (Kindle ed.). Penguin Group. p. 2. ISBN 0-13-236678-9.

[2] In fact, there are mathematical issues regarding quantum field theories still under debate (see e.g. Landau pole), but the predictions extracted from the Standard Model by current methods applicable to current experiments are all self-consistent. For a further discussion see e.g. Chapter 25 of R. Mann (2010). *An Introduction to Particle Physics and the Standard Model.* CRC Press. ISBN 978-1-4200-8298-2.

[3] Sean Carroll, Ph.D., Cal Tech, 2007, The Teaching Company, *Dark Matter, Dark Energy: The Dark Side of the Universe*, Guidebook Part 2 page 59, Accessed Oct. 7, 2013, "...Standard Model of Particle Physics: The modern theory of elementary particles and their interactions ... It does not, strictly speaking, include gravity, although it's often convenient to include gravitons among the known particles of nature..."

[4] S.L. Glashow (1961). "Partial-symmetries of weak interactions". *Nuclear Physics* **22** (4): 579–588. Bibcode:1961NucPh..22..579G. doi:10.1016/0029-5582(61)90469-2.

[5] S. Weinberg (1967). "A Model of Leptons". *Physical Review Letters* **19** (21): 1264–1266. Bibcode:1967PhRvL..19.1264W. doi:10.1103/PhysRevLett.19.1264.

[6] A. Salam (1968). N. Svartholm, ed. *Elementary Particle Physics: Relativistic Groups and Analyticity*. Eighth Nobel Symposium. Stockholm: Almquist and Wiksell. p. 367.

[7] F. Englert, R. Brout (1964). "Broken Symmetry and the Mass of Gauge Vector Mesons". *Physical Review Letters* **13** (9): 321–323. Bibcode:1964PhRvL..13..321E. doi:10.1103/PhysRevLett.13.321.

[8] P.W. Higgs (1964). "Broken Symmetries and the Masses of Gauge Bosons". *Physical Review Letters* **13** (16): 508–509. Bibcode:1964PhRvL..13..508H. doi:10.1103/PhysRevLett.13.508.

[9] G.S. Guralnik, C.R. Hagen, T.W.B. Kibble (1964). "Global Conservation Laws and Massless Particles". *Physical Review Letters* **13** (20): 585–587. Bibcode:1964PhRvL..13..585G. doi:10.1103/PhysRevLett.13.585.

[10] F.J. Hasert et al. (1973). "Search for elastic muon-neutrino electron scattering". *Physics Letters B* **46** (1): 121. Bibcode:1973PhLB...46..121H. doi:10.1016/0370-2693(73)90494-2.

[11] F.J. Hasert et al. (1973). "Observation of neutrino-like interactions without muon or electron in the Gargamelle neutrino experiment". *Physics Letters B* **46** (1): 138. Bibcode:1973PhLB...46..138H. doi:10.1016/0370-2693(73)90499-1.

[12] F.J. Hasert et al. (1974). "Observation of neutrino-like interactions without muon or electron in the Gargamelle neutrino experiment". *Nuclear Physics B* **73** (1): 1. Bibcode:1974NuPhB..73....1H. doi:10.1016/0550-3213(74)90038-8.

[13] D. Haidt (4 October 2004). "The discovery of the weak neutral currents". *CERN Courier*. Retrieved 8 May 2008.

[14] "Details can be worked out if the situation is simple enough for us to make an approximation, which is almost never, but often we can understand more or less what is happening." from *The Feynman Lectures on Physics*, Vol 1. pp. 2–7

[15] G.S. Guralnik (2009). "The History of the Guralnik, Hagen and Kibble development of the Theory of Spontaneous Symmetry Breaking and Gauge Particles". *International Journal of Modern Physics A* **24** (14): 2601–2627. arXiv:0907.3466. Bibcode:2009IJMPA..24.2601G. doi:10.1142/S0217751X09045431.

[16] B.W. Lee, C. Quigg, H.B. Thacker (1977). "Weak interactions at very high energies: The role of the Higgs-boson mass". *Physical Review D* **16** (5): 1519–1531. Bibcode:1977PhRvD..16.1519L. doi:10.1103/PhysRevD.16.1519.

[17] "Huge $10 billion collider resumes hunt for 'God particle'". CNN. 11 November 2009. Retrieved 2010-05-04.

[18] M. Strassler (10 July 2012). "Higgs Discovery: Is it a Higgs?". Retrieved 2013-08-06.

[19] "CERN experiments observe particle consistent with long-sought Higgs boson". CERN. 4 July 2012. Retrieved 2012-07-04.

[20] "Observation of a New Particle with a Mass of 125 GeV". CERN. 4 July 2012. Retrieved 2012-07-05.

[21] "ATLAS Experiment". ATLAS. 1 January 2006. Retrieved 2012-07-05.

[22] "Confirmed: CERN discovers new particle likely to be the Higgs boson". *YouTube*. Russia Today. 4 July 2012. Retrieved 2013-08-06.

[23] D. Overbye (4 July 2012). "A New Particle Could Be Physics' Holy Grail". *New York Times*. Retrieved 2012-07-04.

[24] "New results indicate that new particle is a Higgs boson". CERN. 14 March 2013. Retrieved 2013-08-06.

[25] S. Braibant, G. Giacomelli, M. Spurio (2009). *Particles and Fundamental Interactions: An Introduction to Particle Physics*. Springer. pp. 313–314. ISBN 978-94-007-2463-1.

[26] http://home.web.cern.ch/about/physics/standard-model Official CERN website

[27] http://www.pha.jhu.edu/~{}dfehling/particle.gif

[28] "BABAR Data in Tension with the Standard Model". SLAC. 31 May 2012. Retrieved 2013-08-06.

[29] BaBar Collaboration (2012). "Evidence for an excess of $B \to D^{(*)} \tau^- \nu\tau$ decays". *Physical Review Letters* **109** (10): 101802. arXiv:1205.5442. Bibcode:2012PhRvL.109j1802L. doi:10.1103/PhysRevLett.109.101802.

[30] "BaBar data hint at cracks in the Standard Model". *e! Science News*. 18 June 2012. Retrieved 2013-08-06.

[31] J. Bagdonaite et al. (2012). "A Stringent Limit on a Drifting Proton-to-Electron Mass Ratio from Alcohol in the Early Universe". *Science* **339** (6115): 46. Bibcode:2013Sci...339...46B. doi:10.1126/science.1224898.

[32] C. Moskowitz (13 December 2012). "Phew! Universe's Constant Has Stayed Constant". Space.com. Retrieved 2012-12-14.

[33] "Particle chameleon caught in the act of changing". CERN. 31 May 2010. Retrieved 2012-07-05.

[34] S. Weinberg (1979). "Baryon and Lepton Non-conserving Processes". *Physical Review Letters* **43** (21): 1566. Bibcode:1979PhRvL..43.1566W. doi:10.1103/PhysRevLett.43.1566.

[35] P. Minkowski (1977). "μ → e γ at a Rate of One Out of 10^9 Muon Decays?". *Physics Letters B* **67** (4): 421. Bibcode:1977PhLB...67..421M. doi:10.1016/0370-2693(77)90435-X.

[36] R. N. Mohapatra, G. Senjanovic (1980). "Neutrino Mass and Spontaneous Parity Nonconservation". *Physical Review Letters* **44** (14): 912–915. Bibcode:1980PhRvL..44..912M. doi:10.1103/PhysRevLett.44.912.

[37] M. Gell-Mann, P. Ramond and R. Slansky (1979). F. van Nieuwenhuizen and D. Z. Freedman, ed. *Supergravity*. North Holland. pp. 315–321. ISBN 0-444-85438-X.

[38] Salvio, Strumia (2014-03-17). "Agravity". *JHEP 1406 (2014) 080*. arXiv:1403.4226. Bibcode:2014JHEP...06..080S. doi:10.1007/JHEP06(2014)080.

1.1.11   Further reading

- R. Oerter (2006). *The Theory of Almost Everything: The Standard Model, the Unsung Triumph of Modern Physics*. Plume.

- B.A. Schumm (2004). *Deep Down Things: The Breathtaking Beauty of Particle Physics*. Johns Hopkins University Press. ISBN 0-8018-7971-X.

- "The Standard Model of Particle Physics Interactive Graphic".

Introductory textbooks

- I. Aitchison, A. Hey (2003). *Gauge Theories in Particle Physics: A Practical Introduction*. Institute of Physics. ISBN 978-0-585-44550-2.

- W. Greiner, B. Müller (2000). *Gauge Theory of Weak Interactions*. Springer. ISBN 3-540-67672-4.

- G.D. Coughlan, J.E. Dodd, B.M. Gripaios (2006). *The Ideas of Particle Physics: An Introduction for Scientists*. Cambridge University Press.

- D.J. Griffiths (1987). *Introduction to Elementary Particles*. John Wiley & Sons. ISBN 0-471-60386-4.

- G.L. Kane (1987). *Modern Elementary Particle Physics*. Perseus Books. ISBN 0-201-11749-5.

Advanced textbooks

- T.P. Cheng, L.F. Li (2006). *Gauge theory of elementary particle physics*. Oxford University Press. ISBN 0-19-851961-3. Highlights the gauge theory aspects of the Standard Model.

- J.F. Donoghue, E. Golowich, B.R. Holstein (1994). *Dynamics of the Standard Model*. Cambridge University Press. ISBN 978-0-521-47652-2. Highlights dynamical and phenomenological aspects of the Standard Model.

- L. O'Raifeartaigh (1988). *Group structure of gauge theories*. Cambridge University Press. ISBN 0-521-34785-8.

- Nagashima Y. Elementary Particle Physics: Foundations of the Standard Model, Volume 2. (Wiley 2013) 920 рапуы

- Schwartz, M.D. Quantum Field Theory and the Standard Model (Cambridge University Press 2013) 952 pages

- Langacker P. The standard model and beyond. (CRC Press, 2010) 670 pages Highlights group-theoretical aspects of the Standard Model.

Journal articles

- E.S. Abers, B.W. Lee (1973). "Gauge theories". *Physics Reports* **9**: 1–141. Bibcode:1973PhR.....9....1A. doi:10.1016/0370-1573(73)90027-6.

- M. Baak et al. (2012). "The Electroweak Fit of the Standard Model after the Discovery of a New Boson at the LHC". *The European Physical Journal C* **72** (11). arXiv:1209.2716. Bibcode:2012EPJC...72.2205B. doi:10.1140/epjc/s10052-012-2205-9.

- Y. Hayato et al. (1999). "Search for Proton Decay through $p \to \nu K^+$ in a Large Water Cherenkov Detector". *Physical Review Letters* **83** (8): 1529. arXiv:hep-ex/9904020. Bibcode:1999PhRvL..83.1529H. doi:10.1103/PhysRevLett.83.1529.

- S.F. Novaes (2000). "Standard Model: An Introduction". arXiv:hep-ph/0001283 [hep-ph].

- D.P. Roy (1999). "Basic Constituents of Matter and their Interactions — A Progress Report". arXiv:hep-ph/9912523 [hep-ph].

- F. Wilczek (2004). "The Universe Is A Strange Place". *Nuclear Physics B - Proceedings Supplements* **134**: 3. arXiv:astro-ph/0401347. Bibcode:2004NuPhS.134....3W. doi:10.1016/j.nuclphysbps.2004.08.001.

1.1.12 External links

- "The Standard Model explained in Detail by CERN's John Ellis" omega tau podcast.

- "LHC sees hint of lightweight Higgs boson" "New Scientist".

- "Standard Model may be found incomplete," *New Scientist*.

- "Observation of the Top Quark" at Fermilab.

- "The Standard Model Lagrangian." After electroweak symmetry breaking, with no explicit Higgs boson.

- "Standard Model Lagrangian" with explicit Higgs terms. PDF, PostScript, and LaTeX versions.

- "The particle adventure." Web tutorial.

- Nobes, Matthew (2002) "Introduction to the Standard Model of Particle Physics" on Kuro5hin: Part 1, Part 2, Part 3a, Part 3b.

- "The Standard Model" The Standard Model on the CERN web site explains how the basic building blocks of matter interact, governed by four fundamental forces.

1.2 List of particles

This is a list of the different types of particles found or believed to exist in the whole of the universe. For individual lists of the different particles, see the list below.

1.2.1 Elementary particles

Main article: Elementary particle

Elementary particles are particles with no measurable internal structure; that is, they are not composed of other particles. They are the fundamental objects of quantum field theory. Many families and sub-families of elementary particles exist. Elementary particles are classified according to their spin. Fermions have half-integer spin while bosons have integer spin. All the particles of the Standard Model have been experimentally observed, recently including the Higgs boson.[1][2]

Fermions

Main article: Fermion

Fermions are one of the two fundamental classes of particles, the other being bosons. Fermion particles are described by Fermi–Dirac statistics and have quantum numbers described by the Pauli exclusion principle. They include the quarks and leptons, as well as any composite particles consisting of an odd number of these, such as all baryons and many atoms and nuclei.

Fermions have half-integer spin; for all known elementary fermions this is $1/2$. All known fermions, except neutrinos, are also Dirac fermions; that is, each known fermion has its own distinct antiparticle. It is not known whether the neutrino is a Dirac fermion or a Majorana fermion.[3] Fermions are the basic building blocks of all matter. They are classified according to whether they interact via the color force or not. In the Standard Model, there are 12 types of elementary fermions: six quarks and six leptons.

Quarks Main article: Quark

Quarks are the fundamental constituents of hadrons and interact via the strong interaction. Quarks are the only known carriers of fractional charge, but because they combine in groups of three (baryons) or in groups of two with antiquarks (mesons), only integer charge is observed in nature. Their respective antiparticles are the antiquarks, which are identical except for the fact that they carry the opposite electric charge (for example the up quark carries charge $+2/3$, while the up antiquark carries charge $-2/3$), color charge, and baryon number. There are six flavors of quarks; the three positively charged quarks are called "up-type quarks" and the three negatively charged quarks are called "down-type quarks".

Leptons Main article: Leptons

Leptons do not interact via the strong interaction. Their respective antiparticles are the antileptons which are identical, except for the fact that they carry the opposite electric charge and lepton number. The antiparticle of an electron is an antielectron, which is nearly always called a "positron" for historical reasons. There are six leptons in total; the three charged leptons are called "electron-like leptons", while the neutral leptons are called "neutrinos". Neutrinos are known to oscillate, so that neutrinos of definite flavor do not have definite mass, rather they exist in a superposition of mass eigenstates. The hypothetical heavy right-handed neutrino, called a "sterile neutrino", has been left off the list.

Bosons

Main article: Boson

Bosons are one of the two fundamental classes of particles, the other being fermions. Bosons are characterized by Bose–Einstein statistics and all have integer spins. Bosons may be either elementary, like photons and gluons, or composite, like mesons.

The fundamental forces of nature are mediated by gauge bosons, and mass is believed to be created by the Higgs field. According to the Standard Model the elementary bosons are:

The graviton is added to the list although it is not predicted by the Standard Model, but by other theories in the framework of quantum field theory. Furthermore, gravity is non-renormalizable. There are a total of eight independent gluons. The Higgs boson is postulated by the electroweak theory primarily to explain the origin of particle masses. In a process known as the "Higgs mechanism", the Higgs boson and the other gauge bosons in the Standard Model acquire mass via spontaneous symmetry breaking of the SU(2) gauge symmetry. The Minimal Supersymmetric Standard Model (MSSM) predicts several Higgs bosons. A new particle expected to be the Higgs boson was observed at the CERN/LHC on March 14, 2013, around the energy of 126.5GeV with an accuracy of close to five sigma (99.9999%, which is accepted as definitive). The Higgs mechanism giving mass to other particles has not been observed yet.

Hypothetical particles

Supersymmetric theories predict the existence of more particles, none of which have been confirmed experimentally as of 2014:

Note: just as the photon, Z boson and $W^\pm$ bosons are superpositions of the B^0, W^0, W^1, and W^2 fields – the photino, zino, and wino$^\pm$ are superpositions of the bino^0, wino^0,

wino^1, and wino^2 by definition.

No matter if one uses the original gauginos or this superpositions as a basis, the only predicted physical particles are neutralinos and charginos as a superposition of them together with the Higgsinos.

Other theories predict the existence of additional bosons:

Mirror particles are predicted by theories that restore parity symmetry.

"Magnetic monopole" is a generic name for particles with non-zero magnetic charge. They are predicted by some GUTs.

"Tachyon" is a generic name for hypothetical particles that travel faster than the speed of light and have an imaginary rest mass.

Preons were suggested as subparticles of quarks and leptons, but modern collider experiments have all but ruled out their existence.

Kaluza–Klein towers of particles are predicted by some models of extra dimensions. The extra-dimensional momentum is manifested as extra mass in four-dimensional spacetime.

1.2.2   Composite particles

Hadrons

Main article: Hadron

Hadrons are defined as strongly interacting composite particles. Hadrons are either:

- Composite fermions, in which case they are called baryons.

- Composite bosons, in which case they are called mesons.

Quark models, first proposed in 1964 independently by Murray Gell-Mann and George Zweig (who called quarks "aces"), describe the known hadrons as composed of valence quarks and/or antiquarks, tightly bound by the color force, which is mediated by gluons. A "sea" of virtual quark-antiquark pairs is also present in each hadron.

Baryons   See also: List of baryons

Ordinary baryons (composite fermions) contain three valence quarks or three valence antiquarks each.

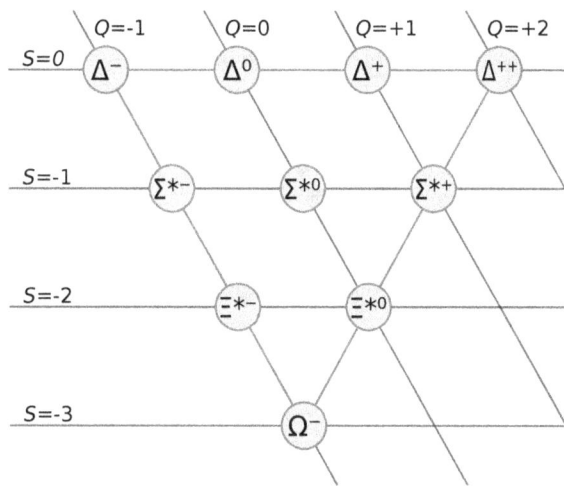

A combination of three u, d or s-quarks with a total spin of $^3\!/_2$ form the so-called "baryon decuplet".

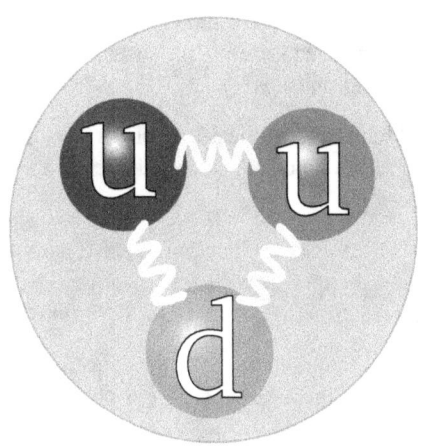

Proton quark structure: 2 up quarks and 1 down quark. The gluon tubes or flux tubes are now known to be Y shaped.

- Nucleons are the fermionic constituents of normal atomic nuclei:

  - Protons, composed of two up and one down quark (uud)

  - Neutrons, composed of two down and one up quark (ddu)

- Hyperons, such as the Λ, Σ, Ξ, and Ω particles, which contain one or more strange quarks, are short-lived and heavier than nucleons. Although not normally present in atomic nuclei, they can appear in short-lived hypernuclei.

- A number of charmed and bottom baryons have also been observed.

Some hints at the existence of exotic baryons have been found recently; however, negative results have also been reported. Their existence is uncertain.

- Pentaquarks consist of four valence quarks and one valence antiquark.

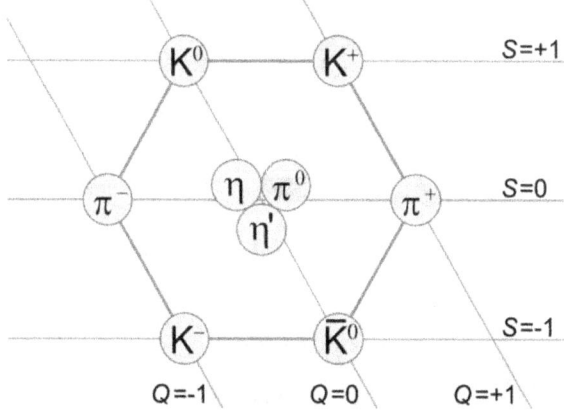

Mesons of spin 0 form a nonet

Mesons   See also: List of mesons

Ordinary mesons are made up of a valence quark and a valence antiquark. Because mesons have spin of 0 or 1 and are not themselves elementary particles, they are "composite" bosons. Examples of mesons include the pion, kaon, and the J/ψ. In quantum hydrodynamic models, mesons mediate the residual strong force between nucleons.

At one time or another, positive signatures have been reported for all of the following exotic mesons but their existences have yet to be confirmed.

- A tetraquark consists of two valence quarks and two valence antiquarks;

- A glueball is a bound state of gluons with no valence quarks;

- Hybrid mesons consist of one or more valence quark-antiquark pairs and one or more real gluons.

Atomic nuclei

Atomic nuclei consist of protons and neutrons. Each type of nucleus contains a specific number of protons and a specific number of neutrons, and is called a "nuclide" or "isotope".

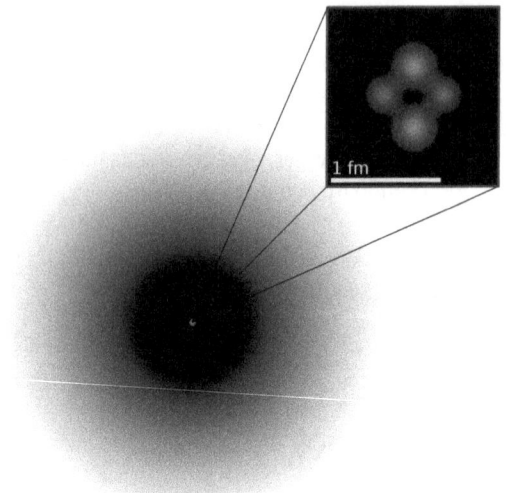

1 Å = 100,000 fm

A semi-accurate depiction of the helium atom. In the nucleus, the protons are in red and neutrons are in purple. In reality, the nucleus is also spherically symmetrical.

Nuclear reactions can change one nuclide into another. See table of nuclides for a complete list of isotopes.

Atoms

Atoms are the smallest neutral particles into which matter can be divided by chemical reactions. An atom consists of a small, heavy nucleus surrounded by a relatively large, light cloud of electrons. Each type of atom corresponds to a specific chemical element. To date, 118 elements have been discovered, while only the elements 1-112,114, and 116 have received official names.

The atomic nucleus consists of protons and neutrons. Protons and neutrons are, in turn, made of quarks.

Molecules

Molecules are the smallest particles into which a non-elemental substance can be divided while maintaining the physical properties of the substance. Each type of molecule corresponds to a specific chemical compound. Molecules are a composite of two or more atoms. See list of compounds for a list of molecules.

1.2.3 Condensed matter

The field equations of condensed matter physics are remarkably similar to those of high energy particle physics. As a result, much of the theory of particle physics applies to condensed matter physics as well; in particular, there are a selection of field excitations, called quasi-particles, that can be created and explored. These include:

- Phonons are vibrational modes in a crystal lattice.

- Excitons are bound states of an electron and a hole.

- Plasmons are coherent excitations of a plasma.

- Polaritons are mixtures of photons with other quasi-particles.

- Polarons are moving, charged (quasi-) particles that are surrounded by ions in a material.

- Magnons are coherent excitations of electron spins in a material.

1.2.4 Other

- An anyon is a generalization of fermion and boson in two-dimensional systems like sheets of graphene that obeys braid statistics.

- A plekton is a theoretical kind of particle discussed as a generalization of the braid statistics of the anyon to dimension > 2.

- A WIMP (weakly interacting massive particle) is any one of a number of particles that might explain dark matter (such as the neutralino or the axion).

- The pomeron, used to explain the elastic scattering of hadrons and the location of Regge poles in Regge theory.

- The skyrmion, a topological solution of the pion field, used to model the low-energy properties of the nucleon, such as the axial vector current coupling and the mass.

- A genon is a particle existing in a closed timelike world line where spacetime is curled as in a Frank Tipler or Ronald Mallett time machine.

- A goldstone boson is a massless excitation of a field that has been spontaneously broken. The pions are quasi-goldstone bosons (quasi- because they are not exactly massless) of the broken chiral isospin symmetry of quantum chromodynamics.

- A goldstino is a goldstone fermion produced by the spontaneous breaking of supersymmetry.

- An instanton is a field configuration which is a local minimum of the Euclidean action. Instantons are used in nonperturbative calculations of tunneling rates.

- A dyon is a hypothetical particle with both electric and magnetic charges.

- A geon is an electromagnetic or gravitational wave which is held together in a confined region by the gravitational attraction of its own field energy.

- An inflaton is the generic name for an unidentified scalar particle responsible for the cosmic inflation.

- A spurion is the name given to a "particle" inserted mathematically into an isospin-violating decay in order to analyze it as though it conserved isospin.

- What is called "true muonium", a bound state of a muon and an antimuon, is a theoretical exotic atom which has never been observed.

1.2.5 Classification by speed

- A tardyon or bradyon travels slower than light and has a non-zero rest mass.

- A luxon travels at the speed of light and has no rest mass.

- A tachyon (mentioned above) is a hypothetical particle that travels faster than the speed of light and has an imaginary rest mass.

1.2.6 See also

- Acceleron

- List of baryons

- List of compounds for a list of molecules.

- List of fictional elements, materials, isotopes and atomic particles

- List of mesons

- Periodic table for an overview of atoms.

- Standard Model for the current theory of these particles.

- Table of nuclides

- Timeline of particle discoveries

1.2.7 References

[1] Observation of a new boson at a mass of 125 GeV with the CMS experiment at the LHC (2013). *arXiv:1207.7235.*

[2] Observation of a new particle in the search for the Standard Model Higgs boson with the ATLAS detector at the LHC (2012). *arXiv:1207.7214.*

[3] B. Kayser, *Two Questions About Neutrinos*, arXiv:1012. 4469v1 [hep-ph] (2010).

[4] R. Maartens (2004). *Brane-World Gravity* (PDF). *Living Reviews in Relativity* **7**. p. 7. Also available in web format at http://www.livingreviews.org/lrr-2004-7.

- C. Amsler *et al.* (Particle Data Group) (2008). "Review of Particle Physics". *Physics Letters B* **667** (1–5): 1. Bibcode:2008PhLB..667....1P. doi:10.1016/j.physletb.2008.07.018. *(All information on this list, and more, can be found in the extensive, biannually-updated review by the Particle Data Group)*

Chapter 2

Quarks

2.1 Quark

This article is about the particle. For other uses, see Quark (disambiguation).

A **quark** (/ˈkwɔrk/ or /ˈkwɑrk/) is an elementary particle and a fundamental constituent of matter. Quarks combine to form composite particles called hadrons, the most stable of which are protons and neutrons, the components of atomic nuclei.[1] Due to a phenomenon known as *color confinement*, quarks are never directly observed or found in isolation; they can be found only within hadrons, such as baryons (of which protons and neutrons are examples), and mesons.[2][3] For this reason, much of what is known about quarks has been drawn from observations of the hadrons themselves.

Quarks have various intrinsic properties, including electric charge, mass, color charge and spin. Quarks are the only elementary particles in the Standard Model of particle physics to experience all four fundamental interactions, also known as *fundamental forces* (electromagnetism, gravitation, strong interaction, and weak interaction), as well as the only known particles whose electric charges are not integer multiples of the elementary charge.

There are six types of quarks, known as *flavors*: up, down, strange, charm, top, and bottom.[4] Up and down quarks have the lowest masses of all quarks. The heavier quarks rapidly change into up and down quarks through a process of particle decay: the transformation from a higher mass state to a lower mass state. Because of this, up and down quarks are generally stable and the most common in the universe, whereas strange, charm, bottom, and top quarks can only be produced in high energy collisions (such as those involving cosmic rays and in particle accelerators). For every quark flavor there is a corresponding type of antiparticle, known as an *antiquark*, that differs from the quark only in that some of its properties have equal magnitude but opposite sign.

The quark model was independently proposed by physicists Murray Gell-Mann and George Zweig in 1964.[5] Quarks were introduced as parts of an ordering scheme for hadrons, and there was little evidence for their physical existence until deep inelastic scattering experiments at the Stanford Linear Accelerator Center in 1968.[6][7] Accelerator experiments have provided evidence for all six flavors. The top quark was the last to be discovered at Fermilab in 1995.[5]

2.1.1 Classification

See also: Standard Model

The Standard Model is the theoretical framework describ-

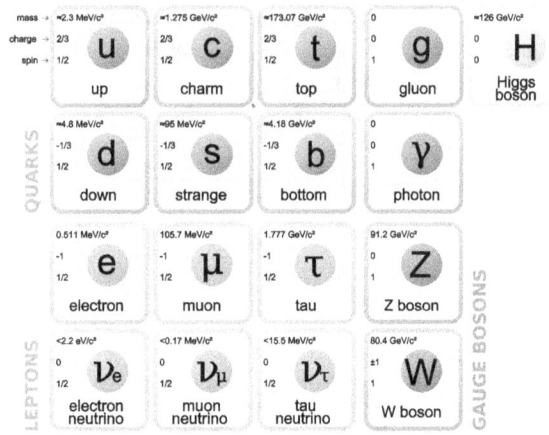

Six of the particles in the Standard Model are quarks (shown in purple). Each of the first three columns forms a generation of matter.

ing all the currently known elementary particles. This model contains six flavors of quarks (q), named up (u), down (d), strange (s), charm (c), bottom (b), and top (t).[4] Antiparticles of quarks are called *antiquarks*, and are denoted by a bar over the symbol for the corresponding quark, such as u for an up antiquark. As with antimatter in general, antiquarks have the same mass, mean lifetime, and spin as their respective quarks, but the electric charge and other charges have the opposite sign.[8]

Quarks are spin-$\frac{1}{2}$ particles, implying that they are fermions according to the spin-statistics theorem. They are subject to the Pauli exclusion principle, which states that no two identical fermions can simultaneously occupy the same quantum state. This is in contrast to bosons (particles with integer spin), any number of which can be in the same state.[9] Unlike leptons, quarks possess color charge, which causes them to engage in the strong interaction. The resulting attraction between different quarks causes the formation of composite particles known as *hadrons* (see "Strong interaction and color charge" below).

The quarks which determine the quantum numbers of hadrons are called *valence quarks*; apart from these, any hadron may contain an indefinite number of virtual (or *sea*) quarks, antiquarks, and gluons which do not influence its quantum numbers.[10] There are two families of hadrons: baryons, with three valence quarks, and mesons, with a valence quark and an antiquark.[11] The most common baryons are the proton and the neutron, the building blocks of the atomic nucleus.[12] A great number of hadrons are known (see list of baryons and list of mesons), most of them differentiated by their quark content and the properties these constituent quarks confer. The existence of "exotic" hadrons with more valence quarks, such as tetraquarks (qqqq) and pentaquarks (qqqqq), has been conjectured[13] but not proven.[nb 1][13][14] However, on 13 July 2015, the LHCb collaboration at CERN reported results consistent with pentaquark states.[15]

Elementary fermions are grouped into three generations, each comprising two leptons and two quarks. The first generation includes up and down quarks, the second strange and charm quarks, and the third bottom and top quarks. All searches for a fourth generation of quarks and other elementary fermions have failed,[16] and there is strong indirect evidence that no more than three generations exist.[nb 2][17] Particles in higher generations generally have greater mass and less stability, causing them to decay into lower-generation particles by means of weak interactions. Only first-generation (up and down) quarks occur commonly in nature. Heavier quarks can only be created in high-energy collisions (such as in those involving cosmic rays), and decay quickly; however, they are thought to have been present during the first fractions of a second after the Big Bang, when the universe was in an extremely hot and dense phase (the quark epoch). Studies of heavier quarks are conducted in artificially created conditions, such as in particle accelerators.[18]

Having electric charge, mass, color charge, and flavor, quarks are the only known elementary particles that engage in all four fundamental interactions of contemporary physics: electromagnetism, gravitation, strong interaction, and weak interaction.[12] Gravitation is too weak to be relevant to individual particle interactions except at extremes of energy (Planck energy) and distance scales (Planck distance). However, since no successful quantum theory of gravity exists, gravitation is not described by the Standard Model.

See the table of properties below for a more complete overview of the six quark flavors' properties.

2.1.2 History

Murray Gell-Mann at TED in 2007. Gell-Mann and George Zweig proposed the quark model in 1964.

The quark model was independently proposed by physicists Murray Gell-Mann[19] (pictured) and George Zweig[20][21] in 1964.[5] The proposal came shortly after Gell-Mann's 1961 formulation of a particle classification system known as the *Eightfold Way*—or, in more technical terms, SU(3) flavor symmetry.[22] Physicist Yuval Ne'eman had independently developed a scheme similar to the Eightfold Way in the same year.[23][24]

At the time of the quark theory's inception, the "particle zoo" included, amongst other particles, a multitude of hadrons. Gell-Mann and Zweig posited that they were not elementary particles, but were instead composed of combinations of quarks and antiquarks. Their model involved three flavors of quarks, up, down, and strange, to which they ascribed properties such as spin and electric charge.[19][20][21] The initial reaction of the physics community to the proposal was mixed. There was particular contention about whether the quark was a physical entity or a mere abstraction used to explain concepts that were not fully understood at the time.[25]

In less than a year, extensions to the Gell-Mann–Zweig model were proposed. Sheldon Lee Glashow and James Bjorken predicted the existence of a fourth flavor of quark, which they called *charm*. The addition was proposed because it allowed for a better description of the weak interaction (the mechanism that allows quarks to decay), equalized the number of known quarks with the number of known leptons, and implied a mass formula that correctly reproduced the masses of the known mesons.[26]

In 1968, deep inelastic scattering experiments at the Stanford Linear Accelerator Center (SLAC) showed that the proton contained much smaller, point-like objects and was therefore not an elementary particle.[6][7][27] Physicists were reluctant to firmly identify these objects with quarks at the time, instead calling them "partons"—a term coined by Richard Feynman.[28][29][30] The objects that were observed at SLAC would later be identified as up and down quarks as the other flavors were discovered.[31] Nevertheless, "parton" remains in use as a collective term for the constituents of hadrons (quarks, antiquarks, and gluons).

The strange quark's existence was indirectly validated by SLAC's scattering experiments: not only was it a necessary component of Gell-Mann and Zweig's three-quark model, but it provided an explanation for the kaon (K) and pion (π) hadrons discovered in cosmic rays in 1947.[32]

In a 1970 paper, Glashow, John Iliopoulos and Luciano Maiani presented further reasoning for the existence of the as-yet undiscovered charm quark.[33][34] The number of supposed quark flavors grew to the current six in 1973, when Makoto Kobayashi and Toshihide Maskawa noted that the experimental observation of CP violation[nb 3][35] could be explained if there were another pair of quarks.

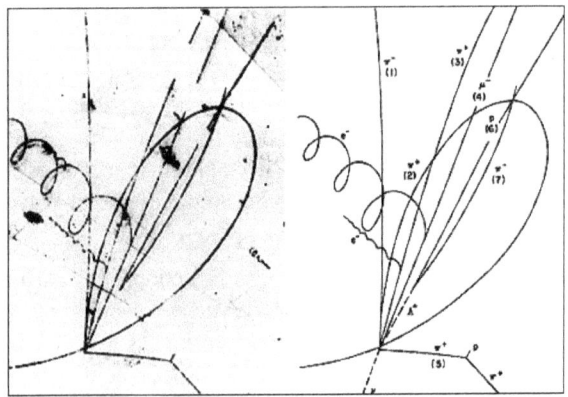

Photograph of the event that led to the discovery of the $\Sigma++$ c baryon, at the Brookhaven National Laboratory in 1974

Charm quarks were produced almost simultaneously by two teams in November 1974 (see November Revolution)—one

at SLAC under Burton Richter, and one at Brookhaven National Laboratory under Samuel Ting. The charm quarks were observed bound with charm antiquarks in mesons. The two parties had assigned the discovered meson two different symbols, J and ψ; thus, it became formally known as the J/ψ meson. The discovery finally convinced the physics community of the quark model's validity.[30]

In the following years a number of suggestions appeared for extending the quark model to six quarks. Of these, the 1975 paper by Haim Harari[36] was the first to coin the terms *top* and *bottom* for the additional quarks.[37]

In 1977, the bottom quark was observed by a team at Fermilab led by Leon Lederman.[38][39] This was a strong indicator of the top quark's existence: without the top quark, the bottom quark would have been without a partner. However, it was not until 1995 that the top quark was finally observed, also by the CDF[40] and DØ[41] teams at Fermilab.[5] It had a mass much larger than had been previously expected,[42] almost as large as that of a gold atom.[43]

2.1.3 Etymology

For some time, Gell-Mann was undecided on an actual spelling for the term he intended to coin, until he found the word *quark* in James Joyce's book *Finnegans Wake*:

> Three quarks for Muster Mark!
> Sure he has not got much of a bark
> And sure any he has it's all beside the mark.
> — James Joyce, *Finnegans Wake*[44]

Gell-Mann went into further detail regarding the name of the quark in his book *The Quark and the Jaguar*:[45]

> In 1963, when I assigned the name "quark" to the fundamental constituents of the nucleon, I had the sound first, without the spelling, which could have been "kwork". Then, in one of my occasional perusals of *Finnegans Wake*, by James Joyce, I came across the word "quark" in the phrase "Three quarks for Muster Mark". Since "quark" (meaning, for one thing, the cry of the gull) was clearly intended to rhyme with "Mark", as well as "bark" and other such words, I had to find an excuse to pronounce it as "kwork". But the book represents the dream of a publican named Humphrey Chimpden Earwicker. Words in the text are typically drawn from several sources at once, like the "portmanteau" words in "Through the Looking-Glass". From time to time, phrases occur in the book that are partially

determined by calls for drinks at the bar. I argued, therefore, that perhaps one of the multiple sources of the cry "Three quarks for Muster Mark" might be "Three quarts for Mister Mark", in which case the pronunciation "kwork" would not be totally unjustified. In any case, the number three fitted perfectly the way quarks occur in nature.

Zweig preferred the name *ace* for the particle he had theorized, but Gell-Mann's terminology came to prominence once the quark model had been commonly accepted.[46]

The quark flavors were given their names for several reasons. The up and down quarks are named after the up and down components of isospin, which they carry.[47] Strange quarks were given their name because they were discovered to be components of the strange particles discovered in cosmic rays years before the quark model was proposed; these particles were deemed "strange" because they had unusually long lifetimes.[48] Glashow, who coproposed charm quark with Bjorken, is quoted as saying, "We called our construct the 'charmed quark', for we were fascinated and pleased by the symmetry it brought to the subnuclear world."[49] The names "bottom" and "top", coined by Harari, were chosen because they are "logical partners for up and down quarks".[36][37][48] In the past, bottom and top quarks were sometimes referred to as "beauty" and "truth" respectively, but these names have somewhat fallen out of use.[50] While "truth" never did catch on, accelerator complexes devoted to massive production of bottom quarks are sometimes called "beauty factories".[51]

2.1.4 Properties

Electric charge

See also: Electric charge

Quarks have fractional electric charge values – either $\frac{1}{3}$ or $\frac{2}{3}$ times the elementary charge (e), depending on flavor. Up, charm, and top quarks (collectively referred to as *up-type quarks*) have a charge of $+\frac{2}{3}$ e, while down, strange, and bottom quarks (*down-type quarks*) have $-\frac{1}{3}$ e. Antiquarks have the opposite charge to their corresponding quarks; up-type antiquarks have charges of $-\frac{2}{3}$ e and down-type antiquarks have charges of $+\frac{1}{3}$ e. Since the electric charge of a hadron is the sum of the charges of the constituent quarks, all hadrons have integer charges: the combination of three quarks (baryons), three antiquarks (antibaryons), or a quark and an antiquark (mesons) always results in integer charges.[52] For example, the hadron constituents of atomic nuclei, neutrons and protons, have charges of 0 e and +1 e respectively; the neutron is composed of two down quarks and one up quark, and the proton of two up quarks and one down quark.[12]

Spin

See also: Spin (physics)

Spin is an intrinsic property of elementary particles, and its direction is an important degree of freedom. It is sometimes visualized as the rotation of an object around its own axis (hence the name "spin"), though this notion is somewhat misguided at subatomic scales because elementary particles are believed to be point-like.[53]

Spin can be represented by a vector whose length is measured in units of the reduced Planck constant $\hbar$ (pronounced "h bar"). For quarks, a measurement of the spin vector component along any axis can only yield the values $+\hbar/2$ or $-\hbar/2$; for this reason quarks are classified as spin-$\frac{1}{2}$ particles.[54] The component of spin along a given axis – by convention the z axis – is often denoted by an up arrow ↑ for the value $+\frac{1}{2}$ and down arrow ↓ for the value $-\frac{1}{2}$, placed after the symbol for flavor. For example, an up quark with a spin of $+\frac{1}{2}$ along the z axis is denoted by u↑.[55]

Weak interaction

Main article: Weak interaction

A quark of one flavor can transform into a quark of another

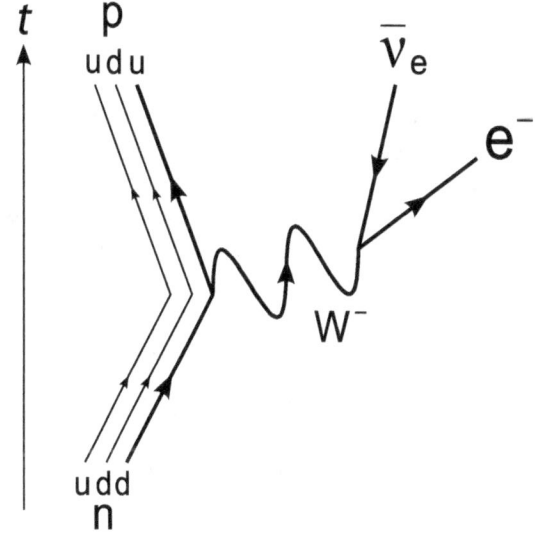

Feynman diagram of beta decay with time flowing upwards. The CKM matrix (discussed below) encodes the probability of this and other quark decays.

flavor only through the weak interaction, one of the four fundamental interactions in particle physics. By absorbing or emitting a W boson, any up-type quark (up, charm, and top quarks) can change into any down-type quark (down, strange, and bottom quarks) and vice versa. This flavor transformation mechanism causes the radioactive process of beta decay, in which a neutron (n) "splits" into a proton (p), an electron (e−) and an electron antineutrino (ν
e) (see picture). This occurs when one of the down quarks in the neutron (udd) decays into an up quark by emitting a virtual W− boson, transforming the neutron into a proton (uud). The W− boson then decays into an electron and an electron antineutrino.[56]

Both beta decay and the inverse process of *inverse beta decay* are routinely used in medical applications such as positron emission tomography (PET) and in experiments involving neutrino detection.

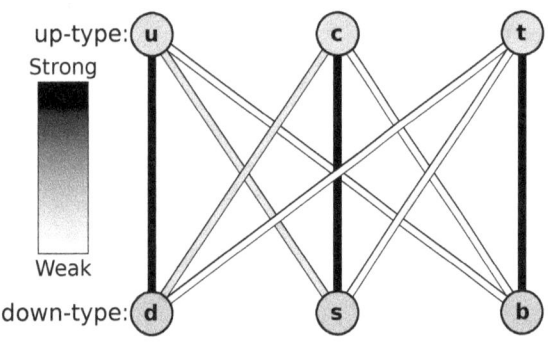

The strengths of the weak interactions between the six quarks. The "intensities" of the lines are determined by the elements of the CKM matrix.

While the process of flavor transformation is the same for all quarks, each quark has a preference to transform into the quark of its own generation. The relative tendencies of all flavor transformations are described by a mathematical table, called the Cabibbo–Kobayashi–Maskawa matrix (CKM matrix). Enforcing unitarity, the approximate magnitudes of the entries of the CKM matrix are:[57]

$$
\begin{bmatrix} |V_{ud}| & |V_{us}| & |V_{ub}| \\ |V_{cd}| & |V_{cs}| & |V_{cb}| \\ |V_{td}| & |V_{ts}| & |V_{tb}| \end{bmatrix} \approx \begin{bmatrix} 0.974 & 0.225 & 0.003 \\ 0.225 & 0.973 & 0.041 \\ 0.009 & 0.040 & 0.999 \end{bmatrix},
$$

where V_{ij} represents the tendency of a quark of flavor i to change into a quark of flavor j (or vice versa).[nb 4]

There exists an equivalent weak interaction matrix for leptons (right side of the W boson on the above beta decay diagram), called the Pontecorvo–Maki–Nakagawa–Sakata matrix (PMNS matrix).[58] Together, the CKM and PMNS matrices describe all flavor transformations, but the links between the two are not yet clear.[59]

Strong interaction and color charge

See also: Color charge and Strong interaction

According to quantum chromodynamics (QCD), quarks possess a property called *color charge*. There are three types of color charge, arbitrarily labeled *blue*, *green*, and *red*.[nb 5] Each of them is complemented by an anticolor – *antiblue*, *antigreen*, and *antired*. Every quark carries a color, while every antiquark carries an anticolor.[60]

The system of attraction and repulsion between quarks charged with different combinations of the three colors is called strong interaction, which is mediated by force carrying particles known as *gluons*; this is discussed at length below. The theory that describes strong interactions is called quantum chromodynamics (QCD). A quark, which will have a single color value, can form a bound system with an antiquark carrying the corresponding anticolor. The result of two attracting quarks will be color neutrality: a quark with color charge ξ plus an antiquark with color charge −ξ will result in a color charge of 0 (or "white" color) and the formation of a meson. This is analogous to the additive color model in basic optics. Similarly, the combination of three quarks, each with different color charges, or three antiquarks, each with anticolor charges, will result in the same "white" color charge and the formation of a baryon or antibaryon.[61]

In modern particle physics, gauge symmetries – a kind of symmetry group – relate interactions between particles (see gauge theories). Color SU(3) (commonly abbreviated to SU(3)$_c$) is the gauge symmetry that relates the color charge in quarks and is the defining symmetry for quantum chromodynamics.[62] Just as the laws of physics are independent of which directions in space are designated x, y, and z, and remain unchanged if the coordinate axes are rotated to a new orientation, the physics of quantum chromodynamics is independent of which directions in three-dimensional color space are identified as blue, red, and green. SU(3)$_c$ color transformations correspond to "rotations" in color space (which, mathematically speaking, is a complex space). Every quark flavor f, each with subtypes fB, fG, fR corresponding to the quark colors,[63] forms a triplet: a three-component quantum field which transforms under the fundamental representation of SU(3)$_c$.[64] The requirement that SU(3)$_c$ should be local – that is, that its transformations be allowed to vary with space and time – determines the properties of the strong interaction, in particular the existence of eight gluon types to act as its force carriers.[62][65]

Mass

See also: Invariant mass

Two terms are used in referring to a quark's mass: *current quark mass* refers to the mass of a quark by itself, while *constituent quark mass* refers to the current quark mass plus the mass of the gluon particle field surrounding the quark.[66] These masses typically have very different values. Most of a hadron's mass comes from the gluons that bind the constituent quarks together, rather than from the quarks themselves. While gluons are inherently massless, they possess energy – more specifically, quantum chromodynamics binding energy (QCBE) – and it is this that contributes so greatly to the overall mass of the hadron (see mass in special relativity). For example, a proton has a mass of approximately 938 MeV/c^2, of which the rest mass of its three valence quarks only contributes about 11 MeV/c^2; much of the remainder can be attributed to the gluons' QCBE.[67][68]

The Standard Model posits that elementary particles derive their masses from the Higgs mechanism, which is related to the Higgs boson. Physicists hope that further research into the reasons for the top quark's large mass of ~173 GeV/c^2, almost the mass of a gold atom,[67][69] might reveal more about the origin of the mass of quarks and other elementary particles.[70]

Table of properties

See also: Flavor (particle physics)

The following table summarizes the key properties of the six quarks. Flavor quantum numbers (isospin (I_3), charm (C), strangeness (S, not to be confused with spin), topness (T), and bottomness (B')) are assigned to certain quark flavors, and denote qualities of quark-based systems and hadrons. The baryon number (B) is +⅓ for all quarks, as baryons are made of three quarks. For antiquarks, the electric charge (Q) and all flavor quantum numbers (B, I_3, C, S, T, and B') are of opposite sign. Mass and total angular momentum (J; equal to spin for point particles) do not change sign for the antiquarks.

J = total angular momentum, *B* = baryon number, *Q* = electric charge, I_3 = isospin, *C* = charm, *S* = strangeness, *T* = topness, *B'* = bottomness.
* Notation such as 4190+180
−60 denotes measurement uncertainty. In the case of the top quark, the first uncertainty is statistical in nature, and the second is systematic.

2.1.5 Interacting quarks

See also: Color confinement and Gluon

As described by quantum chromodynamics, the strong interaction between quarks is mediated by gluons, massless vector gauge bosons. Each gluon carries one color charge and one anticolor charge. In the standard framework of particle interactions (part of a more general formulation known as perturbation theory), gluons are constantly exchanged between quarks through a virtual emission and absorption process. When a gluon is transferred between quarks, a color change occurs in both; for example, if a red quark emits a red–antigreen gluon, it becomes green, and if a green quark absorbs a red–antigreen gluon, it becomes red. Therefore, while each quark's color constantly changes, their strong interaction is preserved.[71][72][73]

Since gluons carry color charge, they themselves are able to emit and absorb other gluons. This causes *asymptotic freedom*: as quarks come closer to each other, the chromodynamic binding force between them weakens.[74] Conversely, as the distance between quarks increases, the binding force strengthens. The color field becomes stressed, much as an elastic band is stressed when stretched, and more gluons of appropriate color are spontaneously created to strengthen the field. Above a certain energy threshold, pairs of quarks and antiquarks are created. These pairs bind with the quarks being separated, causing new hadrons to form. This phenomenon is known as *color confinement*: quarks never appear in isolation.[72][75] This process of hadronization occurs before quarks, formed in a high energy collision, are able to interact in any other way. The only exception is the top quark, which may decay before it hadronizes.[76]

Sea quarks

Hadrons, along with the *valence quarks* (q
v) that contribute to their quantum numbers, contain virtual quark–antiquark (qq) pairs known as *sea quarks* (q
s). Sea quarks form when a gluon of the hadron's color field splits; this process also works in reverse in that the annihilation of two sea quarks produces a gluon. The result is a constant flux of gluon splits and creations colloquially known as "the sea".[77] Sea quarks are much less stable than their valence counterparts, and they typically annihilate each other within the interior of the hadron. Despite this, sea quarks can hadronize into baryonic or mesonic particles under certain circumstances.[78]

Other phases of quark matter

Main article: QCD matter

Under sufficiently extreme conditions, quarks may become deconfined and exist as free particles. In the course of asymptotic freedom, the strong interaction becomes weaker at higher temperatures. Eventually, color confinement would be lost and an extremely hot plasma of freely moving quarks and gluons would be formed. This theoretical phase of matter is called quark–gluon plasma.[81] The exact conditions needed to give rise to this state are unknown and have been the subject of a great deal of speculation and experimentation. A recent estimate puts the needed temperature at $(1.90\pm0.02)\times10^{12}$ kelvin.[82] While a state of entirely free quarks and gluons has never been achieved (despite numerous attempts by CERN in the 1980s and 1990s),[83] recent experiments at the Relativistic Heavy Ion Collider have yielded evidence for liquid-like quark matter exhibiting "nearly perfect" fluid motion.[84]

The quark–gluon plasma would be characterized by a great increase in the number of heavier quark pairs in relation to the number of up and down quark pairs. It is believed that in the period prior to 10^{-6} seconds after the Big Bang (the quark epoch), the universe was filled with quark–gluon plasma, as the temperature was too high for hadrons to be stable.[85]

Given sufficiently high baryon densities and relatively low temperatures – possibly comparable to those found in neutron stars – quark matter is expected to degenerate into a Fermi liquid of weakly interacting quarks. This liquid would be characterized by a condensation of colored quark Cooper pairs, thereby breaking the local $SU(3)_c$ symmetry. Because quark Cooper pairs harbor color charge, such a phase of quark matter would be color superconductive; that is, color charge would be able to pass through it with no resistance.[86]

2.1.6 See also

- Color–flavor locking
- Neutron magnetic moment
- Leptons
- Preons – Hypothetical particles which were once postulated to be subcomponents of quarks and leptons
- Quarkonium – Mesons made of a quark and antiquark of the same flavor
- Quark star – A hypothetical degenerate neutron star with extreme density
- Quark–lepton complementarity – Possible fundamental relation between quarks and leptons

2.1.7 Notes

[1] Several research groups claimed to have proven the existence of tetraquarks and pentaquarks in the early 2000s. While the status of tetraquarks is still under debate, all known pentaquark candidates have previously been established as non-existent.

[2] The main evidence is based on the resonance width of the Z0 boson, which constrains the 4th generation neutrino to have a mass greater than ~45 GeV/c^2. This would be highly contrasting with the other three generations' neutrinos, whose masses cannot exceed 2 MeV/c^2.

[3] CP violation is a phenomenon which causes weak interactions to behave differently when left and right are swapped (P symmetry) and particles are replaced with their corresponding antiparticles (C symmetry).

[4] The actual probability of decay of one quark to another is a complicated function of (amongst other variables) the decaying quark's mass, the masses of the decay products, and the corresponding element of the CKM matrix. This probability is directly proportional (but not equal) to the magnitude squared ($|Vij|^2$) of the corresponding CKM entry.

[5] Despite its name, color charge is not related to the color spectrum of visible light.

2.1.8 References

[1] "Quark (subatomic particle)". *Encyclopædia Britannica*. Retrieved 2008-06-29.

[2] R. Nave. "Confinement of Quarks". *HyperPhysics*. Georgia State University, Department of Physics and Astronomy. Retrieved 2008-06-29.

[3] R. Nave. "Bag Model of Quark Confinement". *HyperPhysics*. Georgia State University, Department of Physics and Astronomy. Retrieved 2008-06-29.

[4] R. Nave. "Quarks". *HyperPhysics*. Georgia State University, Department of Physics and Astronomy. Retrieved 2008-06-29.

[5] B. Carithers, P. Grannis (1995). "Discovery of the Top Quark" (PDF). *Beam Line* (SLAC) **25** (3): 4–16. Retrieved 2008-09-23.

[6] E.D. Bloom et al. (1969). "High-Energy Inelastic e–p Scattering at 6° and 10°". *Physical Review Letters* **23** (16): 930–934. Bibcode:1969PhRvL..23..930B. doi:10.1103/PhysRevLett.23.930.

[7] M. Breidenbach et al. (1969). "Observed Behavior of Highly Inelastic Electron–Proton Scattering". *Physical Review Letters* **23** (16): 935–939. Bibcode:1969PhRvL..23..935B. doi:10.1103/PhysRevLett.23.935.

[8] S.S.M. Wong (1998). *Introductory Nuclear Physics* (2nd ed.). Wiley Interscience. p. 30. ISBN 0-471-23973-9.

[9] K.A. Peacock (2008). *The Quantum Revolution.* Greenwood Publishing Group. p. 125. ISBN 0-313-33448-X.

[10] B. Povh, C. Scholz, K. Rith, F. Zetsche (2008). *Particles and Nuclei.* Springer. p. 98. ISBN 3-540-79367-4.

[11] Section 6.1. in P.C.W. Davies (1979). *The Forces of Nature.* Cambridge University Press. ISBN 0-521-22523-X.

[12] M. Munowitz (2005). *Knowing.* Oxford University Press. p. 35. ISBN 0-19-516737-6.

[13] W.-M. Yao (Particle Data Group) et al. (2006). "Review of Particle Physics: Pentaquark Update" (PDF). *Journal of Physics G* **33** (1): 1–1232. arXiv:astro-ph/0601168. Bibcode:2006JPhG...33....1Y. doi:10.1088/0954-3899/33/1/001.

[14] C. Amsler (Particle Data Group) et al. (2008). "Review of Particle Physics: Pentaquarks" (PDF). *Physics Letters B* **667** (1): 1–1340. Bibcode:2008PhLB..667....1P. doi:10.1016/j.physletb.2008.07.018.
C. Amsler (Particle Data Group) et al. (2008). "Review of Particle Physics: New Charmonium-Like States" (PDF). *Physics Letters B* **667** (1): 1–1340. Bibcode:2008PhLB..667....1P. doi:10.1016/j.physletb.2008.07.018.
E.V. Shuryak (2004). *The QCD Vacuum, Hadrons and Superdense Matter.* World Scientific. p. 59. ISBN 981-238-574-6.

[15] R. Aaij et al. (LHCb collaboration) (2015). "Observation of J/ψp resonances consistent with pentaquark states in Λ0 b→J/ψK−
p decays". *Physical Review Letters* **115** (7). doi:10.1103/PhysRevLett.115.072001.

[16] C. Amsler (Particle Data Group) et al. (2008). "Review of Particle Physics: b′ (4th Generation) Quarks, Searches for" (PDF). *Physics Letters B* **667** (1): 1–1340. Bibcode:2008PhLB..667....1P. doi:10.1016/j.physletb.2008.07.018.
C. Amsler (Particle Data Group) et al. (2008). "Review of Particle Physics: t′ (4th Generation) Quarks, Searches for" (PDF). *Physics Letters B* **667** (1): 1–1340. Bibcode:2008PhLB..667....1P. doi:10.1016/j.physletb.2008.07.018.

[17] D. Decamp; Deschizeaux, B.; Lees, J.-P.; Minard, M.-N.; Crespo, J.M.; Delfino, M.; Fernandez, E.; Martinez, M. et al. (1989). "Determination of the number of light neutrino species". *Physics Letters B* **231** (4): 519. Bibcode:1989PhLB..231..519D. doi:10.1016/0370-2693(89)90704-1.
A. Fisher (1991). "Searching for the Beginning of Time: Cosmic Connection". *Popular Science* **238** (4): 70.
J.D. Barrow (1997) [1994]. "The Singularity and Other Problems". *The Origin of the Universe* (Reprint ed.). Basic Books. ISBN 978-0-465-05314-8.

[18] D.H. Perkins (2003). *Particle Astrophysics.* Oxford University Press. p. 4. ISBN 0-19-850952-9.

[19] M. Gell-Mann (1964). "A Schematic Model of Baryons and Mesons". *Physics Letters* **8** (3): 214–215. Bibcode:1964PhL.....8..214G. doi:10.1016/S0031-9163(64)92001-3.

[20] G. Zweig (1964). "An SU(3) Model for Strong Interaction Symmetry and its Breaking" (PDF). *CERN Report No.8182/TH.401.*

[21] G. Zweig (1964). "An SU(3) Model for Strong Interaction Symmetry and its Breaking: II" (PDF). *CERN Report No.8419/TH.412.*

[22] M. Gell-Mann (2000) [1964]. "The Eightfold Way: A theory of strong interaction symmetry". In M. Gell-Mann, Y. Ne'eman. *The Eightfold Way.* Westview Press. p. 11. ISBN 0-7382-0299-1.
Original: M. Gell-Mann (1961). "The Eightfold Way: A theory of strong interaction symmetry". *Synchrotron Laboratory Report CTSL-20* (California Institute of Technology).

[23] Y. Ne'eman (2000) [1964]. "Derivation of strong interactions from gauge invariance". In M. Gell-Mann, Y. Ne'eman. *The Eightfold Way.* Westview Press. ISBN 0-7382-0299-1.
Original Y. Ne'eman (1961). "Derivation of strong interactions from gauge invariance". *Nuclear Physics* **26** (2): 222. Bibcode:1961NucPh..26..222N. doi:10.1016/0029-5582(61)90134-1.

[24] R.C. Olby, G.N. Cantor (1996). *Companion to the History of Modern Science.* Taylor & Francis. p. 673. ISBN 0-415-14578-3.

[25] A. Pickering (1984). *Constructing Quarks.* University of Chicago Press. pp. 114–125. ISBN 0-226-66799-5.

[26] B.J. Bjorken, S.L. Glashow; Glashow (1964). "Elementary Particles and SU(4)". *Physics Letters* **11** (3): 255–257. Bibcode:1964PhL....11..255B. doi:10.1016/0031-9163(64)90433-0.

[27] J.I. Friedman. "The Road to the Nobel Prize". Hue University. Retrieved 2008-09-29.

[28] R.P. Feynman (1969). "Very High-Energy Collisions of Hadrons". *Physical Review Letters* **23** (24): 1415–1417. Bibcode:1969PhRvL..23.1415F. doi:10.1103/PhysRevLett.23.1415.

[29] S. Kretzer et al. (2004). "CTEQ6 Parton Distributions with Heavy Quark Mass Effects". *Physical Review D* **69** (11): 114005. arXiv:hep-ph/0307022. Bibcode:2004PhRvD..69k4005K. doi:10.1103/PhysRevD.69.114005.

[30] D.J. Griffiths (1987). *Introduction to Elementary Particles.* John Wiley & Sons. p. 42. ISBN 0-471-60386-4.

[31] M.E. Peskin, D.V. Schroeder (1995). *An introduction to quantum field theory.* Addison–Wesley. p. 556. ISBN 0-201-50397-2.

[32] V.V. Ezhela (1996). *Particle physics.* Springer. p. 2. ISBN 1-56396-642-5.

[33] S.L. Glashow, J. Iliopoulos, L. Maiani; Iliopoulos; Maiani (1970). "Weak Interactions with Lepton–Hadron Symmetry". *Physical Review D* **2** (7): 1285–1292. Bibcode:1970PhRvD...2.1285G. doi:10.1103/PhysRevD.2.1285.

[34] D.J. Griffiths (1987). *Introduction to Elementary Particles.* John Wiley & Sons. p. 44. ISBN 0-471-60386-4.

[35] M. Kobayashi, T. Maskawa; Maskawa (1973). "CP-Violation in the Renormalizable Theory of Weak Interaction". *Progress of Theoretical Physics* **49** (2): 652–657. Bibcode:1973PThPh..49..652K. doi:10.1143/PTP.49.652.

[36] H. Harari (1975). "A new quark model for hadrons". *Physics Letters B* **57B** (3): 265. Bibcode:1975PhLB...57..265H. doi:10.1016/0370-2693(75)90072-6.

[37] K.W. Staley (2004). *The Evidence for the Top Quark.* Cambridge University Press. pp. 31–33. ISBN 978-0-521-82710-2.

[38] S.W. Herb et al. (1977). "Observation of a Dimuon Resonance at 9.5 GeV in 400-GeV Proton-Nucleus Collisions". *Physical Review Letters* **39** (5): 252. Bibcode:1977PhRvL..39..252H. doi:10.1103/PhysRevLett.39.252.

[39] M. Bartusiak (1994). *A Positron named Priscilla.* National Academies Press. p. 245. ISBN 0-309-04893-1.

[40] F. Abe (CDF Collaboration) et al. (1995). "Observation of Top Quark Production in pp Collisions with the Collider Detector at Fermilab". *Physical Review Letters* **74** (14): 2626–2631. Bibcode:1995PhRvL..74.2626A. doi:10.1103/PhysRevLett.74.2626. PMID 10057978.

[41] S. Abachi (DØ Collaboration) et al. (1995). "Search for High Mass Top Quark Production in pp Collisions at $\sqrt{s}$ = 1.8 TeV". *Physical Review Letters* **74** (13): 2422–2426. Bibcode:1995PhRvL..74.2422A. doi:10.1103/PhysRevLett.74.2422.

[42] K.W. Staley (2004). *The Evidence for the Top Quark.* Cambridge University Press. p. 144. ISBN 0-521-82710-8.

[43] "New Precision Measurement of Top Quark Mass". Brookhaven National Laboratory News. 2004. Retrieved 2013-11-03.

[44] J. Joyce (1982) [1939]. *Finnegans Wake.* Penguin Books. p. 383. ISBN 0-14-006286-6.

[45] M. Gell-Mann (1995). *The Quark and the Jaguar: Adventures in the Simple and the Complex.* Henry Holt and Co. p. 180. ISBN 978-0-8050-7253-2.

[46] J. Gleick (1992). *Genius: Richard Feynman and modern physics.* Little Brown and Company. p. 390. ISBN 0-316-90316-7.

[47] J.J. Sakurai (1994). S.F Tuan, ed. *Modern Quantum Mechanics* (Revised ed.). Addison–Wesley. p. 376. ISBN 0-201-53929-2.

[48] D.H. Perkins (2000). *Introduction to high energy physics.* Cambridge University Press. p. 8. ISBN 0-521-62196-8.

[49] M. Riordan (1987). *The Hunting of the Quark: A True Story of Modern Physics.* Simon & Schuster. p. 210. ISBN 978-0-671-50466-3.

[50] F. Close (2006). *The New Cosmic Onion.* CRC Press. p. 133. ISBN 1-58488-798-2.

[51] J.T. Volk et al. (1987). "Letter of Intent for a Tevatron Beauty Factory" (PDF). Fermilab Proposal #783.

[52] G. Fraser (2006). *The New Physics for the Twenty-First Century.* Cambridge University Press. p. 91. ISBN 0-521-81600-9.

[53] "The Standard Model of Particle Physics". BBC. 2002. Retrieved 2009-04-19.

[54] F. Close (2006). *The New Cosmic Onion.* CRC Press. pp. 80–90. ISBN 1-58488-798-2.

[55] D. Lincoln (2004). *Understanding the Universe.* World Scientific. p. 116. ISBN 981-238-705-6.

[56] "Weak Interactions". *Virtual Visitor Center.* Stanford Linear Accelerator Center. 2008. Retrieved 2008-09-28.

[57] K. Nakamura et al. (2010). "Review of Particles Physics: The CKM Quark-Mixing Matrix" (PDF). *J. Phys.* G **37** (75021): 150.

[58] Z. Maki, M. Nakagawa, S. Sakata (1962). "Remarks on the Unified Model of Elementary Particles". *Progress of Theoretical Physics* **28** (5): 870. Bibcode:1962PThPh..28..870M. doi:10.1143/PTP.28.870.

[59] B.C. Chauhan, M. Picariello, J. Pulido, E. Torrente-Lujan (2007). "Quark–lepton complementarity, neutrino and standard model data predict θPMNS 13 = 9°+1° −2°". *European Physical Journal* **C50** (3): 573–578. arXiv:hep-ph/0605032. Bibcode:2007EPJC...50..573C. doi:10.1140/epjc/s10052-007-0212-z.

[60] R. Nave. "The Color Force". *HyperPhysics.* Georgia State University, Department of Physics and Astronomy. Retrieved 2009-04-26.

[61] B.A. Schumm (2004). *Deep Down Things*. Johns Hopkins University Press. pp. 131–132. ISBN 0-8018-7971-X. OCLC 55229065.

[62] Part III of M.E. Peskin, D.V. Schroeder (1995). *An Introduction to Quantum Field Theory*. Addison–Wesley. ISBN 0-201-50397-2.

[63] V. Icke (1995). *The force of symmetry*. Cambridge University Press. p. 216. ISBN 0-521-45591-X.

[64] M.Y. Han (2004). *A story of light*. World Scientific. p. 78. ISBN 981-256-034-3.

[65] C. Sutton. "Quantum chromodynamics (physics)". *Encyclopædia Britannica Online*. Retrieved 2009-05-12.

[66] A. Watson (2004). *The Quantum Quark*. Cambridge University Press. pp. 285–286. ISBN 0-521-82907-0.

[67] K.A. Olive *et al.* (Particle Data Group), Chin. Phys. **C38**, 090001 (2014) (URL: http://pdg.lbl.gov)

[68] W. Weise, A.M. Green (1984). *Quarks and Nuclei*. World Scientific. pp. 65–66. ISBN 9971-966-61-1.

[69] D. McMahon (2008). *Quantum Field Theory Demystified*. McGraw–Hill. p. 17. ISBN 0-07-154382-1.

[70] S.G. Roth (2007). *Precision electroweak physics at electron–positron colliders*. Springer. p. VI. ISBN 3-540-35164-7.

[71] R.P. Feynman (1985). *QED: The Strange Theory of Light and Matter* (1st ed.). Princeton University Press. pp. 136–137. ISBN 0-691-08388-6.

[72] M. Veltman (2003). *Facts and Mysteries in Elementary Particle Physics*. World Scientific. pp. 45–47. ISBN 981-238-149-X.

[73] F. Wilczek, B. Devine (2006). *Fantastic Realities*. World Scientific. p. 85. ISBN 981-256-649-X.

[74] F. Wilczek, B. Devine (2006). *Fantastic Realities*. World Scientific. pp. 400ff. ISBN 981-256-649-X.

[75] T. Yulsman (2002). *Origin*. CRC Press. p. 55. ISBN 0-7503-0765-X.

[76] F. Garberson (2008). "Top Quark Mass and Cross Section Results from the Tevatron". arXiv:0808.0273 [hep-ex].

[77] J. Steinberger (2005). *Learning about Particles*. Springer. p. 130. ISBN 3-540-21329-5.

[78] C.-Y. Wong (1994). *Introduction to High-energy Heavy-ion Collisions*. World Scientific. p. 149. ISBN 981-02-0263-6.

[79] S.B. Rüester, V. Werth, M. Buballa, I.A. Shovkovy, D.H. Rischke; Werth; Buballa; Shovkovy; Rischke (2005). "The phase diagram of neutral quark matter: Self-consistent treatment of quark masses". *Physical Review D* **72** (3): 034003. arXiv:hep-ph/0503184. Bibcode:2005PhRvD..72c4004R. doi:10.1103/PhysRevD.72.034004.

[80] M.G. Alford, K. Rajagopal, T. Schaefer, A. Schmitt; Schmitt; Rajagopal; Schäfer (2008). "Color superconductivity in dense quark matter". *Reviews of Modern Physics* **80** (4): 1455–1515. arXiv:0709.4635. Bibcode:2008RvMP...80.1455A. doi:10.1103/RevModPhys.80.1455.

[81] S. Mrowczynski (1998). "Quark–Gluon Plasma". *Acta Physica Polonica B* **29**: 3711. arXiv:nucl-th/9905005. Bibcode:1998AcPPB..29.3711M.

[82] Z. Fodor, S.D. Katz; Katz (2004). "Critical point of QCD at finite T and μ, lattice results for physical quark masses". *Journal of High Energy Physics* **2004** (4): 50. arXiv:hep-lat/0402006. Bibcode:2004JHEP...04..050F. doi:10.1088/1126-6708/2004/04/050.

[83] U. Heinz, M. Jacob (2000). "Evidence for a New State of Matter: An Assessment of the Results from the CERN Lead Beam Programme". arXiv:nucl-th/0002042.

[84] "RHIC Scientists Serve Up "Perfect" Liquid". Brookhaven National Laboratory News. 2005. Retrieved 2009-05-22.

[85] T. Yulsman (2002). *Origins: The Quest for Our Cosmic Roots*. CRC Press. p. 75. ISBN 0-7503-0765-X.

[86] A. Sedrakian, J.W. Clark, M.G. Alford (2007). *Pairing in fermionic systems*. World Scientific. pp. 2–3. ISBN 981-256-907-3.

2.1.9   Further reading

- A. Ali, G. Kramer; Kramer (2011). "JETS and QCD: A historical review of the discovery of the quark and gluon jets and its impact on QCD". *European Physical Journal H* **36** (2): 245. arXiv:1012.2288. Bibcode:2011EPJH...36..245A. doi:10.1140/epjh/e2011-10047-1.

- D.J. Griffiths (2008). *Introduction to Elementary Particles* (2nd ed.). Wiley–VCH. ISBN 3-527-40601-8.

- I.S. Hughes (1985). *Elementary particles* (2nd ed.). Cambridge University Press. ISBN 0-521-26092-2.

- R. Oerter (2005). *The Theory of Almost Everything: The Standard Model, the Unsung Triumph of Modern Physics*. Pi Press. ISBN 0-13-236678-9.

- A. Pickering (1984). *Constructing Quarks: A Sociological History of Particle Physics*. The University of Chicago Press. ISBN 0-226-66799-5.

- B. Povh (1995). *Particles and Nuclei: An Introduction to the Physical Concepts*. Springer–Verlag. ISBN 0-387-59439-6.

- M. Riordan (1987). *The Hunting of the Quark: A true story of modern physics.* Simon & Schuster. ISBN 0-671-64884-5.

- B.A. Schumm (2004). *Deep Down Things: The Breathtaking Beauty of Particle Physics.* Johns Hopkins University Press. ISBN 0-8018-7971-X.

2.1.10 External links

- 1969 Physics Nobel Prize lecture by Murray Gell-Mann

- 1976 Physics Nobel Prize lecture by Burton Richter

- 1976 Physics Nobel Prize lecture by Samuel C.C. Ting

- 2008 Physics Nobel Prize lecture by Makoto Kobayashi

- 2008 Physics Nobel Prize lecture by Toshihide Maskawa

- The Top Quark And The Higgs Particle by T.A. Heppenheimer – A description of CERN's experiment to count the families of quarks.

- Bowley, Roger; Copeland, Ed. "Quarks". *Sixty Symbols*. Brady Haran for the University of Nottingham.

2.2 Up quark

The **up quark** or **u quark** (symbol: u) is the lightest of all quarks, a type of elementary particle, and a major constituent of matter. It, along with the down quark, forms the neutrons (one up quark, two down quarks) and protons (two up quarks, one down quark) of atomic nuclei. It is part of the first generation of matter, has an electric charge of $+\frac{2}{3}\,e$ and a bare mass of 1.8–3.0 MeV/c^2. Like all quarks, the up quark is an elementary fermion with spin-$\frac{1}{2}$, and experiences all four fundamental interactions: gravitation, electromagnetism, weak interactions, and strong interactions. The antiparticle of the up quark is the **up antiquark** (sometimes called *antiup quark* or simply *antiup*), which differs from it only in that some of its properties have equal magnitude but opposite sign.

Its existence (along with that of the down and strange quarks) was postulated in 1964 by Murray Gell-Mann and George Zweig to explain the *Eightfold Way* classification scheme of hadrons. The up quark was first observed by experiments at the Stanford Linear Accelerator Center in 1968.

2.2.1 History

In the beginnings of particle physics (first half of the 20th century), hadrons such as protons, neutrons and pions were thought to be elementary particles. However, as new hadrons were discovered, the 'particle zoo' grew from a few particles in the early 1930s and 1940s to several dozens of them in the 1950s. The relationships between each of them were unclear until 1961, when Murray Gell-Mann[2] and Yuval Ne'eman[3] (independently of each other) proposed a hadron classification scheme called the *Eightfold Way*, or in more technical terms, SU(3) flavor symmetry.

This classification scheme organized the hadrons into isospin multiplets, but the physical basis behind it was still unclear. In 1964, Gell-Mann[4] and George Zweig[5][6] (independently of each other) proposed the quark model, then consisting only of up, down, and strange quarks.[7] However, while the quark model explained the Eightfold Way, no direct evidence of the existence of quarks was found until 1968 at the Stanford Linear Accelerator Center.[8][9] Deep inelastic scattering experiments indicated that protons had substructure, and that protons made of three more-fundamental particles explained the data (thus confirming the quark model).[10]

At first people were reluctant to describe the three bodies as quarks, instead preferring Richard Feynman's parton description,[11][12][13] but over time the quark theory became accepted (see *November Revolution*).[14]

2.2.2 Mass

Despite being extremely common, the bare mass of the up quark is not well determined, but probably lies between 1.8 and 3.0 MeV/c^2.[1] Lattice QCD calculations give a more precise value: 2.01±0.14 MeV/c^2.[15]

When found in mesons (particles made of one quark and one antiquark) or baryons (particles made of three quarks), the 'effective mass' (or 'dressed' mass) of quarks becomes greater because of the binding energy caused by the gluon field between each quark (see mass–energy equivalence). The bare mass of up quarks is so light, it cannot be straightforwardly calculated because relativistic effects have to be taken into account.

2.2.3 See also

- Down quark

- Isospin

- Quark model

- Quantum Mechanics

2.2.4 References

[1] J. Beringer *et al.* (Particle Data Group) (2012). "PDGLive Particle Summary 'Quarks (u, d, s, c, b, t, b', t', Free)'" (PDF). Particle Data Group. Retrieved 2013-02-21.

[2] M. Gell-Mann (2000) [1964]. "The Eightfold Way: A theory of strong interaction symmetry". In M. Gell-Mann, Y. Ne'eman. *The Eightfold Way*. Westview Press. p. 11. ISBN 0-7382-0299-1.
Original: M. Gell-Mann (1961). "The Eightfold Way: A theory of strong interaction symmetry". *Synchrotron Laboratory Report CTSL-20* (California Institute of Technology)

[3] Y. Ne'eman (2000) [1964]. "Derivation of strong interactions from gauge invariance". In M. Gell-Mann, Y. Ne'eman. *The Eightfold Way*. Westview Press. ISBN 0-7382-0299-1.
Original Y. Ne'eman (1961). "Derivation of strong interactions from gauge invariance". *Nuclear Physics* **26** (2): 222. Bibcode:1961NucPh..26..222N. doi:10.1016/0029-5582(61)90134-1.

[4] M. Gell-Mann (1964). "A Schematic Model of Baryons and Mesons". *Physics Letters* **8** (3): 214–215. Bibcode:1964PhL.....8..214G. doi:10.1016/S0031-9163(64)92001-3.

[5] G. Zweig (1964). "An SU(3) Model for Strong Interaction Symmetry and its Breaking". *CERN Report No.8181/Th 8419*.

[6] G. Zweig (1964). "An SU(3) Model for Strong Interaction Symmetry and its Breaking: II". *CERN Report No.8419/Th 8412*.

[7] B. Carithers, P. Grannis (1995). "Discovery of the Top Quark" (PDF). *Beam Line* (SLAC) **25** (3): 4–16. Retrieved 2008-09-23.

[8] E. D. Bloom; Coward, D.; Destaebler, H.; Drees, J.; Miller, G.; Mo, L.; Taylor, R.; Breidenbach, M. et al. (1969). "High-Energy Inclastic *e–p* Scattering at 6° and 10°". *Physical Review Letters* **23** (16): 930–934. Bibcode:1969PhRvL..23..930B. doi:10.1103/PhysRevLett.23.930.

[9] M. Breidenbach; Friedman, J.; Kendall, H.; Bloom, E.; Coward, D.; Destaebler, H.; Drees, J.; Mo, L.; Taylor, R. et al. (1969). "Observed Behavior of Highly Inelastic Electron–Proton Scattering". *Physical Review Letters* **23** (16): 935–939. Bibcode:1969PhRvL..23..935B. doi:10.1103/PhysRevLett.23.935.

[10] J. I. Friedman. "The Road to the Nobel Prize". Hue University. Retrieved 2008-09-29.

[11] R. P. Feynman (1969). "Very High-Energy Collisions of Hadrons". *Physical Review Letters* **23** (24): 1415–1417. Bibcode:1969PhRvL..23.1415F. doi:10.1103/PhysRevLett.23.1415.

[12] S. Kretzer; Lai, H.; Olness, Fredrick; Tung, W. et al. (2004). "CTEQ6 Parton Distributions with Heavy Quark Mass Effects". *Physical Review D* **69** (11): 114005. arXiv:hep-ph/0307022. Bibcode:2004PhRvD..69k4005K. doi:10.1103/PhysRevD.69.114005.

[13] D. J. Griffiths (1987). *Introduction to Elementary Particles*. John Wiley & Sons. p. 42. ISBN 0-471-60386-4.

[14] M. E. Peskin, D. V. Schroeder (1995). *An introduction to quantum field theory*. Addison–Wesley. p. 556. ISBN 0-201-50397-2.

[15] Cho, Adrian (April 2010). "Mass of the Common Quark Finally Nailed Down". Science Magazine.

2.2.5 Further reading

- A. Ali, G. Kramer; Kramer (2011). "JETS and QCD: A historical review of the discovery of the quark and gluon jets and its impact on QCD". *European Physical Journal H* **36** (2): 245. arXiv:1012.2288. Bibcode:2011EPJH...36..245A. doi:10.1140/epjh/e2011-10047-1.

- R. Nave. "Quarks". *HyperPhysics*. Georgia State University, Department of Physics and Astronomy. Retrieved 2008-06-29.

- A. Pickering (1984). *Constructing Quarks*. University of Chicago Press. pp. 114–125. ISBN 0-226-66799-5.

2.3 Down quark

The **down quark** or **d quark** (symbol: d) is the second-lightest of all quarks, a type of elementary particle, and a major constituent of matter. Together with the up quark, it forms the neutrons (one up quark, two down quarks) and protons (two up quarks, one down quark) of atomic nuclei. It is part of the first generation of matter, has an electric charge of $-\frac{1}{3} e$ and a bare mass of 4.8+0.5 −0.3 MeV/c^2.[1] Like all quarks, the down quark is an elementary fermion with spin-$\frac{1}{2}$, and experiences all four fundamental interactions: gravitation, electromagnetism, weak interactions, and strong interactions. The antiparticle of the down quark is the **down antiquark** (sometimes called *antidown quark* or simply *antidown*), which differs from it only in that some of its properties have equal magnitude but opposite sign.

Its existence (along with that of the up and strange quarks) was postulated in 1964 by Murray Gell-Mann and George Zweig to explain the *Eightfold Way* classification scheme of hadrons. The down quark was first observed by experiments at the Stanford Linear Accelerator Center in 1968.

2.3.1   History

In the beginnings of particle physics (first half of the 20th century), hadrons such as protons, neutrons, and pions were thought to be elementary particles. However, as new hadrons were discovered, the 'particle zoo' grew from a few particles in the early 1930s and 1940s to several dozens of them in the 1950s. The relationships between each of them was unclear until 1961, when Murray Gell-Mann[2] and Yuval Ne'eman[3] (independently of each other) proposed a hadron classification scheme called the *Eightfold Way*, or in more technical terms, SU(3) flavor symmetry.

This classification scheme organized the hadrons into isospin multiplets, but the physical basis behind it was still unclear. In 1964, Gell-Mann[4] and George Zweig[5][6] (independently of each other) proposed the quark model, then consisting only of up, down, and strange quarks.[7] However, while the quark model explained the Eightfold Way, no direct evidence of the existence of quarks was found until 1968 at the Stanford Linear Accelerator Center.[8][9] Deep inelastic scattering experiments indicated that protons had substructure, and that protons made of three more-fundamental particles explained the data (thus confirming the quark model).[10]

At first people were reluctant to identify the three-bodies as quarks, instead preferring Richard Feynman's parton description,[11][12][13] but over time the quark theory became accepted (see *November Revolution*).[14]

2.3.2   Mass

Despite being extremely common, the bare mass of the down quark is not well determined, but probably lies between 4.5 and $5.3 \cdot 10^0$ MeV/c^2.[1] Lattice QCD calculations give a more precise value: 4.79 ± 0.16 MeV/c^2.[15]

When found in mesons (particles made of one quark and one antiquark) or baryons (particles made of three quarks), the 'effective mass' (or 'dressed' mass) of quarks becomes greater because of the binding energy caused by the gluon field between quarks (see mass–energy equivalence). For example, the effective mass of down quarks in a proton is around 330 MeV/c^2. Because the bare mass of down quarks is so small, it cannot be straightforwardly calculated because relativistic effects have to be taken into account.

2.3.3   See also

- Up quark
- Isospin
- Quark model

2.3.4   References

[1] J. Beringer (Particle Data Group) et al. (2013). "PDGLive Particle Summary 'Quarks (u, d, s, c, b, t, b′, t′, Free)'" (PDF). Particle Data Group. Retrieved 2013-07-23.

[2] M. Gell-Mann (2000) [1964]. "The Eightfold Way: A theory of strong interaction symmetry". In M. Gell-Mann, Y. Ne'eman. *The Eightfold Way*. Westview Press. p. 11. ISBN 0-7382-0299-1.
Original: M. Gell-Mann (1961). "The Eightfold Way: A theory of strong interaction symmetry". *Synchrotron Laboratory Report CTSL-20* (California Institute of Technology).

[3] Y. Ne'eman (2000) [1964]. "Derivation of strong interactions from gauge invariance". In M. Gell-Mann, Y. Ne'eman. *The Eightfold Way*. Westview Press. ISBN 0-7382-0299-1.
Original Y. Ne'eman (1961). "Derivation of strong interactions from gauge invariance". *Nuclear Physics* **26** (2): 222. Bibcode:1961NucPh..26..222N. doi:10.1016/0029-5582(61)90134-1.

[4] M. Gell-Mann (1964). "A Schematic Model of Baryons and Mesons". *Physics Letters* **8** (3): 214–215. Bibcode:1964PhL.....8..214G. doi:10.1016/S0031-9163(64)92001-3.

[5] G. Zweig (1964). "An SU(3) Model for Strong Interaction Symmetry and its Breaking". *CERN Report No.8181/Th 8419*.

[6] G. Zweig (1964). "An SU(3) Model for Strong Interaction Symmetry and its Breaking: II". *CERN Report No.8419/Th 8412*.

[7] B. Carithers, P. Grannis (1995). "Discovery of the Top Quark" (PDF). *Beam Line* (SLAC) **25** (3): 4–16. Retrieved 2008-09-23.

[8] E. D. Bloom; Coward, D.; Destaebler, H.; Drees, J.; Miller, G.; Mo, L.; Taylor, R.; Breidenbach, M. et al. (1969). "High-Energy Inelastic *e–p* Scattering at 6° and 10°". *Physical Review Letters* **23** (16): 930–934. Bibcode:1969PhRvL..23..930B. doi:10.1103/PhysRevLett.23.930.

[9] M. Breidenbach; Friedman, J.; Kendall, H.; Bloom, E.; Coward, D.; Destaebler, H.; Drees, J.; Mo, L.; Taylor, R. et al. (1969). "Observed Behavior of Highly Inelastic Electron–Proton Scattering". *Physical Review Letters* **23** (16): 935–939. Bibcode:1969PhRvL..23..935B. doi:10.1103/PhysRevLett.23.935.

[10] J. I. Friedman. "The Road to the Nobel Prize". Hue University. Retrieved 2008-09-29.

[11] R. P. Feynman (1969). "Very High-Energy Collisions of Hadrons". *Physical Review Letters* **23** (24): 1415–1417. Bibcode:1969PhRvL..23.1415F. doi:10.1103/PhysRevLett.23.1415.

[12] S. Kretzer; Lai, H.; Olness, Fredrick; Tung, W. et al. (2004). "CTEQ6 Parton Distributions with Heavy Quark Mass Effects". *Physical Review D* **69** (11): 114005. arXiv:hep-ph/0307022. Bibcode:2004PhRvD..69k4005K. doi:10.1103/PhysRevD.69.114005.

[13] D. J. Griffiths (1987). *Introduction to Elementary Particles.* John Wiley & Sons. p. 42. ISBN 0-471-60386-4.

[14] M. E. Peskin, D. V. Schroeder (1995). *An introduction to quantum field theory.* Addison–Wesley. p. 556. ISBN 0-201-50397-2.

[15] Cho, Adrian (April 2010). "Mass of the Common Quark Finally Nailed Down". Science Magazine.

2.3.5 Further reading

- A. Ali, G. Kramer; Kramer (2011). "JETS and QCD: A historical review of the discovery of the quark and gluon jets and its impact on QCD". *European Physical Journal H* **36** (2): 245. arXiv:1012.2288. Bibcode:2011EPJH...36..245A. doi:10.1140/epjh/e2011-10047-1.

- R. Nave. "Quarks". *HyperPhysics*. Georgia State University, Department of Physics and Astronomy. Retrieved 2008-06-29.

- A. Pickering (1984). *Constructing Quarks.* University of Chicago Press. pp. 114–125. ISBN 0-226-66799-5.

2.4 Charm quark

The **charm quark** or **c quark** (from its symbol, c) is the third most massive of all quarks, a type of elementary particle. Charm quarks are found in hadrons, which are subatomic particles made of quarks. Example of hadrons containing charm quarks include the J/ψ meson (J/ψ), D mesons (D), charmed Sigma baryons (Σ c), and other charmed particles.

It, along with the strange quark is part of the second generation of matter, and has an electric charge of $+\frac{2}{3}$ e and a bare mass of $1.29 + 0.05$ -0.11 GeV/c^2.[1] Like all quarks, the charm quark is an elementary fermion with spin-$\frac{1}{2}$, and experiences all four fundamental interactions: gravitation, electromagnetism, weak interactions, and strong interactions. The antiparticle of the charm quark is the **charm antiquark** (sometimes called *anticharm quark* or simply *anticharm*), which differs from it only in that some of its properties have equal magnitude but opposite sign.

The existence of a fourth quark had been speculated by a number of authors around 1964 (for instance by James Bjorken and Sheldon Glashow[4]), but its prediction is usually credited to Sheldon Glashow, John Iliopoulos and Luciano Maiani in 1970 (see GIM mechanism).[5] The first charmed particle (a particle containing a charm quark) to be discovered was the J/ψ meson. It was discovered by a team at the Stanford Linear Accelerator Center (SLAC), led by Burton Richter,[6] and one at the Brookhaven National Laboratory (BNL), led by Samuel Ting.[7]

The 1974 discovery of the J/ψ (and thus the charm quark) ushered in a series of breakthroughs which are collectively known as the *November Revolution*.

2.4.1 Hadrons containing charm quarks

Main articles: List of baryons and list of mesons

Some of the hadrons containing charm quarks include:

- D mesons contain a charm quark (or its antiparticle) and an up or down quark.

- D
s mesons contain a charm quark and a strange quark.

- There are many charmonium states, for example the J/ψ particle. These consist of a charm quark and its antiparticle.

- Charmed baryons have been observed, and are named in analogy with strange baryons (e.g. Λ+ c).

2.4.2 See also

- Quark model

2.4.3 Notes

[1] K. Nakamura *et al.* (Particle Data Group) (2011). "PDGLive Particle Summary 'Quarks (u, d, s, c, b, t, b′, t′, Free)'" (PDF). Particle Data Group. Retrieved 2011-08-08.

[2] Carl Rod Nave. "Transformation of Quark Flavors by the Weak Interaction". Retrieved 2010-12-06. The c quark has about 5% probability of decaying into a d quark instead of an s quark.

[3] K. Nakamura et al. (2010). "Review of Particles Physics: The CKM Quark-Mixing Matrix" (PDF). *J. Phys.* G **37** (75021): 150.

[4] B.J. Bjorken, S.L. Glashow; Glashow (1964). "Elementary particles and SU(4)". *Physics Letters* **11** (3): 255–257. Bibcode:1964PhL....11..255B. doi:10.1016/0031-9163(64)90433-0.

[5] S.L. Glashow, J. Iliopoulos, L. Maiani; Iliopoulos; Maiani (1970). "Weak Interactions with Lepton–Hadron Symmetry". *Physical Review D* **2** (7): 1285–1292. Bibcode:1970PhRvD...2.1285G. doi:10.1103/PhysRevD.2.1285.

[6] J.-E. Augustin; Boyarski, A.; Breidenbach, M.; Bulos, F.; Dakin, J.; Feldman, G.; Fischer, G.; Fryberger, D.; Hanson, G.; Jean-Marie, B.; Larsen, R.; Lüth, V.; Lynch, H.; Lyon, D.; Morehouse, C.; Paterson, J.; Perl, M.; Richter, B.; Rapidis, P.; Schwitters, R.; Tanenbaum, W.; Vannucci, F.; Abrams, G.; Briggs, D.; Chinowsky, W.; Friedberg, C.; Goldhaber, G.; Hollebeek, R.; Kadyk, J.; Lulu, B. (1974). "Discovery of a Narrow Resonance in e^+e^- Annihilation". *Physical Review Letters* **33** (23): 1406. Bibcode:1974PhRvL..33.1406A. doi:10.1103/PhysRevLett.33.1406.

[7] J.J. Aubert et al. (1974). "Experimental Observation of a Heavy Particle *J*". *Physical Review Letters* **33** (23): 1404. Bibcode:1974PhRvL..33.1404A. doi:10.1103/PhysRevLett.33.1404.

2.4.4 Further reading

- R. Nave. "Quarks". *HyperPhysics*. Georgia State University, Department of Physics and Astronomy. Retrieved 2008-06-29.

- A. Pickering (1984). *Constructing Quarks*. University of Chicago Press. pp. 114–125. ISBN 0-226-66799-5.

2.5 Strange quark

The **strange quark** or **s quark** (from its symbol, *s*) is the third-lightest of all quarks, a type of elementary particle. Strange quarks are found in subatomic particles called hadrons. Example of hadrons containing strange quarks include kaons (K), strange D mesons (D
s), Sigma baryons (Σ), and other strange particles.

Along with the charm quark, it is part of the second generation of matter, and has an electric charge of $-\frac{1}{3}\,e$ and a bare mass of 95+5
−5 MeV/c^2.[1] Like all quarks, the strange quark is an elementary fermion with spin-$\frac{1}{2}$, and experiences all four fundamental interactions: gravitation, electromagnetism, weak interactions, and strong interactions. The antiparticle of the strange quark is the **strange antiquark** (sometimes called *antistrange quark* or simply *antistrange*), which differs from it only in that some of its properties have equal magnitude but opposite sign.

The first strange particle (a particle containing a strange quark) was discovered in 1947 (kaons), but the existence of the strange quark itself (and that of the up and down quarks) was only postulated in 1964 by Murray Gell-Mann and George Zweig to explain the *Eightfold Way* classification scheme of hadrons. The first evidence for the existence of quarks came in 1968, in deep inelastic scattering experiments at the Stanford Linear Accelerator Center. These experiments confirmed the existence of up and down quarks, and by extension, strange quarks, as they were required to explain the Eightfold Way.

2.5.1 History

In the beginnings of particle physics (first half of the 20th century), hadrons such as protons, neutron and pions were thought to be elementary particles. However, new hadrons were discovered, the 'particle zoo' grew from a few particles in the early 1930s and 1940s to several dozens of them in the 1950s. However some particles were much longer lived than others; most particles decayed through the strong interaction and had lifetimes of around 10^{-23} seconds. But when they decayed through the weak interactions, they had lifetimes of around 10^{-10} seconds to decay. While studying these decays Murray Gell-Mann (in 1953)[2][3] and Kazuhiko Nishijima (in 1955)[4] developed the concept of *strangeness* (which Nishijima called *eta-charge*, after the eta meson (η)) which explained the 'strangeness' of the longer-lived particles. The Gell-Mann–Nishijima formula is the result of these efforts to understand strange decays.

However, the relationships between each particles and the physical basis behind the strangeness property was still unclear. In 1961, Gell-Mann[5] and Yuval Ne'eman[6] (independently of each other) proposed a hadron classification scheme called the *Eightfold Way*, or in more technical terms, SU(3) flavor symmetry. This ordered hadrons into isospin multiplets. The physical basis behind both isospin and strangeness was only explained in 1964, when Gell-Mann[7] and George Zweig[8][9] (independently of each other) proposed the quark model, then consisting only of up, down, and strange quarks.[10] Up and down quarks were the carriers of isospin, while the strange quark carried strangeness. While the quark model explained the Eightfold Way, no direct evidence of the existence of quarks was found until 1968 at the Stanford Linear Accelerator Center.[11][12] Deep inelastic scattering experiments indicated that protons had substructure, and that protons made of three more-fundamental particles explained the data (thus confirming the quark model).[13]

At first people were reluctant to identify the three-bodies as quarks, instead preferring Richard Feynman's parton description,[14][15][16] but over time the quark theory became accepted (see *November Revolution*).[17]

2.5.2 See also

- Quark model

- Strange matter

- Strangeness production

- Strangelet

- Strange star

2.5.3 References

[1] J. Beringer *et al.* (Particle Data Group) (2012). "PDGLive Particle Summary 'Quarks (u, d, s, c, b, t, b′, t′, Free)'" (PDF). Particle Data Group. Retrieved 2012-11-30.

[2] M. Gell-Mann (1953). "Isotopic Spin and New Unstable Particles". *Physical Review* **92** (3): 833. Bibcode:1953PhRv...92..833G. doi:10.1103/PhysRev.92.833.

[3] G. Johnson (2000). *Strange Beauty: Murray Gell-Mann and the Revolution in Twentieth-Century Physics*. Random House. p. 119. ISBN 0-679-43764-9. By the end of the summer... [Gell-Mann] completed his first paper, "Isotopic Spin and Curious Particles" and send it of to *Physical Review*. The editors hated the title, so he amended it to "Strange Particles". They wouldn't go for that either—never mind that almost everybody used the term—suggesting instead "Isotopic Spin and New Unstable Particles".

[4] K. Nishijima, Kazuhiko (1955). "Charge Independence Theory of V Particles". *Progress of Theoretical Physics* **13** (3): 285. Bibcode:1955PThPh..13..285N. doi:10.1143/PTP.13.285.

[5] M. Gell-Mann (2000) [1964]. "The Eightfold Way: A theory of strong interaction symmetry". In M. Gell-Mann, Y. Ne'eman. *The Eightfold Way*. Westview Press. p. 11. ISBN 0-7382-0299-1.
Original: M. Gell-Mann (1961). "The Eightfold Way: A theory of strong interaction symmetry". *Synchrotron Laboratory Report CTSL-20* (California Institute of Technology)

[6] Y. Ne'eman (2000) [1964]. "Derivation of strong interactions from gauge invariance". In M. Gell-Mann, Y. Ne'eman. *The Eightfold Way*. Westview Press. ISBN 0-7382-0299-1.
Original Y. Ne'eman (1961). "Derivation of strong interactions from gauge invariance". *Nuclear Physics* **26** (2): 222. Bibcode:1961NucPh..26..222N. doi:10.1016/0029-5582(61)90134-1.

[7] M. Gell-Mann (1964). "A Schematic Model of Baryons and Mesons". *Physics Letters* **8** (3): 214–215. Bibcode:1964PhL.....8..214G. doi:10.1016/S0031-9163(64)92001-3.

[8] G. Zweig (1964). "An SU(3) Model for Strong Interaction Symmetry and its Breaking". *CERN Report No.8181/Th 8419*.

[9] G. Zweig (1964). "An SU(3) Model for Strong Interaction Symmetry and its Breaking: II". *CERN Report No.8419/Th 8412*.

[10] B. Carithers, P. Grannis (1995). "Discovery of the Top Quark" (PDF). *Beam Line* (SLAC) **25** (3): 4–16. Retrieved 2008-09-23.

[11] E. D. Bloom; Coward, D.; Destaebler, H.; Drees, J.; Miller, G.; Mo, L.; Taylor, R.; Breidenbach, M. et al. (1969). "High-Energy Inelastic *e–p* Scattering at 6° and 10°". *Physical Review Letters* **23** (16): 930–934. Bibcode:1969PhRvL..23..930B. doi:10.1103/PhysRevLett.23.930.

[12] M. Breidenbach; Friedman, J.; Kendall, H.; Bloom, E.; Coward, D.; Destaebler, H.; Drees, J.; Mo, L.; Taylor, R. et al. (1969). "Observed Behavior of Highly Inelastic Electron–Proton Scattering". *Physical Review Letters* **23** (16): 935–939. Bibcode:1969PhRvL..23..935B. doi:10.1103/PhysRevLett.23.935.

[13] J. I. Friedman. "The Road to the Nobel Prize". Hue University. Retrieved 2008-09-29.

[14] R. P. Feynman (1969). "Very High-Energy Collisions of Hadrons". *Physical Review Letters* **23** (24): 1415–1417. Bibcode:1969PhRvL..23.1415F. doi:10.1103/PhysRevLett.23.1415.

[15] S. Kretzer; Lai, H.; Olness, Fredrick; Tung, W. et al. (2004). "CTEQ6 Parton Distributions with Heavy Quark Mass Effects". *Physical Review D* **69** (11): 114005. arXiv:hep-th/0307022. Bibcode:2004PhRvD..69k4005K. doi:10.1103/PhysRevD.69.114005.

[16] D. J. Griffiths (1987). *Introduction to Elementary Particles*. John Wiley & Sons. p. 42. ISBN 0-471-60386-4.

[17] M. E. Peskin, D. V. Schroeder (1995). *An introduction to quantum field theory*. Addison–Wesley. p. 556. ISBN 0-201-50397-2.

2.5.4 Further reading

- R. Nave. "Quarks". *HyperPhysics*. Georgia State University, Department of Physics and Astronomy. Retrieved 2008-06-29.

- A. Pickering (1984). *Constructing Quarks*. University of Chicago Press. pp. 114–125. ISBN 0-226-66799-5.

2.6 Top quark

The **top quark**, also known as the **t quark** (symbol: t) or **truth quark**, is an elementary particle and a fundamental constituent of matter. Like all quarks, the top quark is an elementary fermion with spin-$\frac{1}{2}$, and experiences all four fundamental interactions: gravitation, electromagnetism, weak interactions, and strong interactions. It has an electric charge of $+\frac{2}{3}$ e,[2] and is the most massive of all observed elementary particles. It has a mass of 173.34 ± 0.27 (stat) ± 0.71 (syst)10^0 GeV/c^2,[1] which is about the same mass as an atom of tungsten. The antiparticle of the top quark is the **top antiquark** (symbol: t, sometimes called *antitop quark* or simply *antitop*), which differs from it only in that some of its properties have equal magnitude but opposite sign.

The top quark interacts primarily by the strong interaction, but can only decay through the weak force. It decays to a W boson and either a bottom quark (most frequently), a strange quark, or, on the rarest of occasions, a down quark. The Standard Model predicts its mean lifetime to be roughly 5×10^{-25} s.[3] This is about a twentieth of the timescale for strong interactions, and therefore it does not form hadrons, giving physicists a unique opportunity to study a "bare" quark (all other quarks hadronize, meaning that they combine with other quarks to form hadrons, and can only be observed as such). Because it is so massive, the properties of the top quark allow predictions to be made of the mass of the Higgs boson under certain extensions of the Standard Model (see Mass and coupling to the Higgs boson below). As such, it is extensively studied as a means to discriminate between competing theories.

Its existence (and that of the bottom quark) was postulated in 1973 by Makoto Kobayashi and Toshihide Maskawa to explain the observed CP violations in kaon decay,[4] and was discovered in 1995 by the CDF[5] and DØ[6] experiments at Fermilab. Kobayashi and Maskawa won the 2008 Nobel Prize in Physics for the prediction of the top and bottom quark, which together form the third generation of quarks.[7]

2.6.1 History

In 1973, Makoto Kobayashi and Toshihide Maskawa predicted the existence of a third generation of quarks to explain observed CP violations in kaon decay.[4] The names top and bottom were introduced by Haim Harari in 1975,[8][9] to match the names of the first generation of quarks (up and down) reflecting the fact that the two were the 'up' and 'down' component of a weak isospin doublet.[10] The top quark was sometimes called *truth quark* in the past, but over time *top quark* became the predominant use.[11]

The proposal of Kobayashi and Maskawa heavily relied on the GIM mechanism put forward by Sheldon Lee Glashow, John Iliopoulos and Luciano Maiani,[12] which predicted the existence of the then still unobserved charm quark. When in November 1974 teams at Brookhaven National Laboratory (BNL) and the Stanford Linear Accelerator Center (SLAC) simultaneously announced the discovery of the J/ψ meson, it was soon after identified as a bound state of the missing charm quark with its antiquark. This discovery allowed the GIM mechanism to become part of the Standard Model.[13] With the acceptance of the GIM mechanism, Kobayashi and Maskawa's prediction also gained in credibility. Their case was further strengthened by the discovery of the tau by Martin Lewis Perl's team at SLAC between 1974 and 1978.[14] This announced a third generation of leptons, breaking the new symmetry between leptons and quarks introduced by the GIM mechanism. Restoration of the symmetry implied the existence of a fifth and sixth quark.

It was in fact not long until a fifth quark, the bottom, was discovered by the E288 experiment team, led by Leon Lederman at Fermilab in 1977.[15][16][17] This strongly suggested that there must also be a sixth quark, the top, to complete the pair. It was known that this quark would be heavier than the bottom, requiring more energy to create in particle collisions, but the general expectation was that the sixth quark would soon be found. However, it took another 18 years before the existence of the top was confirmed.[18]

Early searches for the top quark at SLAC and DESY (in Hamburg) came up empty-handed. When, in the early eighties, the Super Proton Synchrotron (SPS) at CERN discovered the W boson and the Z boson, it was again felt that the discovery of the top was imminent. As the SPS gained competition from the Tevatron at Fermilab there was still no sign of the missing particle, and it was announced by the group at CERN that the top mass must be at least 41 GeV/c^2. After a race between CERN and Fermilab to discover the top, the accelerator at CERN reached its limits without creating a single top, pushing the lower bound on its mass up to 77 GeV/c^2.[18]

The Tevatron was (until the start of LHC operation at CERN in 2009) the only hadron collider powerful enough to produce top quarks. In order to be able to confirm a future discovery, a second detector, the DØ detector, was added to the complex (in addition to the Collider Detector at Fermilab (CDF) already present). In October 1992, the two groups found their first hint of the top, with a single creation event that appeared to contain the top. In the following years, more evidence was collected and on April 22, 1994, the CDF group submitted their paper presenting tentative evidence for the existence of a top quark with a mass of about 175 GeV/c^2. In the meantime, DØ had found no more evidence than the suggestive event in 1992. A year later, on March 2, 1995, after having gathered more evidence and a reanalysis of the DØ data (who had been searching for a much lighter top), the two groups jointly reported the discovery of the top with a certainty of 99.9998% at a mass of 176±18 GeV/c^2.[5][6][18]

In the years leading up to the top quark discovery, it was realized that certain precision measurements of the electroweak vector boson masses and couplings are very sensitive to the value of the top quark mass. These effects become much larger for higher values of the top mass and therefore could indirectly see the top quark even if it could not be directly produced in any experiment at the time. The largest effect from the top quark mass was on the T parameter and by 1994 the precision of these indirect measurements had led to a prediction of the top quark mass to be between 145 GeV/c^2 and 185 GeV/c^2. It is the development of techniques that ultimately allowed such precision calculations that led to Gerardus 't Hooft and Martinus Veltman winning the Nobel Prize in physics in 1999.[19][20]

2.6.2 Properties

- At the final Tevatron energy of 1.96 TeV, top–antitop pairs were produced with a cross section of about 7 picobarns (pb).[21] The Standard Model prediction (at next-to-leading order with m_t = 175 GeV/c^2) is 6.7–7.5 pb.

- The W bosons from top quark decays carry polarization from the parent particle, hence pose themselves as a unique probe to top polarization.

- In the Standard Model, the top quark is predicted to have a spin quantum number of $^1/_2$ and electric charge $+^2/_3$. A first measurement of the top quark charge has been published, resulting in approximately 90% confidence limit that the top quark charge is indeed $+^2/_3$.[22]

2.6.3 Production

Because top quarks are very massive, large amounts of energy are needed to create one. The only way to achieve such high energies is through high energy collisions. These occur naturally in the Earth's upper atmosphere as cosmic rays collide with particles in the air, or can be created in a particle accelerator. In 2011, after the Tevatron ceased operations, the Large Hadron Collider at CERN became the only accelerator that generates a beam of sufficient energy to produce top quarks, with a center-of-mass energy of 7 TeV.

There are multiple processes that can lead to the production of a top quark. The most common is production of a top–antitop pair via strong interactions. In a collision, a highly energetic gluon is created, which subsequently decays into a top and antitop. This process was responsible for the majority of the top events at Tevatron and was the process observed when the top was first discovered in 1995.[23] It is also possible to produce pairs of top–antitop through the decay of an intermediate photon or Z-boson. However, these processes are predicted to be much rarer and have a virtually identical experimental signature in a hadron collider like Tevatron.

A distinctly different process is the production of single tops via weak interaction. This can happen in two ways (called channels): either an intermediate W-boson decays into a top and antibottom quark ("s-channel") or a bottom quark (probably created in a pair through the decay of a gluon) transforms to top quark by exchanging a W-boson with an up or down quark ("t-channel"). The first evidence for these processes was published by the DØ collaboration in December 2006,[24] and in March 2009 the CDF[25] and DØ[23] collaborations released twin papers with the definitive observation of these processes. The main significance of measuring these production processes is that their frequency is directly proportional to the $| V_{tb} |^2$ component of the CKM matrix.

2.6.4 Decay

The only known way that a top quark can decay is through the weak interaction producing a W-boson and a down-type quark (down, strange, or bottom). Because of its enormous mass, the top quark is extremely short-lived with a predicted lifetime of only 5×10^{-25} s.[3] As a result top quarks do not have time to form hadrons before they decay, as other quarks do. This provides physicists with the unique opportunity to study the behavior of a "bare" quark.

In particular, it is possible to directly determine the branching ratio $\Gamma(W^+b) / \Gamma(W^+q$ (q = b,s,d)). The best current determination of this ratio is 0.91±0.04.[26] Since this ratio is equal to $| V_{tb} |^2$ according to the Standard Model, this gives another way of determining the CKM element $| V_{tb} |$, or in combination with the determination of $| V_{tb} |$ from single top production provides tests for the assumption that the CKM matrix is unitary.[27]

The Standard Model also allows more exotic decays, but only at one loop level, meaning that they are extremely suppressed. In particular, it is possible for a top quark to decay into another up-type quark (an up or a charm) by emitting a photon or a Z-boson.[28] Searches for these exotic decay modes have provided no evidence for their existence in accordance with expectations from the Standard Model. The branching ratios for these decays have been determined to be less than 5.9 in 1,000 for photonic decay and less than 2.1 in 1,000 for Z-boson decay at 95% confidence.[26]

2.6.5   Mass and coupling to the Higgs boson

The Standard Model describes fermion masses through the Higgs mechanism. The Higgs boson has a Yukawa coupling to the left- and right-handed top quarks. After electroweak symmetry breaking (when the Higgs acquires a vacuum expectation value), the left- and right-handed components mix, becoming a mass term.

$$\mathcal{L} = y_t h q u^c \rightarrow \frac{y_t v}{\sqrt{2}}(1 + h^0/v) u u^c$$

The top quark Yukawa coupling has a value of

$$y_t = \sqrt{2} m_t / v \simeq 1$$

where $v = 246$ GeV is the value of the Higgs vacuum expectation value.

Yukawa couplings

See also: Beta function (physics)

In the Standard Model, all of the quark and lepton Yukawa couplings are small compared to the top quark Yukawa coupling. Understanding this hierarchy in the fermion masses is an open problem in theoretical physics. Yukawa couplings are not constants and their values change depending on the energy scale (distance scale) at which they are measured. The dynamics of Yukawa couplings are determined by the renormalization group equation.

One of the prevailing views in particle physics is that the size of the top quark Yukawa coupling is determined by the renormalization group, leading to the "quasi-infrared fixed point."

The Yukawa couplings of the up, down, charm, strange and bottom quarks, are hypothesized to have small values at the extremely high energy scale of grand unification, 10^{15} GeV. They increase in value at lower energy scales, at which the quark masses are generated by the Higgs. The slight growth is due to corrections from the QCD coupling. The corrections from the Yukawa couplings are negligible for the lower mass quarks.

If, however, a quark Yukawa coupling has a large value at very high energies, its Yukawa corrections will evolve and cancel against the QCD corrections. This is known as a (quasi-) infrared fixed point. No matter what the initial starting value of the coupling is, if it is sufficiently large it will reach this fixed point value. The corresponding quark mass is then predicted.

The top quark Yukawa coupling lies very near the infrared fixed point of the Standard Model. The renormalization group equation is:

$$\mu \frac{\partial}{\partial \mu} y_t \approx \frac{y_t}{16\pi^2} \left(\frac{9}{2} y_t^2 - 8g_3^2 - \frac{9}{4} g_2^2 - \frac{17}{20} g_1^2 \right),$$

where g_3 is the color gauge coupling, g_2 is the weak isospin gauge coupling, and g_1 is the weak hypercharge gauge coupling. This equation describes how the Yukawa coupling changes with energy scale μ. Solutions to this equation for large initial values y_t cause the right-hand side of the equation to quickly approach zero, locking y_t to the QCD coupling g_3. The value of the fixed point is fairly precisely determined in the Standard Model, leading to a top quark mass of 230 GeV. However, if there is more than one Higgs doublet, the mass value will be reduced by Higgs mixing angle effects in an unpredicted way.

In the minimal supersymmetric extension of the Standard Model (MSSM), there are two Higgs doublets and the renormalization group equation for the top quark Yukawa coupling is slightly modified:

$$\mu \frac{\partial}{\partial \mu} y_t \approx \frac{y_t}{16\pi^2} \left(6y_t^2 + y_b^2 - \frac{16}{3} g_3^2 - 3g_2^2 - \frac{13}{15} g_1^2 \right),$$

where y_b is the bottom quark Yukawa coupling. This leads to a fixed point where the top mass is smaller, 170–200 GeV. The uncertainty in this prediction arises because the bottom quark Yukawa coupling can be amplified in the MSSM. Some theorists believe this is supporting evidence for the MSSM.

The quasi-infrared fixed point has subsequently formed the basis of top quark condensation theories of electroweak symmetry breaking in which the Higgs boson is composite at *extremely* short distance scales, composed of a pair of top and antitop quarks.

2.6.6   See also

- CDF experiment

- Topness

- Top quark condensate

- Topcolor

- Quark model

2.6.7 References

[1] The ATLAS, CDF, CMS, D0 Collaborations (2014). "First combination of Tevatron and LHC measurements of the top-quark mass". Retrieved 2014-03-19.

[2] S. Willenbrock (2003). "The Standard Model and the Top Quark". In H.B Prosper and B. Danilov (eds.). *Techniques and Concepts of High-Energy Physics XII*. NATO Science Series **123**. Kluwer Academic. pp. 1–41. arXiv:hep-ph/0211067v3. ISBN 1-4020-1590-9.

[3] A. Quadt (2006). "Top quark physics at hadron colliders". *European Physical Journal C* **48** (3): 835–1000. Bibcode:2006EPJC...48..835Q. doi:10.1140/epjc/s2006-02631-6.

[4] M. Kobayashi, T. Maskawa (1973). "*CP*-Violation in the Renormalizable Theory of Weak Interaction". *Progress of Theoretical Physics* **49** (2): 652. Bibcode:1973PThPh..49..652K. doi:10.1143/PTP.49.652.

[5] F. Abe *et al.* (CDF Collaboration) (1995). "Observation of Top Quark Production in pp Collisions with the Collider Detector at Fermilab". *Physical Review Letters* **74** (14): 2626–2631. Bibcode:1995PhRvL..74.2626A. doi:10.1103/PhysRevLett.74.2626. PMID 10057978.

[6] S. Abachi *et al.* (DØ Collaboration) (1995). "Search for High Mass Top Quark Production in pp Collisions at √s = 1.8 TeV". *Physical Review Letters* **74** (13): 2422–2426. Bibcode:1995PhRvL..74.2422A. doi:10.1103/PhysRevLett.74.2422.

[7] "2008 Nobel Prize in Physics". The Nobel Foundation. 2008. Retrieved 2009-09-11.

[8] H. Harari (1975). "A new quark model for hadrons". *Physics Letters B* **57** (3): 265. Bibcode:1975PhLB...57..265H. doi:10.1016/0370-2693(75)90072-6.

[9] K.W. Staley (2004). *The Evidence for the Top Quark*. Cambridge University Press. pp. 31–33. ISBN 978-0-521-82710-2.

[10] D.H. Perkins (2000). *Introduction to high energy physics*. Cambridge University Press. p. 8. ISBN 0-521-62196-8.

[11] F. Close (2006). *The New Cosmic Onion*. CRC Press. p. 133. ISBN 1-58488-798-2.

[12] S.L. Glashow, J. Iliopoulous, L. Maiani (1970). "Weak Interactions with Lepton–Hadron Symmetry". *Physical Review D* **2** (7): 1285–1292. Bibcode:1970PhRvD...2.1285G. doi:10.1103/PhysRevD.2.1285.

[13] A. Pickering (1999). *Constructing Quarks: A Sociological History of Particle Physics*. University of Chicago Press. pp. 253–254. ISBN 978-0-226-66799-7.

[14] M.L. Perl et al. (1975). "Evidence for Anomalous Lepton Production in e+e– Annihilation". *Physical Review Letters* **35** (22): 1489. Bibcode:1975PhRvL..35.1489P. doi:10.1103/PhysRevLett.35.1489.

[15] "Discoveries at Fermilab – Discovery of the Bottom Quark" (Press release). Fermilab. 7 August 1977. Retrieved 2009-07-24.

[16] L.M. Lederman (2005). "Logbook: Bottom Quark". *Symmetry Magazine* **2** (8).

[17] S.W. Herb et al. (1977). "Observation of a Dimuon Resonance at 9.5 GeV in 400-GeV Proton-Nucleus Collisions". *Physical Review Letters* **39** (5): 252. Bibcode:1977PhRvL..39..252H. doi:10.1103/PhysRevLett.39.252.

[18] T.M. Liss, P.L. Tipton (1997). "The Discovery of the Top Quark" (PDF). *Scientific American*: 54–59.

[19] "The Nobel Prize in Physics 1999". The Nobel Foundation. Retrieved 2009-09-10.

[20] "The Nobel Prize in Physics 1999, Press Release" (Press release). The Nobel Foundation. 12 October 1999. Retrieved 2009-09-10.

[21] D. Chakraborty (DØ and CDF collaborations) (2002). *Top quark and W/Z results from the Tevatron* (PDF). Rencontres de Moriond. p. 26.

[22] V.M. Abazov *et al.* (DØ Collaboration) (2007). "Experimental discrimination between charge 2e/3 top quark and charge 4e/3 exotic quark production scenarios". *Physical Review Letters* **98** (4): 041801. arXiv:hep-ex/0608044. Bibcode:2007PhRvL..98d1801A. doi:10.1103/PhysRevLett.98.041801. PMID 17358756.

[23] V.M. Abazov *et al.* (DØ Collaboration) (2009). "Observation of Single Top Quark Production". *Physical Review Letters* **103** (9). arXiv:0903.0850. Bibcode:2009PhRvL.103i2001A. doi:10.1103/PhysRevLett.103.092001.

[24] V.M. Abazov *et al.* (DØ Collaboration) (2007). "Evidence for production of single top quarks and first direct measurement of |V$_{tb}$|". *Physical Review Letters* **98** (18): 181802. arXiv:hep-ex/0612052. Bibcode:2007PhRvL..98r1802A. doi:10.1103/PhysRevLett.98.181802. PMID 17501561.

[25] T. Aaltonen *et al.* (CDF Collaboration) (2009). "First Observation of Electroweak Single Top Quark Production". *Physical Review Letters* **103** (9). arXiv:0903.0885. Bibcode:2009PhRvL.103i2002A. doi:10.1103/PhysRevLett.103.092002.

[26] J. Beringer *et al.* (Particle Data Group) (2012). "PDGLive Particle Summary 'Quarks (u, d, s, c, b, t, b', t', Free)'" (PDF). Particle Data Group. Retrieved 2013-07-23.

[27] V.M. Abazov *et al.* (DØ Collaboration) (2008). "Simultaneous measurement of the ratio B(t→Wb)/B(t→Wq) and the top-quark pair production cross section with the DØ detector at √s = 1.96 TeV". *Physical Review Letters* **100** (19): 192003. arXiv:0801.1326. Bibcode:2008PhRvL.100s2003A. doi:10.1103/PhysRevLett.100.192003.

[28] S. Chekanov *et al.* (ZEUS Collaboration) (2003). "Search for single-top production in ep collisions at HERA". *Physics Letters B* **559** (3–4): 153. arXiv:hep-ex/0302010. Bibcode:2003PhLB..559..153Z.   doi:10.1016/S0370-2693(03)00333-2.

2.6.8   Further reading

- Frank Fiedler; for the D0; CDF Collaborations (June 2005). "Top Quark Production and Properties at the Tevatron". arXiv:hep-ex/0506005 [hep-ex].

- R. Nave. "Quarks". *HyperPhysics*. Georgia State University, Department of Physics and Astronomy. Retrieved 2008-06-29.

- A. Pickering (1984). *Constructing Quarks*. University of Chicago Press. pp. 114–125. ISBN 0-226-66799-5.

2.6.9   External links

- Top quark on arxiv.org
- Tevatron Electroweak Working Group
- Top quark information on Fermilab website
- Logbook pages from CDF and DZero collaborations' top quark discovery
- Scientific American article on the discovery of the top quark
- Public Homepage of Top Quark Analysis Results from DØ Collaboration at Fermilab
- Public Homepage of Top Quark Analysis Results from CDF Collaboration at Fermilab
- Harvard Magazine article about the 1994 top quark discovery
- 1999 Nobel Prize in Physics

2.7   Bottom quark

The **bottom quark** or **b quark**, also known as the **beauty quark**, is a third-generation quark with a charge of $-\frac{1}{3}$ e. Although all quarks are described in a similar way by quantum chromodynamics, the bottom quark's large bare mass (around 4.2 GeV/c^2,[3] a bit more than four times the mass of a proton), combined with low values of the CKM matrix elements V_{ub} and V_{cb}, gives it a distinctive signature that makes it relatively easy to identify experimentally

(using a technique called B-tagging). Because three generations of quark are required for CP violation (see CKM matrix), mesons containing the bottom quark are the easiest particles to use to investigate the phenomenon; such experiments are being performed at the BaBar, Belle and LHCb experiments. The bottom quark is also notable because it is a product in almost all top quark decays, and is a frequent decay product for the Higgs boson.

The bottom quark was theorized in 1973 by physicists Makoto Kobayashi and Toshihide Maskawa to explain CP violation.[1] The name "bottom" was introduced in 1975 by Haim Harari.[4][5] The bottom quark was discovered in 1977 by the Fermilab E288 experiment team led by Leon M. Lederman, when collisions produced bottomonium.[2][6][7] Kobayashi and Maskawa won the 2008 Nobel Prize in Physics for their explanation of CP-violation.[8][9] On its discovery, there were efforts to name the bottom quark "beauty", but "bottom" became the predominant usage.

The bottom quark can decay into either an up quark or charm quark via the weak interaction. Both these decays are suppressed by the CKM matrix, making lifetimes of most bottom particles ($\sim 10^{-12}$ s) somewhat higher than those of charmed particles ($\sim 10^{-13}$ s), but lower than those of strange particles (from $\sim 10^{-10}$ to $\sim 10^{-8}$ s).

2.7.1   Hadrons containing bottom quarks

Main articles: list of baryons and list of mesons

Some of the hadrons containing bottom quarks include:

- B mesons contain a bottom quark (or its antiparticle) and an up or down quark.
- B c and B s mesons contain a bottom quark along with a charm quark or strange quark respectively.
- There are many bottomonium states, for example the Υ meson and χ_b(3P), the first particle discovered in LHC. These consist of a bottom quark and its antiparticle.
- Bottom baryons have been observed, and are named in analogy with strange baryons (e.g. $\Lambda 0$ b).

2.7.2   See also

- Quark model

2.7.3 References

[1] M. Kobayashi; T. Maskawa (1973). "CP-Violation in the Renormalizable Theory of Weak Interaction". *Progress of Theoretical Physics* **49** (2): 652–657. Bibcode:1973PThPh..49..652K. doi:10.1143/PTP.49.652.

[2] "Discoveries at Fermilab – Discovery of the Bottom Quark" (Press release). Fermilab. 7 August 1977. Retrieved 2009-07-24.

[3] J. Beringer (Particle Data Group) et al. (2012). "PDGLive Particle Summary 'Quarks (u, d, s, c, b, t, b′, t′, Free)'" (PDF). Particle Data Group. Retrieved 2012-12-18.

[4] H. Harari (1975). "A new quark model for hadrons". *Physics Letters B* **57** (3): 265. Bibcode:1975PhLB...57..265H. doi:10.1016/0370-2693(75)90072-6.

[5] K.W. Staley (2004). *The Evidence for the Top Quark.* Cambridge University Press. pp. 31–33. ISBN 978-0-521-82710-2.

[6] L.M. Lederman (2005). "Logbook: Bottom Quark". *Symmetry Magazine* **2** (8).

[7] S.W. Herb; Hom, D.; Lederman, L.; Sens, J.; Snyder, H.; Yoh, J.; Appel, J.; Brown, B.; Brown, C.; Innes, W.; Ueno, K.; Yamanouchi, T.; Ito, A.; Jöstlein, H.; Kaplan, D.; Kephart, R. et al. (1977). "Observation of a Dimuon Resonance at 9.5 GeV in 400-GeV Proton-Nucleus Collisions". *Physical Review Letters* **39** (5): 252. Bibcode:1977PhRvL..39..252H. doi:10.1103/PhysRevLett.39.252.

[8] 2008 Physics Nobel Prize lecture by Makoto Kobayashi

[9] 2008 Physics Nobel Prize lecture by Toshihide Maskawa

2.7.4 Further reading

- L. Lederman (1978). "The Upsilon Particle". *Scientific American* **239** (4): 72. doi:10.1038/scientificamerican1078-72.

- R. Nave. "Quarks". *HyperPhysics*. Georgia State University, Department of Physics and Astronomy. Retrieved 2008-06-29.

- A. Pickering (1984). *Constructing Quarks.* University of Chicago Press. pp. 114–125. ISBN 0-226-66799-5.

- J. Yoh (1997). "The Discovery of the b Quark at Fermilab in 1977: The Experiment Coordinator's Story" (PDF). *Proceedings of Twenty Beautiful Years of Bottom Physics*. Fermilab. Retrieved 2009-07-24.

2.7.5 External links

- History of the discovery of the bottom quark / Upsilon meson

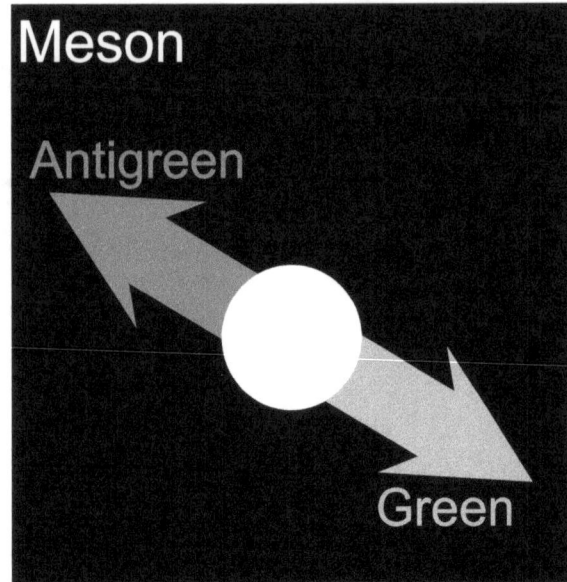

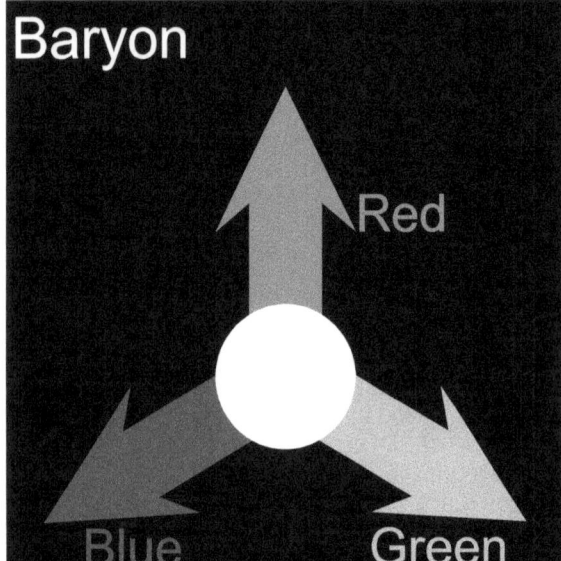

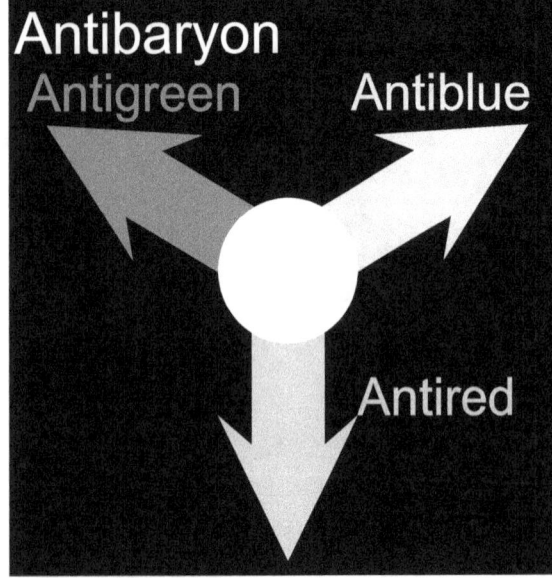

All types of hadrons have zero total color charge.

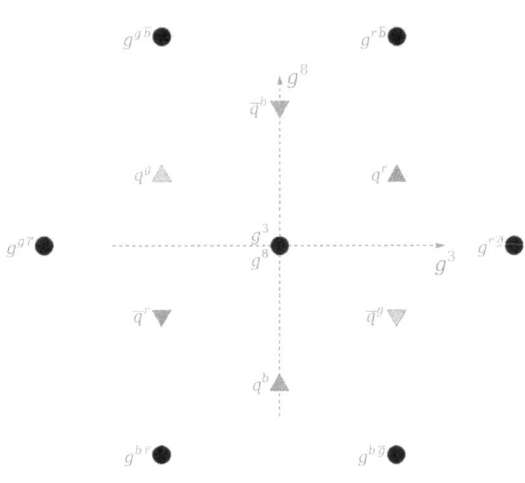

The pattern of strong charges for the three colors of quark, three antiquarks, and eight gluons (with two of zero charge overlapping).

Current quark masses for all six flavors in comparison, as balls of proportional volumes. Proton and electron (red) are shown in bottom left corner for scale

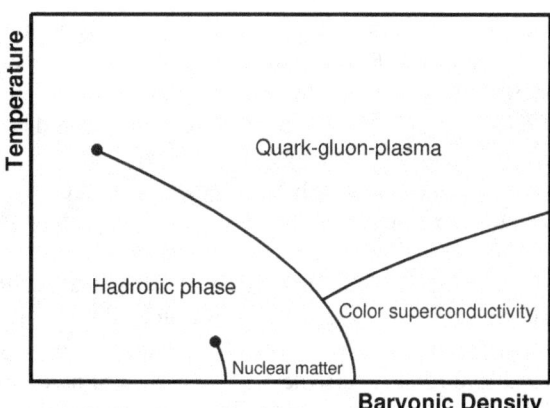

A qualitative rendering of the phase diagram of quark matter. The precise details of the diagram are the subject of ongoing research.[79][80]

Chapter 3

Leptons

3.1 Lepton

For other uses, see Lepton (disambiguation).

A **lepton** is an elementary, half-integer spin (spin $\frac{1}{2}$) particle that does not undergo strong interactions, but is subject to the Pauli exclusion principle.[1] The best known of all leptons is the electron, which is directly tied to all chemical properties. Two main classes of leptons exist: charged leptons (also known as the *electron-like* leptons), and neutral leptons (better known as neutrinos). Charged leptons can combine with other particles to form various composite particles such as atoms and positronium, while neutrinos rarely interact with anything, and are consequently rarely observed.

There are six types of leptons, known as *flavours*, forming three *generations*.[2] The first generation is the *electronic leptons*, comprising the electron (e−) and electron neutrino (ν e); the second is the *muonic leptons*, comprising the muon (μ−) and muon neutrino (ν μ); and the third is the *tauonic leptons*, comprising the tau (τ−) and the tau neutrino (ν τ). Electrons have the least mass of all the charged leptons. The heavier muons and taus will rapidly change into electrons through a process of particle decay: the transformation from a higher mass state to a lower mass state. Thus electrons are stable and the most common charged lepton in the universe, whereas muons and taus can only be produced in high energy collisions (such as those involving cosmic rays and those carried out in particle accelerators).

Leptons have various intrinsic properties, including electric charge, spin, and mass. Unlike quarks however, leptons are not subject to the strong interaction, but they are subject to the other three fundamental interactions: gravitation, electromagnetism (excluding neutrinos, which are electrically neutral), and the weak interaction. For every lepton flavor there is a corresponding type of antiparticle, known as antilepton, that differs from the lepton only in that some of its properties have equal magnitude but opposite sign.

However, according to certain theories, neutrinos may be their own antiparticle, but it is not currently known whether this is the case or not.

The first charged lepton, the electron, was theorized in the mid-19th century by several scientists[3][4][5] and was discovered in 1897 by J. J. Thomson.[6] The next lepton to be observed was the muon, discovered by Carl D. Anderson in 1936, which was classified as a meson at the time.[7] After investigation, it was realized that the muon did not have the expected properties of a meson, but rather behaved like an electron, only with higher mass. It took until 1947 for the concept of "leptons" as a family of particle to be proposed.[8] The first neutrino, the electron neutrino, was proposed by Wolfgang Pauli in 1930 to explain certain characteristics of beta decay.[8] It was first observed in the Cowan–Reines neutrino experiment conducted by Clyde Cowan and Frederick Reines in 1956.[8][9] The muon neutrino was discovered in 1962 by Leon M. Lederman, Melvin Schwartz and Jack Steinberger,[10] and the tau discovered between 1974 and 1977 by Martin Lewis Perl and his colleagues from the Stanford Linear Accelerator Center and Lawrence Berkeley National Laboratory.[11] The tau neutrino remained elusive until July 2000, when the DONUT collaboration from Fermilab announced its discovery.[12][13]

Leptons are an important part of the Standard Model. Electrons are one of the components of atoms, alongside protons and neutrons. Exotic atoms with muons and taus instead of electrons can also be synthesized, as well as lepton–antilepton particles such as positronium.

3.1.1 Etymology

The name *lepton* comes from the Greek λεπτός *leptós*, "fine, small, thin" (neuter form: λεπτόν *leptón*);[14][15] the earliest attested form of the word is the Mycenaean Greek 𐀩𐀡𐀵, *re-po-to*, written in Linear B syllabic script.[16] *Lepton* was first used by physicist Léon Rosenfeld in 1948:[17]

Following a suggestion of Prof. C. Møller,

I adopt — as a pendant to "nucleon" — the denomination "lepton" (from λεπτός, small, thin, delicate) to denote a particle of small mass.

The etymology incorrectly implies that all the leptons are of small mass. When Rosenfeld named them, the only known leptons were electrons and muons, which are in fact of small mass — the mass of an electron (0.511 MeV/c^2)[18] and the mass of a muon (with a value of 105.7 MeV/c^2)[19] are fractions of the mass of the "heavy" proton (938.3 MeV/c^2).[20] However, the mass of the tau (discovered in the mid 1970s) (1777 MeV/c^2)[21] is nearly twice that of the proton, and about 3,500 times that of the electron.

3.1.2 History

See also: Electron § Discovery, Muon § History and Tau (particle) § History

The first lepton identified was the electron, discov-

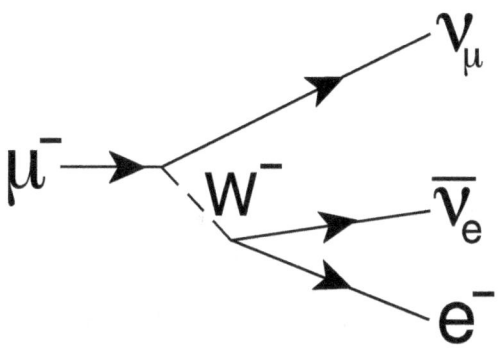

A muon transmutes into a muon neutrino by emitting a W− boson. The W− boson subsequently decays into an electron and an electron antineutrino.

ered by J.J. Thomson and his team of British physicists in 1897.[22][23] Then in 1930 Wolfgang Pauli postulated the electron neutrino to preserve conservation of energy, conservation of momentum, and conservation of angular momentum in beta decay.[24] Pauli theorized that an undetected particle was carrying away the difference between the energy, momentum, and angular momentum of the initial and observed final particles. The electron neutrino was simply called the neutrino, as it was not yet known that neutrinos came in different flavours (or different "generations").

Nearly 40 years after the discovery of the electron, the muon was discovered by Carl D. Anderson in 1936. Due to its mass, it was initially categorized as a meson rather than a lepton.[25] It later became clear that the muon was much more similar to the electron than to mesons, as muons do not undergo the strong interaction, and thus the muon was

reclassified: electrons, muons, and the (electron) neutrino were grouped into a new group of particles – the leptons. In 1962 Leon M. Lederman, Melvin Schwartz and Jack Steinberger showed that more than one type of neutrino exists by first detecting interactions of the muon neutrino, which earned them the 1988 Nobel Prize, although by then the different flavours of neutrino had already been theorized.[26]

The tau was first detected in a series of experiments between 1974 and 1977 by Martin Lewis Perl with his colleagues at the SLAC LBL group.[27] Like the electron and the muon, it too was expected to have an associated neutrino. The first evidence for tau neutrinos came from the observation of "missing" energy and momentum in tau decay, analogous to the "missing" energy and momentum in beta decay leading to the discovery of the electron neutrino. The first detection of tau neutrino interactions was announced in 2000 by the DONUT collaboration at Fermilab, making it the latest particle of the Standard Model to have been directly observed,[28] apart from the Higgs boson, which probably has been discovered in 2012.

Although all present data is consistent with three generations of leptons, some particle physicists are searching for a fourth generation. The current lower limit on the mass of such a fourth charged lepton is 100.8 GeV/c^2,[29] while its associated neutrino would have a mass of at least 45.0 GeV/c^2.[30]

3.1.3 Properties

Spin and chirality

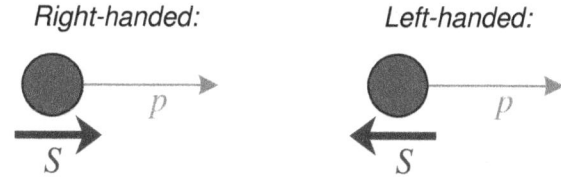

Left-handed and right-handed helicities

Leptons are spin-$\frac{1}{2}$ particles. The spin-statistics theorem thus implies that they are fermions and thus that they are subject to the Pauli exclusion principle; no two leptons of the same species can be in exactly the same state at the same time. Furthermore, it means that a lepton can have only two possible spin states, namely up or down.

A closely related property is chirality, which in turn is closely related to a more easily visualized property called helicity. The helicity of a particle is the direction of its spin relative to its momentum; particles with spin in the same direction as their momentum are called *right-handed* and otherwise they are called *left-handed*. When a particle is mass-less, the direction of its momentum relative to its spin is frame independent, while for massive particles it is possible to 'overtake' the particle by a Lorentz transformation flipping the helicity. Chirality is a technical property (defined through the transformation behaviour under the Poincaré group) that agrees with helicity for (approximately) massless particles and is still well defined for massive particles.

In many quantum field theories—such as quantum electrodynamics and quantum chromodynamics—left and right-handed fermions are identical. However in the Standard Model left-handed and right-handed fermions are treated asymmetrically. Only left-handed fermions participate in the weak interaction, while there are no right-handed neutrinos. This is an example of parity violation. In the literature left-handed fields are often denoted by a capital L subscript (e.g. e–L) and right-handed fields are denoted by a capital R subscript.

Electromagnetic interaction

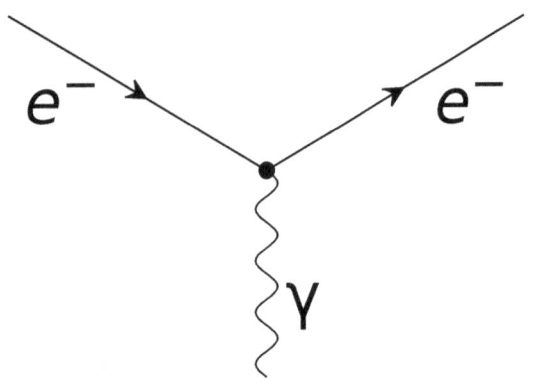

Lepton–photon interaction

One of the most prominent properties of leptons is their electric charge, Q. The electric charge determines the strength of their electromagnetic interactions. It determines the strength of the electric field generated by the particle (see Coulomb's law) and how strongly the particle reacts to an external electric or magnetic field (see Lorentz force). Each generation contains one lepton with $Q = -e$ (conventionally the charge of a particle is expressed in units of the elementary charge) and one lepton with zero electric charge. The lepton with electric charge is commonly simply referred to as a 'charged lepton' while the neutral lepton is called a neutrino. For example the first generation consists of the electron e– with a negative electric charge and the electrically neutral electron neutrino ν
e.

In the language of quantum field theory the electromagnetic

interaction of the charged leptons is expressed by the fact that the particles interact with the quantum of the electromagnetic field, the photon. The Feynman diagram of the electron-photon interaction is shown on the right.

Because leptons possess an intrinsic rotation in the form of their spin, charged leptons generate a magnetic field. The size of their magnetic dipole moment μ is given by,

$$\mu = g\frac{Q\hbar}{4m},$$

where m is the mass of the lepton and g is the so-called g-factor for the lepton. First order approximation quantum mechanics predicts that the g-factor is 2 for all leptons. However, higher order quantum effects caused by loops in Feynman diagrams introduce corrections to this value. These corrections, referred to as the anomalous magnetic dipole moment, are very sensitive to the details of a quantum field theory model and thus provide the opportunity for precision tests of the standard model. The theoretical and measured values for the electron anomalous magnetic dipole moment are within agreement within eight significant figures.[31]

Weak Interaction

In the Standard Model the left-handed charged lepton and the left-handed neutrino are arranged in doublet (ν
eL, e–L) that transforms in the spinor representation ($T = \frac{1}{2}$) of the weak isospin SU(2) gauge symmetry. This means that these particles are eigenstates of the isospin projection T_3 with eigenvalues $\frac{1}{2}$ and $-\frac{1}{2}$ respectively. In the meantime, the right-handed charged lepton transforms as a weak isospin scalar ($T = 0$) and thus does not participate in the weak interaction, while there is no right-handed neutrino at all.

The Higgs mechanism recombines the gauge fields of the weak isospin SU(2) and the weak hypercharge U(1) symmetries to three massive vector bosons (W+, W−, Z0) mediating the weak interaction, and one massless vector boson, the photon, responsible for the electromagnetic interaction. The electric charge Q can be calculated from the isospin projection T_3 and weak hypercharge YW through the Gell-Mann–Nishijima formula,

$$Q = T_3 + Y\text{W}/2$$

To recover the observed electric charges for all particles the left-handed weak isospin doublet (ν
eL, e–L) must thus have YW = −1, while the right-handed isospin scalar e−
R must have YW = −2. The interaction of the leptons with

the massive weak interaction vector bosons is shown in the figure on the left.

Mass

In the Standard Model each lepton starts out with no intrinsic mass. The charged leptons (i.e. the electron, muon, and tau) obtain an effective mass through interaction with the Higgs field, but the neutrinos remain massless. For technical reasons the masslessness of the neutrinos implies that there is no mixing of the different generations of charged leptons as there is for quarks. This is in close agreement with current experimental observations.[32]

However, it is known from experiments – most prominently from observed neutrino oscillations[33] – that neutrinos do in fact have some very small mass, probably less than 2 eV/c^2.[34] This implies the existence of physics beyond the Standard Model. The currently most favoured extension is the so-called seesaw mechanism, which would explain both why the left-handed neutrinos are so light compared to the corresponding charged leptons, and why we have not yet seen any right-handed neutrinos.

Leptonic numbers

Main article: Lepton number

The members of each generation's weak isospin doublet are assigned leptonic numbers that are conserved under the Standard Model.[35] Electrons and electron neutrinos have an *electronic number* of $L_e = 1$, while muons and muon neutrinos have a *muonic number* of $L\mu = 1$, while tau particles and tau neutrinos have a *tauonic number* of $L\tau = 1$. The antileptons have their respective generation's leptonic numbers of −1.

Conservation of the leptonic numbers means that the number of leptons of the same type remains the same, when particles interact. This implies that leptons and antileptons must be created in pairs of a single generation. For example, the following processes are allowed under conservation of leptonic numbers:

$$\begin{pmatrix} \nu_e \\ e^- \end{pmatrix}, \begin{pmatrix} \nu_\mu \\ \mu^- \end{pmatrix}, \begin{pmatrix} \nu_\tau \\ \tau^- \end{pmatrix}$$

Each generation forms a weak isospin doublet.

e− + e+ → γ + γ,

τ− + τ+ → Z0 + Z0,

but not these:

γ → e− + μ+,

W− → e− + ν

τ,

Z0 → μ− + τ+.

However, neutrino oscillations are known to violate the conservation of the individual leptonic numbers. Such a violation is considered to be smoking gun evidence for physics beyond the Standard Model. A much stronger conservation law is the conservation of the total number of leptons (L), conserved even in the case of neutrino oscillations, but even it is still violated by a tiny amount by the chiral anomaly.

3.1.4 Universality

The coupling of the leptons to gauge bosons are flavour-independent (i.e., the interactions between leptons and gauge bosons are the same for all leptons).[35] This property is called *lepton universality* and has been tested in measurements of the tau and muon lifetimes and of Z boson partial decay widths, particularly at the Stanford Linear Collider (SLC) and Large Electron-Positron Collider (LEP) experiments.[36]:241–243[37]:138

The decay rate (Γ) of muons through the process μ− → e− + ν

e + ν

μ is approximately given by an expression of the form (see muon decay for more details)[35]

$$\Gamma \left(\mu^- \to e^- + \bar{\nu}_e + \nu_\mu \right) = K_1 G_F^2 m_\mu^5,$$

where K_1 is some constant, and G_F is the Fermi coupling constant. The decay rate of tau particles through the process τ− → e− + ν

e + ν

τ is given by an expression of the same form[35]

$$\Gamma \left(\tau^- \to e^- + \bar{\nu}_e + \nu_\tau \right) = K_2 G_F^2 m_\tau^5,$$

where K_2 is some constant. Muon–Tauon universality implies that $K_1 = K_2$. On the other hand, electron–muon universality implies[35]

$$\Gamma \left(\tau^- \to e^- + \bar{\nu}_e + \nu_\tau \right) = \Gamma \left(\tau^- \to \mu^- + \bar{\nu}_\mu + \nu_\tau \right).$$

This explains why the branching ratios for the electronic mode (17.85%) and muonic (17.36%) mode of tau decay are equal (within error).[21]

Universality also accounts for the ratio of muon and tau lifetimes. The lifetime of a lepton (τ_l) is related to the decay rate by[35]

$$\tau_l = \frac{B\left(l^- \to e^- + \bar{\nu}_e + \nu_l\right)}{\Gamma\left(l^- \to e^- + \bar{\nu}_e + \nu_l\right)},$$

where $B(\text{x} \to \text{y})$ and $\Gamma(\text{x} \to \text{y})$ denotes the branching ratios and the resonance width of the process x $\to$ y.

The ratio of tau and muon lifetime is thus given by[35]

$$\frac{\tau_\tau}{\tau_\mu} = \frac{B\left(\tau^- \to e^- + \bar{\nu}_e + \nu_\tau\right)}{B\left(\mu^- \to e^- + \bar{\nu}_e + \nu_\mu\right)} \left(\frac{m_\mu}{m_\tau}\right)^5.$$

Using the values of the 2008 *Review of Particle Physics* for the branching ratios of muons[19] and tau[21] yields a lifetime ratio of ~1.29×10^{-7}, comparable to the measured lifetime ratio of ~1.32×10^{-7}. The difference is due to K_1 and K_2 not actually being constants; they depend on the mass of leptons.

3.1.5   Table of leptons

3.1.6   See also

- Koide formula

- List of particles

- Preons – hypothetical particles which were once postulated to be subcomponents of quarks and leptons

3.1.7   Notes

[1] "Lepton (physics)". *Encyclopædia Britannica*. Retrieved 2010-09-29.

[2] R. Nave. "Leptons". *HyperPhysics*. Georgia State University, Department of Physics and Astronomy. Retrieved 2010-09-29.

[3] W.V. Farrar (1969). "Richard Laming and the Coal-Gas Industry, with His Views on the Structure of Matter". *Annals of Science* **25** (3): 243–254. doi:10.1080/00033796900200141.

[4] T. Arabatzis (2006). *Representing Electrons: A Biographical Approach to Theoretical Entities*. University of Chicago Press. pp. 70–74. ISBN 0-226-02421-0.

[5] J.Z. Buchwald, A. Warwick (2001). *Histories of the Electron: The Birth of Microphysics*. MIT Press. pp. 195–203. ISBN 0-262-52424-4.

[6] J.J. Thomson (1897). "Cathode Rays". *Philosophical Magazine* **44** (269): 293. doi:10.1080/14786449708621070.

[7] S.H. Neddermeyer, C.D. Anderson; Anderson (1937). "Note on the Nature of Cosmic-Ray Particles". *Physical Review* **51** (10): 884–886. Bibcode:1937PhRv...51..884N. doi:10.1103/PhysRev.51.884.

[8] "The Reines-Cowan Experiments: Detecting the Poltergeist" (PDF). *Los Alamos Science* **25**: 3. 1997. Retrieved 2010-02-10.

[9] F. Reines, C.L. Cowan, Jr.; Cowan (1956). "The Neutrino". *Nature* **178** (4531): 446. Bibcode:1956Natur.178..446R. doi:10.1038/178446a0.

[10] G. Danby; Gaillard, J-M.; Goulianos, K.; Lederman, L.; Mistry, N.; Schwartz, M.; Steinberger, J. et al. (1962). "Observation of high-energy neutrino reactions and the existence of two kinds of neutrinos". *Physical Review Letters* **9**: 36. Bibcode:1962PhRvL...9...36D. doi:10.1103/PhysRevLett.9.36.

[11] M.L. Perl; Abrams, G.; Boyarski, A.; Breidenbach, M.; Briggs, D.; Bulos, F.; Chinowsky, W.; Dakin, J.; Feldman, G.; Friedberg, C.; Fryberger, D.; Goldhaber, G.; Hanson, G.; Heile, F.; Jean-Marie, B.; Kadyk, J.; Larsen, R.; Litke, A.; Lüke, D.; Lulu, B.; Lüth, V.; Lyon, D.; Morehouse, C.; Paterson, J.; Pierre, F.; Pun, T.; Rapidis, P.; Richter, B.; Sadoulet, B. et al. (1975). "Evidence for Anomalous Lepton Production in e+e− Annihilation". *Physical Review Letters* **35** (22): 1489. Bibcode:1975PhRvL..35.1489P. doi:10.1103/PhysRevLett.35.1489.

[12] "Physicists Find First Direct Evidence for Tau Neutrino at Fermilab" (Press release). Fermilab. 20 July 2000.

[13] K. Kodama *et al.* (DONUT Collaboration); Kodama; Ushida; Andreopoulos; Saoulidou; Tzanakos; Yager; Baller; Boehnlein; Freeman; Lundberg; Morfin; Rameika; Yun; Song; Yoon; Chung; Berghaus; Kubantsev; Reay; Sidwell; Stanton; Yoshida; Aoki; Hara; Rhee; Ciampa; Erickson; Graham et al. (2001). "Observation of tau neutrino interactions". *Physics Letters B* **504** (3): 218. arXiv:hep-ex/0012035. Bibcode:2001PhLB..504..218D. doi:10.1016/S0370-2693(01)00307-0.

[14] "lepton". *Online Etymology Dictionary*.

[15] λεπτός. Liddell, Henry George; Scott, Robert; *A Greek-English Lexicon* at the Perseus Project.

[16] Found on the KN L 693 and PY Un 1322 tablets. "The Linear B word re-po-to". Palaeolexicon. Word study tool of ancient languages. Raymoure, K.A. "re-po-to". *Minoan Linear A & Mycenaean Linear B*. Deaditerranean. "KN 693 L (103)". "PY 1322 Un + fr. (Cii)". *DĀMOS: Database of Mycenaean at Oslo*. University of Oslo.

[17] L. Rosenfeld (1948)

[18] C. Amsler *et al.* (2008): Particle listings – e–

[19] C. Amsler *et al.* (2008): Particle listings – μ–

[20] C. Amsler *et al.* (2008): Particle listings – p+

[21] C. Amsler *et al.* (2008): Particle listings – τ–

[22] S. Weinberg (2003)

[23] R. Wilson (1997)

[24] K. Riesselmann (2007)

[25] S.H. Neddermeyer, C.D. Anderson (1937)

[26] I.V. Anicin (2005)

[27] M.L. Perl et al. (1975)

[28] K. Kodama (2001)

[29] C. Amsler *et al.* (2008) Heavy Charged Leptons Searches

[30] C. Amsler *et al.* (2008) Searches for Heavy Neutral Leptons

[31] M.E. Peskin, D.V. Schroeder (1995), p. 197

[32] M.E. Peskin, D.V. Schroeder (1995), p. 27

[33] Y. Fukuda *et al.* (1998)

[34] C.Amsler et al. (2008): Particle listings – Neutrino properties

[35] B.R. Martin, G. Shaw (1992)

[36] J. P. Cumalat (1993). *Physics in Collision 12*. Atlantica Séguier Frontières. ISBN 978-2-86332-129-4.

[37] G Fraser (1 January 1998). *The Particle Century*. CRC Press. ISBN 978-1-4200-5033-2.

[38] J. Peltoniemi, J. Sarkamo (2005)

3.1.8 References

- C. Amsler *et al.* (Particle Data Group); Amsler; Doser; Antonelli; Asner; Babu; Baer; Band; Barnett; Bergren; Beringer; Bernardi; Bertl; Bichsel; Biebel; Bloch; Blucher; Blusk; Cahn; Carena; Caso; Ceccucci; Chakraborty; Chen; Chivukula; Cowan; Dahl; d'Ambrosio; Damour et al. (2008). "Review of Particle Physics". *Physics Letters B* **667**: 1. Bibcode:2008PhLB..667....1P. doi:10.1016/j.physletb.2008.07.018.

- I.V. Anicin (2005). "The Neutrino – Its Past, Present and Future". *SFIN (Institute of Physics, Belgrade) year XV, Series A: Conferences, No. A2 (2002) 3–59*: 3172. arXiv:physics/0503172. Bibcode:2005physics...3172A.

- Y.Fukuda; Hayakawa, T.; Ichihara, E.; Inoue, K.; Ishihara, K.; Ishino, H.; Itow, Y.; Kajita, T. et al. (1998). "Evidence for Oscillation of Atmospheric Neutrinos". *Physical Review Letters* **81** (8): 1562–1567. arXiv:hep-ex/9807003. Bibcode:1998PhRvL..81.1562F. doi:10.1103/PhysRevLett.81.1562.

- K. Kodama; Ushida, N.; Andreopoulos, C.; Saoulidou, N.; Tzanakos, G.; Yager, P.; Baller, B.; Boehnlein, D.; Freeman, W.; Lundberg, B.; Morfin, J.; Rameika, R.; Yun, J.C.; Song, J.S.; Yoon, C.S.; Chung, S.H.; Berghaus, P.; Kubantsev, M.; Reay, N.W.; Sidwell, R.; Stanton, N.; Yoshida, S.; Aoki, S.; Hara, T.; Rhee, J.T.; Ciampa, D.; Erickson, C.; Graham, M.; Heller, K. et al. (2001). "Observation of tau neutrino interactions". *Physics Letters B* **504** (3): 218. arXiv:hep-ex/0012035. Bibcode:2001PhLB..504..218D. doi:10.1016/S0370-2693(01)00307-0.

- B.R. Martin, G. Shaw (1992). "Chapter 2 – Leptons, quarks and hadrons". *Particle Physics*. John Wiley & Sons. pp. 23–47. ISBN 0-471-92358-3.

- S.H. Neddermeyer, C.D. Anderson; Anderson (1937). "Note on the Nature of Cosmic-Ray Particles". *Physical Review* **51** (10): 884–886. Bibcode:1937PhRv...51..884N. doi:10.1103/PhysRev.51.884.

- J. Peltoniemi, J. Sarkamo (2005). "Laboratory measurements and limits for neutrino properties". *The Ultimate Neutrino Page*. Retrieved 2008-11-07.

- M.L. Perl; Abrams, G.; Boyarski, A.; Breidenbach, M.; Briggs, D.; Bulos, F.; Chinowsky, W.; Dakin, J. et al. (1975). "Evidence for Anomalous Lepton Production in e+–e− Annihilation". *Physical Review Letters* **35** (22): 1489–1492. Bibcode:1975PhRvL..35.1489P. doi:10.1103/PhysRevLett.35.1489.

- M.E. Peskin, D.V. Schroeder (1995). *Introduction to Quantum Field Theory*. Westview Press. ISBN 0-201-50397-2.

- K. Riesselmann (2007). "Logbook: Neutrino Invention". *Symmetry Magazine* **4** (2).

- L. Rosenfeld (1948). *Nuclear Forces*. Interscience Publishers. p. xvii.

- R. Shankar (1994). "Chapter 2 – Rotational Invariance and Angular Momentum". *Principles of Quantum Mechanics* (2nd ed.). Springer. pp. 305–352. ISBN 978-0-306-44790-7.

- S. Weinberg (2003). *The Discovery of Subatomic Particles*. Cambridge University Press. ISBN 0-521-82351-X.

- R. Wilson (1997). *Astronomy Through the Ages: The Story of the Human Attempt to Understand the Universe*. CRC Press. p. 138. ISBN 0-7484-0748-0.

3.1.9   External links

- Particle Data Group homepage. The PDG compiles authoritative information on particle properties.

- Leptons, a summary of leptons from *Hyperphysics*.

3.2   Neutrino

For other uses, see Neutrino (disambiguation).

A **neutrino** (/nuːˈtriːnoʊ/ or /njuːˈtriːnoʊ/, in Italian [neuˈtrino]) is an electrically neutral elementary particle[4] with half-integer spin. The neutrino (meaning "little neutral one" in Italian) is denoted by the Greek letter ν (*nu*). All evidence suggests that neutrinos have mass but that their masses are tiny, even compared to other subatomic particles. They are the only identified candidate for dark matter, specifically hot dark matter.[5]

Neutrinos are leptons, along with the charged electrons, muons, and taus, and come in three flavors: electron neutrinos (ν_e), muon neutrinos (ν_μ), and tau neutrinos (ν_τ). Each flavor is also associated with an antiparticle, called an "antineutrino", which also has no electric charge and half-integer spin. Neutrinos are produced in a way that conserves lepton number; i.e., for every electron neutrino produced, a positron (anti-electron) is produced, and for every electron antineutrino produced, an electron is produced as well.

Neutrinos do not carry any electric charge, which means that they are not affected by the electromagnetic force that acts on charged particles, and are leptons, so they are not affected by the strong force that acts on particles inside atomic nuclei. Neutrinos are therefore affected only by the weak subatomic force and by gravity. The weak force is a very short-range interaction, and gravity is extremely weak on the subatomic scale. Thus, neutrinos typically pass through normal matter unimpeded and undetected.

Neutrinos can be created in several ways, including in certain types of radioactive decay, in nuclear reactions such as those that take place in the Sun, in nuclear reactors, when cosmic rays hit atoms and in supernovas. The majority of neutrinos in the vicinity of the earth are from nuclear reactions in the Sun. In fact, about 65 billion (6.5×10^{10}) solar neutrinos per second pass through every square centimeter perpendicular to the direction of the Sun in the region of the Earth.[6]

Neutrinos are now understood to oscillate between different flavors in flight. That is, an electron neutrino produced in a beta decay reaction may arrive in a detector as a muon or tau neutrino. This oscillation requires that the different neutrino flavors have different masses, although these masses have been shown to be tiny. From cosmological measurements, we know that the sum of the three neutrino masses must be less than one millionth that of the electron.[7]

3.2.1   History

Pauli's proposal

The neutrino[nb 1] was postulated first by Wolfgang Pauli in 1930 to explain how beta decay could conserve energy, momentum, and angular momentum (spin). In contrast to Niels Bohr, who proposed a statistical version of the conservation laws to explain the event, Pauli hypothesized an undetected particle that he called a "neutron" in keeping with convention employed for naming both the proton and the electron, which in 1930 were known to be respective products for alpha and beta decay. He considered that the new particle was emitted from the nucleus together with the electron or beta particle in the process of beta decay.[8][nb 2]

James Chadwick discovered a much more massive nuclear particle in 1932 and also named it a neutron, leaving two kinds of particles with the same name. Pauli earlier had used the term "neutron" for both the particle that conserved energy in beta decay, and a presumed neutral particle in the nucleus.[nb 3] The word "neutrino" entered the international vocabulary through Enrico Fermi, who used it during a conference in Paris in July 1932 and at the Solvay Conference in October 1933, where also Pauli employed it. The name (the Italian equivalent of "little neutral one") was jokingly coined by Edoardo Amaldi during a conversation with Fermi at the Institute of physics of via Panisperna in Rome, in order to distinguish this light neutral particle from Chadwick's neutron. [9]

In Fermi's theory of beta decay, Chadwick's large neutral particle could decay to a proton, electron, and the smaller neutral particle (flavored as an electron antineutrino):

$$n^0 \rightarrow p^+ + e^- + \nu_e$$

Fermi's paper, written in 1934, unified Pauli's neutrino with

Paul Dirac's positron and Werner Heisenberg's neutron–proton model and gave a solid theoretical basis for future experimental work. However, the journal Nature rejected Fermi's paper, saying that the theory was "too remote from reality". He submitted the paper to an Italian journal, which accepted it, but the general lack of interest in his theory at that early date caused him to switch to experimental physics.[10][11]

Nevertheless, even in 1934 there were hints that Bohr's idea that the energy conservation laws were not followed, was incorrect. At the Solvay conference of 1934, the first measurements of the energy spectra of beta decay were reported, and these spectra were found to impose a strict limit on the energy of electrons from each type of beta decay. Such a limit was not expected if the conservation of energy was not upheld, in which case any amount of energy would be expected to be statistically available in at least a few decays. The natural explanation of the beta decay spectrum as first measured in 1934 was that only a limited (and conserved) amount of energy was available, and a new particle was sometimes taking a varying fraction of this limited energy, leaving the rest for the beta particle. Pauli made use of the occasion to publicly emphasize that the still-undetected "neutrino" must be an actual particle.

Direct detection

Clyde Cowan conducting the neutrino experiment c. 1956

In 1942 Wang Ganchang first proposed the use of beta capture to experimentally detect neutrinos.[12] In the 20 July 1956 issue of *Science*, Clyde Cowan, Frederick Reines, F. B. Harrison, H. W. Kruse, and A. D. McGuire published

confirmation that they had detected the neutrino,[13][14] a result that was rewarded almost forty years later with the 1995 Nobel Prize.[15]

In this experiment, now known as the Cowan–Reines neutrino experiment, antineutrinos created in a nuclear reactor by beta decay reacted with protons to produce neutrons and positrons:

$$\nu$$
$$e + p+ \rightarrow n0 + e+$$

The positron quickly finds an electron, and they annihilate each other. The two resulting gamma rays (γ) are detectable. The neutron can be detected by its capture on an appropriate nucleus, releasing a gamma ray. The coincidence of both events – positron annihilation and neutron capture – gives a unique signature of an antineutrino interaction.

Neutrino flavor

The antineutrino discovered by Cowan and Reines is the antiparticle of the electron neutrino. In 1962, Leon M. Lederman, Melvin Schwartz and Jack Steinberger showed that more than one type of neutrino exists by first detecting interactions of the muon neutrino (already hypothesised with the name *neutretto*),[16] which earned them the 1988 Nobel Prize in Physics. When the third type of lepton, the tau, was discovered in 1975 at the Stanford Linear Accelerator Center, it too was expected to have an associated neutrino (the tau neutrino). First evidence for this third neutrino type came from the observation of missing energy and momentum in tau decays analogous to the beta decay leading to the discovery of the electron neutrino. The first detection of tau neutrino interactions was announced in summer of 2000 by the DONUT collaboration at Fermilab; its existence had already been inferred by both theoretical consistency and experimental data from the Large Electron–Positron Collider.

Solar neutrino problem

Main article: Solar neutrino problem

Starting in the late 1960s, several experiments found that the number of electron neutrinos arriving from the Sun was between one third and one half the number predicted by the Standard Solar Model. This discrepancy, which became known as the solar neutrino problem, remained unresolved for some thirty years. It was resolved by discovery of neutrino oscillation and mass. (The Standard Model of particle physics had assumed that neutrinos are massless and cannot change flavor. However, if neutrinos had mass, they could change flavor, or *oscillate* between flavors).

Oscillation

A practical method for investigating neutrino oscillations was first suggested by Bruno Pontecorvo in 1957 using an analogy with kaon oscillations; over the subsequent 10 years he developed the mathematical formalism and the modern formulation of vacuum oscillations. In 1985 Stanislav Mikheyev and Alexei Smirnov (expanding on 1978 work by Lincoln Wolfenstein) noted that flavor oscillations can be modified when neutrinos propagate through matter. This so-called Mikheyev–Smirnov–Wolfenstein effect (MSW effect) is important to understand because many neutrinos emitted by fusion in the Sun pass through the dense matter in the solar core (where essentially all solar fusion takes place) on their way to detectors on Earth.

Starting in 1998, experiments began to show that solar and atmospheric neutrinos change flavors (see Super-Kamiokande and Sudbury Neutrino Observatory). This resolved the solar neutrino problem: the electron neutrinos produced in the Sun had partly changed into other flavors which the experiments could not detect.

Although individual experiments, such as the set of solar neutrino experiments, are consistent with non-oscillatory mechanisms of neutrino flavor conversion, taken altogether, neutrino experiments imply the existence of neutrino oscillations. Especially relevant in this context are the reactor experiment KamLAND and the accelerator experiments such as MINOS. The KamLAND experiment has indeed identified oscillations as the neutrino flavor conversion mechanism involved in the solar electron neutrinos. Similarly MINOS confirms the oscillation of atmospheric neutrinos and gives a better determination of the mass squared splitting.[17]

Supernova neutrinos

See also: Supernova Early Warning System

Raymond Davis, Jr. and Masatoshi Koshiba were jointly awarded the 2002 Nobel Prize in Physics; Davis for his pioneer work on cosmic neutrinos and Koshiba for the first real time observation of supernova neutrinos. The detection of solar neutrinos, and of neutrinos of the SN 1987A supernova in 1987 marked the beginning of neutrino astronomy. In an average supernova, approximately 10^{57} (an Octodecillion) neutrinos are released.

3.2.2   Properties and reactions

The neutrino has half-integer spin ($\hbar/2$) and is therefore a fermion. Neutrinos interact primarily through the weak force. The discovery of neutrino flavor oscillations implies that neutrinos have mass. The existence of a neutrino mass strongly suggests the existence of a tiny neutrino magnetic moment[18] of the order of 10^{-19} μB, allowing the possibility that neutrinos may interact electromagnetically as well. An experiment done by C. S. Wu at Columbia University showed that neutrinos always have left-handed chirality.[19] It is very hard to uniquely identify neutrino interactions among the natural background of radioactivity. For this reason, in early experiments a special reaction channel was chosen to facilitate the identification: the interaction of an antineutrino with one of the hydrogen nuclei in the water molecules. A hydrogen nucleus is a single proton, so simultaneous nuclear interactions, which would occur within a heavier nucleus, don't need to be considered for the detection experiment. Within a cubic metre of water placed right outside a nuclear reactor, only relatively few such interactions can be recorded, but the setup is now used for measuring the reactor's plutonium production rate.

Mikheyev–Smirnov–Wolfenstein effect

Main article: Mikheyev–Smirnov–Wolfenstein effect

Neutrinos traveling through matter, in general, undergo a process analogous to light traveling through a transparent material. This process is not directly observable because it does not produce ionizing radiation, but gives rise to the MSW effect. Only a small fraction of the neutrino's energy is transferred to the material.

Nuclear reactions

Neutrinos can interact with a nucleus, changing it to another nucleus. This process is used in radiochemical neutrino detectors. In this case, the energy levels and spin states within the target nucleus have to be taken into account to estimate the probability for an interaction. In general the interaction probability increases with the number of neutrons and protons within a nucleus.

Induced fission

Very much like neutrons do in nuclear reactors, neutrinos can induce fission reactions within heavy nuclei.[20] So far, this reaction has not been measured in a laboratory, but is predicted to happen within stars and supernovae. The process affects the abundance of isotopes seen in the universe.[21] Neutrino fission of deuterium nuclei has been observed in the Sudbury Neutrino Observatory, which uses a heavy water detector.

Types

There are three known types (*flavors*) of neutrinos: electron neutrino ν
e, muon neutrino ν
μ and tau neutrino ν
τ, named after their partner leptons in the Standard Model (see table at right). The current best measurement of the number of neutrino types comes from observing the decay of the Z boson. This particle can decay into any light neutrino and its antineutrino, and the more types of light neutrinos[nb 4] available, the shorter the lifetime of the Z boson. Measurements of the Z lifetime have shown that the number of light neutrino types is 3.[18] The correspondence between the six quarks in the Standard Model and the six leptons, among them the three neutrinos, suggests to physicists' intuition that there should be exactly three types of neutrino. However, actual proof that there are only three kinds of neutrinos remains an elusive goal of particle physics.

The possibility of *sterile* neutrinos—relatively light neutrinos which do not participate in the weak interaction but which could be created through flavor oscillation (see below)—is unaffected by these Z-boson-based measurements, and the existence of such particles is in fact hinted by experimental data from the LSND experiment. However, the currently running MiniBooNE experiment suggested, until recently, that sterile neutrinos are not required to explain the experimental data,[22] although the latest research into this area is on-going and anomalies in the MiniBooNE data may allow for exotic neutrino types, including sterile neutrinos.[23] A recent re-analysis of reference electron spectra data from the Institut Laue-Langevin[24] has also hinted at a fourth, sterile neutrino.[25]

Recently analyzed data from the Wilkinson Microwave Anisotropy Probe of the cosmic background radiation is compatible with either three or four types of neutrinos. It is hoped that the addition of two more years of data from the probe will resolve this uncertainty.[26]

Antineutrinos

Antineutrinos, the antiparticles of neutrinos, are neutral particles produced in nuclear beta decay. These are emitted during beta particle emissions, in which a neutron decays into a proton, electron, and antineutrino. They have a spin of ½, and are part of the lepton family of particles. All antineutrinos observed thus far possess right-handed helicity (i.e. only one of the two possible spin states has ever been seen), while neutrinos are left-handed. Antineutrinos, like neutrinos, interact with other matter only through the gravitational and weak forces, making them very difficult to detect experimentally. Neutrino oscillation experiments indicate that antineutrinos have mass, but beta decay experiments constrain that mass to be very small. A neutrino–antineutrino interaction has been suggested in attempts to form a composite photon with the neutrino theory of light.

Because antineutrinos and neutrinos are neutral particles, it is possible that they are actually the same particle. Particles that have this property are known as Majorana particles. Majorana neutrinos have the property that the neutrino and antineutrino could be distinguished only by chirality; what experiments observe as a difference between the neutrino and antineutrino could simply be due to one particle with two possible chiralities. If neutrinos are indeed Majorana particles, neutrinoless double beta decay, as well as a range of other lepton number violating phenomena, would be allowed. Several experiments have been and are being conducted to search for this process.

Researchers around the world have begun to investigate the possibility of using antineutrinos for reactor monitoring in the context of preventing the proliferation of nuclear weapons.[27][28][29]

Antineutrinos were first detected as a result of their interaction with protons in a large tank of water. This was installed next to a nuclear reactor as a controllable source of the antineutrinos. (See: Cowan–Reines neutrino experiment)

Only antineutrinos, not neutrinos, take part in the Glashow resonance.

Flavor oscillations

Main article: Neutrino oscillation

Neutrinos are most often created or detected with a well defined flavor (electron, muon, tau). However, in a phenomenon known as neutrino flavor oscillation, neutrinos are able to oscillate among the three available flavors while they propagate through space. Specifically, this occurs because the neutrino flavor eigenstates are not the same as the neutrino mass eigenstates (simply called 1, 2, 3). This allows for a neutrino that was produced as an electron neutrino at a given location to have a calculable probability to be detected as either a muon or tau neutrino after it has traveled to another location. This quantum mechanical effect was first hinted by the discrepancy between the number of electron neutrinos detected from the Sun's core failing to match the expected numbers, dubbed as the "solar neutrino problem". In the Standard Model the existence of flavor oscillations implies nonzero differences between the neutrino masses, because the amount of mixing between neutrino flavors at a given time depends on the differences between their squared masses. There are other possibilities in which neutrino can oscillate even if they are massless. If Lorentz invariance is not an exact symmetry, neutrinos can experience Lorentz-violating oscillations.[30]

It is possible that the neutrino and antineutrino are in fact the same particle, a hypothesis first proposed by the Italian physicist Ettore Majorana. The neutrino could transform into an antineutrino (and vice versa) by flipping the orientation of its spin state.[31]

This change in spin would require the neutrino and antineutrino to have nonzero mass, and therefore travel slower than light, because such a spin flip, caused only by a change in point of view, can take place only if inertial frames of reference exist that move faster than the particle: such a particle has a spin of one orientation when seen from a frame which moves slower than the particle, but the opposite spin when observed from a frame that moves faster than the particle.

On July 19, 2013 the results from the T2K experiment presented at the European Physical Society Conference on High Energy Physics in Stockholm, Sweden, confirmed neutrino oscillation theory.[32][33]

Speed

Main article: Measurements of neutrino speed

Before neutrinos were found to oscillate, they were generally assumed to be massless, propagating at the speed of light. According to the theory of special relativity, the question of neutrino velocity is closely related to their mass. If neutrinos are massless, they must travel at the speed of light. However, if they have mass, they cannot reach the speed of light.

Also some Lorentz-violating variants of quantum gravity might allow faster-than-light neutrinos. A comprehensive framework for Lorentz violations is the Standard-Model Extension (SME).

In the early 1980s, first measurements of neutrino speed were done using pulsed pion beams (produced by pulsed proton beams hitting a target). The pions decayed producing neutrinos, and the neutrino interactions observed within a time window in a detector at a distance were consistent with the speed of light. This measurement was repeated in 2007 using the MINOS detectors, which found the speed of 3 GeV neutrinos to be, at the 99% confidence level, in the range between $0.999976\,c$ and $1.000126\,c$. The central value of $1.000051c$ is higher than the speed of light but is also consistent with a velocity of exactly c or even slightly

less. This measurement set an upper bound on the mass of the muon neutrino of 50 MeV at 99% confidence.[34][35] After the detectors for the project were upgraded in 2012, MINOS refined their initial result and found agreement with the speed of light, with the difference in the arrival time of neutrinos and light of -0.0006% ($\pm 0.0012\%$).[36]

A similar observation was made, on a much larger scale, with supernova 1987A (SN 1987A). 10-MeV antineutrinos from the supernova were detected within a time window that was consistent with the speed of light for the neutrinos. Currently, the question of whether or not neutrinos have mass cannot be decided; their speed is (as yet) indistinguishable from the speed of light.

In September 2011, the OPERA collaboration released calculations showing velocities of 17-GeV and 28-GeV neutrinos exceeding the speed of light in their experiments (see Faster-than-light neutrino anomaly). In November 2011, OPERA repeated its experiment with changes so that the speed could be determined individually for each detected neutrino. The results showed the same faster-than-light speed. However, in February 2012 reports came out that the results may have been caused by a loose fiber optic cable attached to one of the atomic clocks which measured the departure and arrival times of the neutrinos. An independent recreation of the experiment in the same laboratory by ICARUS found no discernible difference between the speed of a neutrino and the speed of light.[37] In June 2012, CERN announced that new measurements conducted by four Gran Sasso experiments (OPERA, ICARUS, Borexino and LVD) found agreement between the speed of light and the speed of neutrinos, finally refuting the initial OPERA result.[38]

Mass

The Standard Model of particle physics assumed that neutrinos are massless. However the experimentally established phenomenon of neutrino oscillation, which mixes neutrino flavour states with neutrino mass states (analogously to CKM mixing), requires neutrinos to have nonzero masses.[22] Massive neutrinos were originally conceived by Bruno Pontecorvo in the 1950s. Enhancing the basic framework to accommodate their mass is straightforward by adding a right-handed Lagrangian. This can be done in two ways. If, like other fundamental Standard Model particles, mass is generated by the Dirac mechanism, then the framework would require an SU(2) singlet. This particle would have no other Standard Model interactions (apart from the Yukawa interactions with the neutral component of the Higgs doublet), so is called a sterile neutrino. Or, mass can be generated by the Majorana mechanism, which would require the neutrino and antineutrino to be the same particle.

The strongest upper limit on the masses of neutrinos comes from cosmology: the Big Bang model predicts that there is a fixed ratio between the number of neutrinos and the number of photons in the cosmic microwave background. If the total energy of all three types of neutrinos exceeded an average of 50 eV per neutrino, there would be so much mass in the universe that it would collapse.[39] This limit can be circumvented by assuming that the neutrino is unstable; however, there are limits within the Standard Model that make this difficult. A much more stringent constraint comes from a careful analysis of cosmological data, such as the cosmic microwave background radiation, galaxy surveys, and the Lyman-alpha forest. These indicate that the summed masses of the three neutrinos must be less than 0.3 eV.[40]

In 1998, research results at the Super-Kamiokande neutrino detector determined that neutrinos can oscillate from one flavor to another, which requires that they must have a nonzero mass.[41] While this shows that neutrinos have mass, the absolute neutrino mass scale is still not known. This is because neutrino oscillations are sensitive only to the difference in the squares of the masses.[42] The best estimate of the difference in the squares of the masses of mass eigenstates 1 and 2 was published by KamLAND in 2005: $\Delta m^2_{21} = 0.000079$ eV^2.[43] In 2006, the MINOS experiment measured oscillations from an intense muon neutrino beam, determining the difference in the squares of the masses between neutrino mass eigenstates 2 and 3. The initial results indicate $|\Delta m^2_{32}| = 0.0027$ eV^2, consistent with previous results from Super-Kamiokande.[44] Since $|\Delta m^2_{32}|$ is the difference of two squared masses, at least one of them has to have a value which is at least the square root of this value. Thus, there exists at least one neutrino mass eigenstate with a mass of at least 0.04 eV.[45]

In 2009, lensing data of a galaxy cluster were analyzed to predict a neutrino mass of about 1.5 eV.[46] This surprisingly high value requires that the three neutrino masses be nearly equal, with neutrino oscillations of order meV. The masses lie below the Mainz-Troitsk upper bound of 2.2 eV for the electron antineutrino.[47] The latter will be tested in 2015 in the KATRIN experiment, that searches for a mass between 0.2 eV and 2 eV.

A number of efforts are under way to directly determine the absolute neutrino mass scale in laboratory experiments. The methods applied involve nuclear beta decay (KATRIN and MARE).

On 31 May 2010, OPERA researchers observed the first tau neutrino candidate event in a muon neutrino beam, the first time this transformation in neutrinos had been observed, providing further evidence that they have mass.[48]

In July 2010 the 3-D MegaZ DR7 galaxy survey reported that they had measured a limit of the combined mass of the three neutrino varieties to be less than 0.28 eV.[49] A tighter upper bound yet for this sum of masses, 0.23 eV, was reported in March 2013 by the Planck collaboration,[50] whereas a February 2014 result estimates the sum as 0.320 ± 0.081 eV based on discrepancies between the cosmological consequences implied by Planck's detailed measurements of the Cosmic Microwave Background and predictions arising from observing other phenomena, combined with the assumption that neutrinos are responsible for the observed weaker gravitational lensing than would be expected from massless neutrinos.[51]

If the neutrino is a Majorana particle, the mass may be calculated by finding the half life of neutrinoless double-beta decay of certain nuclei. As of 2015, the lowest upper limit on the Majorana mass of the neutrino has been set by KamLAND-Zen: 0.12–0.25 eV.[52]

Size

Standard Model neutrinos are fundamental point-like particles. An effective size can be defined using their electroweak cross section (apparent size in electroweak interaction). The average electroweak characteristic size is $r^2 = n \times 10^{-33}$ cm^2 ($n \times 1$ nanobarn), where $n = 3.2$ for electron neutrino, $n = 1.7$ for muon neutrino and $n = 1.0$ for tau neutrino; it depends on no other properties than mass.[53] However, this is best understood as being relevant only to probability of scattering. Since the neutrino does not interact electromagnetically, and is defined quantum mechanically by a wavefunction, it does not have a size in the same sense as everyday objects.[54] Furthermore, processes that produce neutrinos impart such high energies to them that they travel at almost the speed of light. Nevertheless, neutrinos are fermions, and thus obey the Pauli exclusion principle, i.e. that increasing their density forces them into progressively higher momentum states.

Chirality

Experimental results show that (nearly) all produced and observed neutrinos have left-handed helicities (spins antiparallel to momenta), and all antineutrinos have right-handed helicities, within the margin of error. In the massless limit, it means that only one of two possible chiralities is observed for either particle. These are the only chiralities included in the Standard Model of particle interactions.

It is possible that their counterparts (right-handed neutrinos and left-handed antineutrinos) simply do not exist. If they do, their properties are substantially different from observable neutrinos and antineutrinos. It is theorized that they are either very heavy (on the order of GUT scale—see *Seesaw mechanism*), do not participate in weak interaction (so-called sterile neutrinos), or both.

The existence of nonzero neutrino masses somewhat complicates the situation. Neutrinos are produced in weak interactions as chirality eigenstates. However, chirality of a massive particle is not a constant of motion; helicity is, but the chirality operator does not share eigenstates with the helicity operator. Free neutrinos propagate as mixtures of left- and right-handed helicity states, with mixing amplitudes on the order of mv/E. This does not significantly affect the experiments, because neutrinos involved are nearly always ultrarelativistic, and thus mixing amplitudes are vanishingly small. For example, most solar neutrinos have energies on the order of 100 keV–1 MeV, so the fraction of neutrinos with "wrong" helicity among them cannot exceed 10^{-10}.[55][56]

3.2.3   Sources

Artificial

Reactor neutrinos   Nuclear reactors are the major source of human-generated neutrinos. Antineutrinos are made in the beta-decay of neutron-rich daughter fragments in the fission process. Generally, the four main isotopes contributing to the antineutrino flux are 235U, 238U, 239Pu and 241Pu (i.e. via the antineutrinos emitted during beta-minus decay of their respective fission fragments). The average nuclear fission releases about 200 MeV of energy, of which roughly 4.5% (or about 9 MeV)[57] is radiated away as antineutrinos. For a typical nuclear reactor with a thermal power of 4000 MW, meaning that the core produces this much heat, and an electrical power generation of 1300 MW, the total power production from fissioning atoms is actually 4185 MW, of which 185 MW is radiated away as antineutrino radiation and never appears in the engineering. This is to say, 185 MW of fission energy is *lost* from this reactor and does not appear as heat available to run turbines, since antineutrinos penetrate all building materials practically without interaction.[nb 5]

The antineutrino energy spectrum depends on the degree to which the fuel is burned (plutonium-239 fission antineutrinos on average have slightly more energy than those from uranium-235 fission), but in general, the *detectable* antineutrinos from fission have a peak energy between about 3.5 and 4 MeV, with a maximum energy of about 10 MeV.[58] There is no established experimental method to measure the flux of low-energy antineutrinos. Only antineutrinos with an energy above threshold of 1.8 MeV can be uniquely

identified (see *neutrino detection* below). An estimated 3% of all antineutrinos from a nuclear reactor carry an energy above this threshold. Thus, an average nuclear power plant may generate over 10^{20} antineutrinos per second above this threshold, but also a much larger number (97%/3% = ~30 times this number) below the energy threshold, which cannot be seen with present detector technology.

Accelerator neutrinos   Some particle accelerators have been used to make neutrino beams. The technique is to collide protons with a fixed target, producing charged pions or kaons. These unstable particles are then magnetically focused into a long tunnel where they decay while in flight. Because of the relativistic boost of the decaying particle, the neutrinos are produced as a beam rather than isotropically. Efforts to construct an accelerator facility where neutrinos are produced through muon decays are ongoing.[59] Such a setup is generally known as a neutrino factory.

Nuclear bombs   Nuclear bombs also produce very large quantities of neutrinos. Fred Reines and Clyde Cowan considered the detection of neutrinos from a bomb prior to their search for reactor neutrinos; a fission reactor was recommended as a better alternative by Los Alamos physics division leader J.M.B. Kellogg.[60] Fission bombs produce antineutrinos (from the fission process), and fusion bombs produce both neutrinos (from the fusion process) and antineutrinos (from the initiating fission explosion).

Geologic

Main article: Geoneutrino

Neutrinos are part of the natural background radiation. In particular, the decay chains of 238U and 232Th isotopes, as well as 40K, include beta decays which emit antineutrinos. These so-called geoneutrinos can provide valuable information on the Earth's interior. A first indication for geoneutrinos was found by the KamLAND experiment in 2005. KamLAND's main background in the geoneutrino measurement are the antineutrinos coming from reactors. Several future experiments aim at improving the geoneutrino measurement and these will necessarily have to be far away from reactors.

Atmospheric

Atmospheric neutrinos result from the interaction of cosmic rays with atomic nuclei in the Earth's atmosphere, creating showers of particles, many of which are unstable and produce neutrinos when they decay. A collaboration of particle

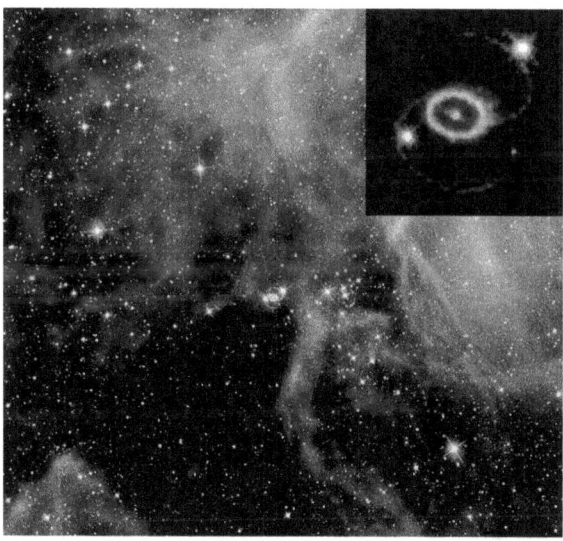

Solar neutrinos (proton–proton chain) in the Standard Solar Model

SN 1987A

physicists from Tata Institute of Fundamental Research (India), Osaka City University (Japan) and Durham University (UK) recorded the first cosmic ray neutrino interaction in an underground laboratory in Kolar Gold Fields in India in 1965.

Solar

Solar neutrinos originate from the nuclear fusion powering the Sun and other stars. The details of the operation of the Sun are explained by the Standard Solar Model. In short: when four protons fuse to become one helium nucleus, two of them have to convert into neutrons, and each such conversion releases one electron neutrino.

The Sun sends enormous numbers of neutrinos in all directions. Each second, about 65 billion (6.5×10^{10}) solar neutrinos pass through every square centimeter on the part of the Earth that faces the Sun.[6] Since neutrinos are insignificantly absorbed by the mass of the Earth, the surface area on the side of the Earth opposite the Sun receives about the same number of neutrinos as the side facing the Sun.

Supernovae

In 1966 Colgate and White[61] calculated that neutrinos carry away most of the gravitational energy released by the collapse of massive stars, events now categorized as Type Ib and Ic and Type II supernovae. When such stars collapse, matter densities at the core becomes so high (10^{17} kg/m^3) that the degeneracy of electrons is not enough to prevent protons and electrons from combining to form a neutron and an electron neutrino. A second and more important neutrino source is the thermal energy (100 billion kelvins) of the newly formed neutron core, which is dissipated via the formation of neutrino–antineutrino pairs of all flavors.[62]

Colgate and White's theory of supernova neutrino production was confirmed in 1987, when neutrinos from supernova 1987A were detected. The water-based detectors Kamiokande II and IMB detected 11 and 8 antineutrinos of thermal origin,[62] respectively, while the scintillator-based Baksan detector found 5 neutrinos (lepton number = 1) of either thermal or electron-capture origin, in a burst lasting less than 13 seconds. The neutrino signal from the supernova arrived at earth several hours before the arrival of the first electromagnetic radiation, as expected from the evident fact that the latter emerges along with the shock wave. The exceptionally feeble interaction with normal matter allowed the neutrinos to pass through the churning mass of the exploding star, while the electromagnetic photons were slowed.

Because neutrinos interact so little with matter, it is thought that a supernova's neutrino emissions carry information about the innermost regions of the explosion. Much of the *visible* light comes from the decay of radioactive elements produced by the supernova shock wave, and even light from the explosion itself is scattered by dense and turbulent gases, and thus delayed. The neutrino burst is expected to reach Earth before any electromagnetic waves, including visible light, gamma rays or radio waves. The exact time delay depends on the velocity of the shock wave and on the thickness of the outer layer of the star. For a Type II supernova, astronomers expect the neutrino flood to be released seconds after the stellar core collapse, while the first electromagnetic signal may emerge hours later, after the explosion shock wave has had time to reach the surface of the star. The SNEWS project uses a network of neutrino detectors to monitor the sky for candidate supernova events; the neutrino signal will provide a useful advance warning of a star exploding in the Milky Way.

Although neutrinos pass through the outer gases of a supernova without scattering, they provide information about the deeper supernova core with evidence that here, even neutrinos scatter to a significant extent. In a supernova core the densities are those of a neutron star (which is expected to be formed in this type of supernova),[63] becoming large enough to influence the duration of the neutrino signal by delaying some neutrinos. The length of the neutrino signal from SN 1987A, some 13 seconds, was far longer than it would take in theory for neutrinos to pass directly through the neutrino-generating core of a supernova, expected to be only 32 kilometers in diameter SN 1987A. The number of neutrinos counted was also consistent with a total neutrino energy of 2.2×10^{46} joules, which was estimated to be nearly all of the total energy of the supernova.[64]

Supernova remnants

The energy of supernova neutrinos ranges from a few to several tens of MeV. However, the sites where cosmic rays are accelerated are expected to produce neutrinos that are at least one million times more energetic, produced from turbulent gaseous environments left over by supernova explosions: the supernova remnants. The origin of the cosmic rays was attributed to supernovas by Walter Baade and Fritz Zwicky; this hypothesis was refined by Vitaly L. Ginzburg and Sergei I. Syrovatsky who attributed the origin to supernova remnants, and supported their claim by the crucial remark, that the cosmic ray losses of the Milky Way is compensated, if the efficiency of acceleration in supernova remnants is about 10 percent. Ginzburg and Syrovatskii's hypothesis is supported by the specific mechanism of "shock wave acceleration" happening in supernova remnants, which is consistent with the original theoretical picture drawn by Enrico Fermi, and is receiving support from observational data. The very-high-energy neutrinos are still to be seen, but this branch of neutrino astronomy is just in its infancy. The main existing or forthcoming experiments that aim at observing very-high-energy neutrinos from our galaxy are Baikal, AMANDA, IceCube, ANTARES, NEMO and Nestor. Related information is provided by very-high-energy gamma ray observatories, such as VERITAS, HESS and MAGIC. Indeed, the collisions of cosmic rays are supposed to produce charged pions, whose decay give the neutrinos, and also neutral pions, whose decay give gamma rays: the environment of a supernova remnant is transparent to both types of radiation.

Still-higher-energy neutrinos, resulting from the interactions of extragalactic cosmic rays, could be observed with the Pierre Auger Observatory or with the dedicated experiment named ANITA.

Big Bang

Main article: Cosmic neutrino background

It is thought that, just like the cosmic microwave background radiation left over from the Big Bang, there is a background of low-energy neutrinos in our Universe. In the 1980s it was proposed that these may be the explanation for the dark matter thought to exist in the universe. Neutrinos have one important advantage over most other dark matter candidates: it is known that they exist. However, this idea also has serious problems.

From particle experiments, it is known that neutrinos are very light. This means that they easily move at speeds close to the speed of light. For this reason, dark matter made from neutrinos is termed "hot dark matter". The problem is that being fast moving, the neutrinos would tend to have spread out evenly in the universe before cosmological expansion made them cold enough to congregate in clumps. This would cause the part of dark matter made of neutrinos to be smeared out and unable to cause the large galactic structures that we see.

Further, these same galaxies and groups of galaxies appear to be surrounded by dark matter that is not fast enough to escape from those galaxies. Presumably this matter provided the gravitational nucleus for formation. This implies that neutrinos cannot make up a significant part of the total amount of dark matter.

From cosmological arguments, relic background neutrinos are estimated to have density of 56 of each type per cubic centimeter and temperature 1.9 K (1.7×10^{-4} eV) if they are massless, much colder if their mass exceeds 0.001 eV. Although their density is quite high, they have not yet been observed in the laboratory, as their energy is below thresholds of most detection methods, and due to extremely low neutrino interaction cross-sections at sub-eV energies. In contrast, boron-8 solar neutrinos—which are emitted with a higher energy—have been detected definitively despite having a space density that is lower than that of relic neutrinos by some 6 orders of magnitude.

3.2.4 Detection

Main article: Neutrino detector

Neutrinos cannot be detected directly, because they do not ionize the materials they are passing through (they do not carry electric charge and other proposed effects, like the MSW effect, do not produce traceable radiation). A unique reaction to identify antineutrinos, sometimes referred to as inverse beta decay, as applied by Reines and Cowan (see below), requires a very large detector in order to detect a significant number of neutrinos. All detection methods require the neutrinos to carry a minimum threshold energy. So far, there is no detection method for low-energy neutrinos, in the sense that potential neutrino interactions (for example by the MSW effect) cannot be uniquely distinguished from other causes. Neutrino detectors are often built underground in order to isolate the detector from cosmic rays and other background radiation.

Antineutrinos were first detected in the 1950s near a nuclear reactor. Reines and Cowan used two targets containing a solution of cadmium chloride in water. Two scintillation detectors were placed next to the cadmium targets. Antineutrinos with an energy above the threshold of 1.8 MeV caused charged current interactions with the protons in the water, producing positrons and neutrons. This is very much like β+ decay, where energy is used to convert a proton into a neutron, a positron (e+) and an electron neutrino (ν e) is emitted:

From known β+ decay:

Energy + p → n + e+ + ν
e

In the Cowan and Reines experiment, instead of an outgoing neutrino, you have an incoming antineutrino (ν e) from a nuclear reactor:

Energy (>1.8 MeV) + p + ν
e → n + e+

The resulting positron annihilation with electrons in the detector material created photons with an energy of about 0.5 MeV. Pairs of photons in coincidence could be detected by the two scintillation detectors above and below the target. The neutrons were captured by cadmium nuclei resulting in gamma rays of about 8 MeV that were detected a few microseconds after the photons from a positron annihilation event.

Since then, various detection methods have been used. Super Kamiokande is a large volume of water surrounded by photomultiplier tubes that watch for the Cherenkov radiation emitted when an incoming neutrino creates an electron or muon in the water. The Sudbury Neutrino Observatory is similar, but uses heavy water as the detecting medium, which uses the same effects, but also allows the additional reaction any-flavor neutrino photo-dissociation of deuterium, resulting in a free neutron which is then detected from gamma radiation after chlorine-capture. Other detectors have consisted of large volumes of chlorine or gallium which are periodically checked for excesses of argon or germanium, respectively, which are created by electron-neutrinos interacting with the original substance. MINOS

uses a solid plastic scintillator coupled to photomultiplier tubes, while Borexino uses a liquid pseudocumene scintillator also watched by photomultiplier tubes and the proposed NOvA detector will use liquid scintillator watched by avalanche photodiodes. The IceCube Neutrino Observatory uses 1 km^3 of the Antarctic ice sheet near the south pole with photomultiplier tubes distributed throughout the volume.

3.2.5 Motivation for scientific interest

Neutrinos' low mass and neutral charge mean they interact exceedingly weakly with other particles and fields. This feature of weak interaction interests scientists because it means neutrinos can be used to probe environments that other radiation (such as light or radio waves) cannot penetrate.

Using neutrinos as a probe was first proposed in the mid-20th century as a way to detect conditions at the core of the Sun. The solar core cannot be imaged directly because electromagnetic radiation (such as light) is diffused by the great amount and density of matter surrounding the core. On the other hand, neutrinos pass through the Sun with few interactions. Whereas photons emitted from the solar core may require 40,000 years to diffuse to the outer layers of the Sun, neutrinos generated in stellar fusion reactions at the core cross this distance practically unimpeded at nearly the speed of light.[65][66]

Neutrinos are also useful for probing astrophysical sources beyond the Solar System because they are the only known particles that are not significantly attenuated by their travel through the interstellar medium. Optical photons can be obscured or diffused by dust, gas, and background radiation. High-energy cosmic rays, in the form of swift protons and atomic nuclei, are unable to travel more than about 100 megaparsecs due to the Greisen–Zatsepin–Kuzmin limit (GZK cutoff). Neutrinos, in contrast, can travel even greater distances barely attenuated.

The galactic core of the Milky Way is fully obscured by dense gas and numerous bright objects. Neutrinos produced in the galactic core might be measurable by Earth-based neutrino telescopes.

Another important use of the neutrino is in the observation of supernovae, the explosions that end the lives of highly massive stars. The core collapse phase of a supernova is an extremely dense and energetic event. It is so dense that no known particles are able to escape the advancing core front except for neutrinos. Consequently, supernovae are known to release approximately 99% of their radiant energy in a short (10-second) burst of neutrinos.[67] These neutrinos are a very useful probe for core collapse studies.

The rest mass of the neutrino (see above) is an important

test of cosmological and astrophysical theories (see *Dark matter*). The neutrino's significance in probing cosmological phenomena is as great as any other method, and is thus a major focus of study in astrophysical communities.[68]

The study of neutrinos is important in particle physics because neutrinos typically have the lowest mass, and hence are examples of the lowest-energy particles theorized in extensions of the Standard Model of particle physics.

In November 2012 American scientists used a particle accelerator to send a coherent neutrino message through 780 feet of rock. This marks the first use of neutrinos for communication, and future research may permit binary neutrino messages to be sent immense distances through even the densest materials, such as the Earth's core.[69]

3.2.6 See also

- List of neutrino experiments

3.2.7 Notes

[1] More specifically, the electron neutrino.

[2] Niels Bohr was notably opposed to this interpretation of beta decay and was ready to accept that energy, momentum and angular momentum were not conserved quantities.

[3] These events necessitated renaming Pauli's less massive, momentum-conserving particle.

[4] In this context, "light neutrino" means neutrinos with less than half the mass of the Z boson.

[5] Typically about one third of the heat which is deposited in a reactor core is available to be converted to electricity, and a 4000 MW reactor would produce only 2700 MW of actual heat, with the rest being converted to its 1300 MW of electric power production.

3.2.8 References

[1] "Astronomers Accurately Measure the Mass of Neutrinos for the First Time". *scitechdaily.com*. Image credit:NASA, ESA, and J. Lotz, M. Mountain, A. Koekemoer, and the HFF Team (STScI). February 10, 2014. Archived from the original on May 7, 2014. Retrieved May 7, 2014.

[2] Foley, James A. (February 10, 2014). "Mass of Neutrinos Accurately Calculated for First Time, Physicists Report". *natureworldnews.com*. Image credit: . via Wikimedia Commons. Archived from the original on May 7, 2014. Retrieved May 7, 2014.

[3] Battye, Richard A.; Moss, Adam (2014). "Evidence for Massive Neutrinos from Cosmic Microwave Background and Lensing Observations". *Physical Review Letters* **112** (5): 051303. arXiv:1308.5870v2. Bibcode:2014PhRvL.112e1303B. doi:10.1103/PhysRevLett.112.051303. PMID 24580586.

[4] "Neutrino". *Glossary for the Research Perspectives of the Max Planck Society*. Max Planck Gesellschaft. Retrieved 2012-03-27.

[5] Dodelson, Scott; Widrow, Lawrence M. (1994). "Sterile neutrinos as dark matter" **72** (17).

[6] Bahcall, John N.; Serenelli, Aldo M.; Basu, Sarbani (2005). "New Solar Opacities, Abundances, Helioseismology, and Neutrino Fluxes". *The Astrophysical Journal* **621** (1): L85–8. arXiv:astro-ph/0412440. Bibcode:2005ApJ...621L..85B. doi:10.1086/428929.

[7] Olive, K. A. "Sum of Neutrino Masses" (PDF). *Chinese Physics C*.

[8] Brown, Laurie M. (1978). "The idea of the neutrino". *Physics Today* **31** (9): 23–8. Bibcode:1978PhT....31i..23B. doi:10.1063/1.2995181.

[9] E. Amaldi (1984). "From the discovery of the neutron to the discovery of nuclear fission". *Phys. Rep.* **111** (1–4): 306.

[10] F. Close (2010). *Neutrino*. Oxford University Press. ISBN 978-0-19-957459-9.

[11] E. Fermi (1934). "Versuch einer Theorie der β-Strahlen. I". *Zeitschrift für Physik A* **88** (3–4): 161. Bibcode:1934ZPhy...88..161F. doi:10.1007/BF01351864. Translated in F. L. Wilson (1968). "Fermi's Theory of Beta Decay" (PDF). *American Journal of Physics* **36** (12): 1150. Bibcode:1968AmJPh..36.1150W. doi:10.1119/1.1974382.

[12] K.-C. Wang (1942). "A Suggestion on the Detection of the Neutrino". *Physical Review* **61** (1–2): 97. Bibcode:1942PhRv...61...97W. doi:10.1103/PhysRev.61.97.

[13] C. L. Cowan Jr.; F. Reines; F. B. Harrison; H. W. Kruse et al. (1956). "Detection of the Free Neutrino: a Confirmation". *Science* **124** (3212): 103–4. Bibcode:1956Sci...124..103C. doi:10.1126/science.124.3212.103. PMID 17796274.

[14] K. Winter (2000). *Neutrino physics*. Cambridge University Press. p. 38ff. ISBN 978-0-521-65003-8. This source reproduces the 1956 paper.

[15] "The Nobel Prize in Physics 1995". The Nobel Foundation. Retrieved 29 June 2010.

[16] I. V. Anicin (2005). "The Neutrino – Its Past, Present and Future". arXiv:physics/0503172.

[17] M. Maltoni; T. Schwetz; M. Tórtola; J. W. F. Valle (2004). "Status of global fits to neutrino oscillations". *New Journal of Physics* **6** (1): 122. arXiv:hep-ph/0405172. Bibcode:2004NJPh....6..122M. doi:10.1088/1367-2630/6/1/122.

[18] Particle Data Group; Eidelman, S.; Hayes, K. G.; Olive, K. A.; Aguilar-Benitez, M.; Amsler, C.; Asner, D.; Babu, K. S.; Barnett, R. M.; Beringer, J.; Burchat, P. R.; Carone, C. D.; Caso, S.; Conforto, G.; Dahl, O.; d'Ambrosio, G.; Doser, M.; Feng, J. L.; Gherghetta, T.; Gibbons, L.; Goodman, M.; Grab, C.; Groom, D. E.; Gurtu, A.; Hagiwara, K.; Hernández-Rey, J. J.; Hikasa, K.; Honscheid, K.; Jawahery, H. et al. (2004). "Review of Particle Physics". *Physics Letters B* **592**: 1–5. arXiv:astro-ph/0406663. Bibcode:2004PhLB..592....1P. doi:10.1016/j.physletb.2004.06.001.

[19] S.M. Caroll (25 March 2009). "Ada Lovelace Day: Chien-Shiung Wu". *Discover Magazine*. Retrieved 2011-09-23.

[20] Kolbe, E.; Langanke, K.; Fuller, G. M. (2004). "Neutrino-Induced Fission of Neutron-Rich Nuclei". *Physical Review Letters* **92** (11): 111101. arXiv:astro-ph/0308350. Bibcode:2004PhRvL..92k1101K. doi:10.1103/PhysRevLett.92.111101. PMID 15089120.

[21] Kelić, A.; Zinner, N.; Kolbe, E.; Langanke, K.; Schmidt, K.-H. (2005). "Cross sections and fragment distributions from neutrino-induced fission on r-process nuclei". *Physics Letters B* **616** (1–2): 48–58. arXiv:hep-ex/0312045. Bibcode:2005PhLB..616...48K. doi:10.1016/j.physletb.2005.04.074.

[22] Karagiorgi, G.; Aguilar-Arevalo, A.; Conrad, J. M.; Shaevitz, M. H.; Whisnant, K.; Sorel, M.; Barger, V. (2007). "LeptonicCPviolation studies at MiniBooNE in the (3+2) sterile neutrino oscillation hypothesis". *Physical Review D* **75**: 013011. arXiv:hep-ph/0609177. Bibcode:2007PhRvD..75a3011K. doi:10.1103/PhysRevD.75.013011.

[23] M. Alpert (2007). "Dimensional Shortcuts". *Scientific American*. Retrieved 2009-10-31.

[24] Mueller, Th. A.; Lhuillier, D.; Fallot, M.; Letourneau, A.; Cormon, S.; Fechner, M.; Giot, L.; Lasserre, T.; Martino, J.; Mention, G.; Porta, A.; Yermia, F. (2011). "Improved predictions of reactor antineutrino spectra". *Physical Review C* **83** (5): 054615. arXiv:1101.2663. Bibcode:2011PhRvC..83e4615M. doi:10.1103/PhysRevC.83.054615.

[25] Mention, G.; Fechner, M.; Lasserre, Th.; Mueller, Th. A.; Lhuillier, D.; Cribier, M.; Letourneau, A. (2011). "Reactor antineutrino anomaly". *Physical Review D* **83** (7): 073006. arXiv:1101.2755. Bibcode:2011PhRvD..83g3006M. doi:10.1103/PhysRevD.83.073006.

[26] R. Cowen (2 February 2010). "Ancient Dawn's Early Light Refines the Age of the Universe". *Science News*. Retrieved 2010-02-03.

[27] neutrinos.llnl.gov "LLNL/SNL Applied Antineutrino Physics Project. LLNL-WEB-204112". 2006.

[28] apc.univ-paris7.fr "Applied Antineutrino Physics 2007 workshop". 2007.

[29] "New Tool To Monitor Nuclear Reactors Developed". ScienceDaily. 13 March 2008. Retrieved 2008-03-16.

[30] Alan Kostelecký, V.; Mewes, Matthew (2004). "Lorentz andCPTviolation in neutrinos". *Physical Review D* **69**: 016005. arXiv:hep-ph/0309025. Bibcode:2004PhRvD..69a6005A. doi:10.1103/PhysRevD.69.016005.

[31] C. Giunti; C.W. Kim (2007). *Fundamentals of neutrino physics and astrophysics*. Oxford University Press. p. 255. ISBN 0-19-850871-9.

[32] "Neutrino shape-shift points to new physics" *Physics News*, 19 July 2013.

[33] "Neutrino 'flavour' flip confirmed" *BBC News*, 19 July 2013.

[34] Adamson, P.; Andreopoulos, C.; Arms, K. E.; Armstrong, R.; Auty, D. J.; Avvakumov, S.; Ayres, D. S.; Baller, B.; Barish, B.; Barnes, P. D.; Barr, G.; Barrett, W. L.; Beall, E.; Becker, B. R.; Belias, A.; Bergfeld, T.; Bernstein, R. H.; Bhattacharya, D.; Bishai, M.; Blake, A.; Bock, B.; Bock, G. J.; Boehm, J.; Boehnlein, D. J.; Bogert, D.; Border, P. M.; Bower, C.; Buckley-Geer, E.; Cabrera, A. et al. (2007). "Measurement of neutrino velocity with the MINOS detectors and NuMI neutrino beam". *Physical Review D* **76** (7): 072005. arXiv:0706.0437. Bibcode:2007PhRvD..76g2005A. doi:10.1103/PhysRevD.76.072005.

[35] D. Overbye (22 September 2011). "Tiny neutrinos may have broken cosmic speed limit". *New York Times*. That group found, although with less precision, that the neutrino speeds were consistent with the speed of light.

[36] Hesla, Leah (June 8, 2012). "MINOS reports new measurement of neutrino velocity". Fermilab today. Retrieved April 2, 2015.

[37] Antonello, M.; Aprili, P.; Baiboussinov, B.; Baldo Ceolin, M.; Benetti, P.; Calligarich, E.; Canci, N.; Centro, S.; Cesana, A.; Cieślik, K.; Cline, D.B.; Cocco, A.G.; Dabrowska, A.; Dequal, D.; Dermenev, A.; Dolfini, R.; Farnese, C.; Fava, A.; Ferrari, A.; Fiorillo, G.; Gibin, D.; Gigli Berzolari, A.; Gninenko, S.; Guglielmi, A.; Haranczyk, M.; Holeczek, J.; Ivashkin, A.; Kisiel, J.; Kochanek, I. et al. (2012). "Measurement of the neutrino velocity with the ICARUS detector at the CNGS beam". *Physics Letters B* **713**: 17–22. arXiv:1203.3433. Bibcode:2012PhLB..713...17I. doi:10.1016/j.physletb.2012.05.033.

[38] "Neutrinos sent from CERN to Gran Sasso respect the cosmic speed limit, experiments confirm" (Press release). CERN. June 8, 2012. Retrieved April 2, 2015.

[39] Hut, P.; Olive, K.A. (1979). "A cosmological upper limit on the mass of heavy neutrinos". *Physics Letters B* **87** (1–2): 144–6. Bibcode:1979PhLB...87..144H. doi:10.1016/0370-2693(79)90039-X.

[40] Goobar, Ariel; Hannestad, Steen; Mörtsell, Edvard; Tu, Huitzu (2006). "The neutrino mass bound from WMAP 3 year data, the baryon acoustic peak, the SNLS supernovae and the Lyman-α forest". *Journal of Cosmology and Astroparticle Physics* **2006** (6): 019. arXiv:astro-ph/0602155. Bibcode:2006JCAP...06..019G. doi:10.1088/1475-7516/2006/06/019.

[41] Fukuda, Y.; Hayakawa, T.; Ichihara, E.; Inoue, K.; Ishihara, K.; Ishino, H.; Itow, Y.; Kajita, T.; Kameda, J.; Kasuga, S.; Kobayashi, K.; Kobayashi, Y.; Koshio, Y.; Martens, K.; Miura, M.; Nakahata, M.; Nakayama, S.; Okada, A.; Oketa, M.; Okumura, K.; Ota, M.; Sakurai, N.; Shiozawa, M.; Suzuki, Y.; Takeuchi, Y.; Totsuka, Y.; Yamada, S.; Earl, M.; Habig, A. et al. (1998). "Measurements of the Solar Neutrino Flux from Super-Kamiokande's First 300 Days". *Physical Review Letters* **81** (6): 1158. arXiv:hep-ex/9805021. Bibcode:1998PhRvL..81.1158F. doi:10.1103/PhysRevLett.81.1158.

[42] Mohapatra, R N; Antusch, S; Babu, K S; Barenboim, G; Chen, M-C; De Gouvêa, A; De Holanda, P; Dutta, B; Grossman, Y; Joshipura, A; Kayser, B; Kersten, J; Keum, Y Y; King, S F; Langacker, P; Lindner, M; Loinaz, W; Masina, I; Mocioiu, I; Mohanty, S; Murayama, H; Pascoli, S; Petcov, S T; Pilaftsis, A; Ramond, P; Ratz, M; Rodejohann, W; Shrock, R; Takeuchi, T et al. (2007). "Theory of neutrinos: A white paper". *Reports on Progress in Physics* **70** (11): 1757. arXiv:hep-ph/0510213. Bibcode:2007RPPh...70.1757M. doi:10.1088/0034-4885/70/11/R02.

[43] Araki, T.; Eguchi, K.; Enomoto, S.; Furuno, K.; Ichimura, K.; Ikeda, H.; Inoue, K.; Ishihara, K.; Iwamoto, T.; Kawashima, T.; Kishimoto, Y.; Koga, M.; Koseki, Y.; Maeda, T.; Mitsui, T.; Motoki, M.; Nakajima, K.; Ogawa, H.; Owada, K.; Ricol, J.-S.; Shimizu, I.; Shirai, J.; Suekane, F.; Suzuki, A.; Tada, K.; Tajima, O.; Tamae, K.; Tsuda, Y.; Watanabe, H. et al. (2005). "Measurement of Neutrino Oscillation with KamLAND: Evidence of Spectral Distortion". *Physical Review Letters* **94** (8): 081801. arXiv:hep-ex/0406035. Bibcode:2005PhRvL..94h1801A. doi:10.1103/PhysRevLett.94.081801. PMID 15783875.

[44] "MINOS experiment sheds light on mystery of neutrino disappearance" (Press release). Fermilab. 30 March 2006. Retrieved 2007-11-25.

[45] Amsler, C.; Doser, M.; Antonelli, M.; Asner, D.M.; Babu, K.S.; Baer, H.; Band, H.R.; Barnett, R.M.; Bergren, E.; Beringer, J.; Bernardi, G.; Bertl, W.; Bichsel, H.; Biebel, O.; Bloch, P.; Blucher, E.; Blusk, S.; Cahn, R.N.; Carena, M.; Caso, C.; Ceccucci, A.; Chakraborty, D.; Chen, M.-C.; Chivukula, R.S.; Cowan, G.; Dahl, O.; d'Ambrosio, G.; Damour, T.; De Gouvêa, A. et al. (2008). "Review of Particle Physics". *Physics Letters B* **667**: 1. Bibcode:2008PhLB..667....1P. doi:10.1016/j.physletb.2008.07.018.

[46] Nieuwenhuizen, Th. M. (2009). "Do non-relativistic neutrinos constitute the dark matter?". *EPL* **86** (5): 59001. arXiv:0812.4552. Bibcode:2009EL.....8659001N. doi:10.1209/0295-5075/86/59001.

[47] "The most sensitive analysis on the neutrino mass [...] is compatible with a neutrino mass of zero. Considering its uncertainties this value corresponds to an upper limit on the electron neutrino mass of $m < 2.2$ eV/c^2 (95% Confidence Level)" The Mainz Neutrino Mass Experiment

[48] Agafonova, N.; Aleksandrov, A.; Altinok, O.; Ambrosio, M.; Anokhina, A.; Aoki, S.; Ariga, A.; Ariga, T.; Autiero, D.; Badertscher, A.; Bagulya, A.; Bendhabi, A.; Bertolin, A.; Besnier, M.; Bick, D.; Boyarkin, V.; Bozza, C.; Brugière, T.; Brugnera, R.; Brunet, F.; Brunetti, G.; Buontempo, S.; Cazes, A.; Chaussard, L.; Chernyavsky, M.; Chiarella, V.; Chon-Sen, N.; Chukanov, A.; Ciesielski, R. et al. (2010). "Observation of a first ντ candidate event in the OPERA experiment in the CNGS beam". *Physics Letters B* **691** (3): 138–45. arXiv:1006.1623. Bibcode:2010PhLB..691..138A. doi:10.1016/j.physletb.2010.06.022.

[49] Thomas, Shaun A.; Abdalla, Filipe B.; Lahav, Ofer (2010). "Upper Bound of 0.28 eV on Neutrino Masses from the Largest Photometric Redshift Survey". *Physical Review Letters* **105** (3): 031301. arXiv:0911.5291. Bibcode:2010PhRvL.105c1301T. doi:10.1103/PhysRevLett.105.031301. PMID 20867754.

[50] Planck Collaboration, P. A. R.; Ade, P. A. R.; Aghanim, N.; Armitage-Caplan, C.; Arnaud, M.; Ashdown, M.; Atrio-Barandela, F.; Aumont, J.; Baccigalupi, C.; Banday, A. J.; Barreiro, R. B.; Bartlett, J. G.; Battaner, E.; Benabed, K.; Benoît, A.; Benoit-Lévy, A.; Bernard, J.-P.; Bersanelli, M.; Bielewicz, P.; Bobin, J.; Bock, J. J.; Bonaldi, A.; Bond, J. R.; Borrill, J.; Bouchet, F. R.; Bridges, M.; Bucher, M.; Burigana, C.; Butler, R. C. et al. (2013). "Planck 2013 results. XVI. Cosmological parameters". *Astronomy & Astrophysics* **1303**: 5076. arXiv:1303.5076. Bibcode:2013arXiv1303.5076P. doi:10.1051/0004-6361/201321591.

[51] Battye, Richard A.; Moss, Adam (2014). "Evidence for Massive Neutrinos from Cosmic Microwave Background and Lensing Observations". *Physical Review Letters* **112** (5): 051303. arXiv:1308.5870. Bibcode:2014PhRvL.112e1303B. doi:10.1103/PhysRevLett.112.051303. PMID 24580586.

[52] A. Gando et al. (KamLAND-Zen Collaboration) (Feb 7, 2013). "Limit on Neutrinoless ββ Decay of Xe136 from the First Phase of KamLAND-Zen and Comparison with the Positive Claim in Ge76". *Phys. Rev. Lett.* **110**, 062502. Bibcode:2013PhRvL.110f2502G. doi:10.1103/PhysRevLett.110.062502.

[53] Lucio, J. L.; Rosado, A.; Zepeda, A. (1985). "Characteristic size for the neutrino". *Physical Review D* **31** (5): 1091–1096. Bibcode:1985PhRvD..31.1091L. doi:10.1103/PhysRevD.31.1091. PMID 9955801.

[54] Choi, Charles Q. (2 June 2009). "Particles Larger Than Galaxies Fill the Universe?". *National Geographic News.*

[55] B. Kayser (2005). "Neutrino mass, mixing, and flavor change" (PDF). Particle Data Group. Retrieved 2007-11-25.

[56] S.M. Bilenky; C. Giunti (2001). "Lepton Numbers in the framework of Neutrino Mixing". *International Journal of Modern Physics A* **16** (24): 3931–3949. arXiv:hep-ph/0102320. Bibcode:2001IJMPA..16.3931B. doi:10.1142/S0217751X01004967.

[57] "Nuclear Fission and Fusion, and Nuclear Interactions". NLP National Physical Laboratory. 2008. Retrieved 2009-06-25.

[58] A. Bernstein; Wang, Y.; Gratta, G.; West, T. (2002). "Nuclear reactor safeguards and monitoring with antineutrino detectors". *Journal of Applied Physics* **91** (7): 4672. arXiv:nucl-ex/0108001. Bibcode:2002JAP....91.4672B. doi:10.1063/1.1452775.

[59] A. Bandyopadhyay et al. (ISS Physics Working Group) et al. (2007). "Physics at a future Neutrino Factory and super-beam facility". *Reports on Progress in Physics* **72** (10): 6201. arXiv:0710.4947. Bibcode:2009RPPh...72j6201B. doi:10.1088/0034-4885/72/10/106201.

[60] F. Reines; C. Cowan, Jr. (1997). "The Reines-Cowan Experiments: Detecting the Poltergeist" (PDF). *Los Alamos Science* **25**: 3.

[61] S. A. Colgate & R. H. White (1966). "The Hydrodynamic Behavior of Supernova Explosions". *The Astrophysical Journal* **143**: 626. Bibcode:1966ApJ...143..626C. doi:10.1086/148549.

[62] A.K. Mann (1997). *Shadow of a star: The neutrino story of Supernova 1987A*. W. H. Freeman. p. 122. ISBN 0-7167-3097-9.

[63] Products of the 1987A supernova

[64] Diameter of neutrino-generating core, and total neutrino power of SN 1987A

[65] J.N. Bahcall (1989). *Neutrino Astrophysics*. Cambridge University Press. ISBN 0-521-37975-X.

[66] D.R. David Jr. (2003). "Nobel Lecture: A half-century with solar neutrinos". *Reviews of Modern Physics* **75** (3): 10. Bibcode:2003RvMP...75..985D. doi:10.1103/RevModPhys.75.985.

[67] "Physics – Supernova Starting Gun: Neutrinos". Focus.aps.org. 2009-07-17. Retrieved 2012-04-05.

[68] G.B. Gelmini; A. Kusenko; T.J. Weiler (May 2010). "Through Neutrino Eyes". *Scientific American* **302** (5): 38–45. Bibcode:2010SciAm.302e..38G. doi:10.1038/scientificamerican0510-38.

[69] Stancil, D. D.; Adamson, P.; Alania, M.; Aliaga, L. et al. (2012). "Demonstration of Communication Using Neutrinos" (PDF). *Modern Physics Letters A* **27** (12): 1250077. arXiv:1203.2847. Bibcode:2012MPLA...2750077S. doi:10.1142/S0217732312500770. Lay summary – *Popular Science* (March 15, 2012).

3.2.9 Bibliography

- Adam, T.; *et al.* (OPERA collaboration) (2011). "Measurement of the neutrino velocity with the OPERA detector in the CNGS beam". arXiv:1109.4897 [hep-ex].

- Alberico, W. M.; Bilenky, S. M. (2004). "Neutrino Oscillations, Masses And Mixing". *Physics of Particles and Nuclei* **35**: 297–323. arXiv:hep-ph/0306239. Bibcode:2003hep.ph....6239A.

- Bahcall, J. N. (1989). *Neutrino Astrophysics*. Cambridge University Press. ISBN 0-521-35113-8.

- Bumfiel, G. (1 October 2001). "The Milky Way's Hidden Black Hole". *Scientific American*. Retrieved 2010-04-23.

- Close, F. (2010). *Neutrino*. Oxford University Press. ISBN 978-0-19-957459-9.

- Griffiths, D. J. (1987). *Introduction to Elementary Particles*. John Wiley & Sons. ISBN 0-471-60386-4.

- Perkins, D. H. (1999). *Introduction to High Energy Physics*. Cambridge University Press. ISBN 0-521-62196-8.

- Povh, B. (1995). *Particles and Nuclei: An Introduction to the Physical Concepts*. Springer-Verlag. ISBN 0-387-59439-6.

- Riazuddin (2005). "Neutrinos" (PDF). National Center for Physics.

- Schopper, H. F. (1966). *Weak interactions and nuclear beta decay*. North-Holland.

- Tammann, G. A.; Thielemann, F. K.; Trautmann, D. (2003). "Opening new windows in observing the Universe". Europhysics News. Retrieved 2006-06-08.

- Tipler, P.; Llewellyn, R. (2002). *Modern Physics* (4th ed.). W. H. Freeman. ISBN 0-7167-4345-0.

- Tomonaga, S.-I. (1997). *The Story of Spin*. University of Chicago Press.

- Zuber, K. (2003). *Neutrino Physics*. IOP Publishing. ISBN 978-0-7503-0750-5.

3.2.10 External links

- "What's a Neutrino?", Dave Casper (University of California, Irvine)

- Neutrino unbound: On-line review and e-archive on Neutrino Physics and Astrophysics

- Nova: The Ghost Particle: Documentary on US public television from WGBH

- Measuring the density of the earth's core with neutrinos

- Universe submerged in a sea of chilled neutrinos, *New Scientist*, 5 March 2008

- What's a neutrino?

- Search for neutrinoless double beta decay with enriched 76Ge in Gran Sasso 1990–2003

- Neutrino caught in the act of changing from muon-type to tau-type, CERN press release

- Cosmic Weight Gain: A Wispy Particle Bulks Up by George Johnson

- Neutrino 'ghost particle' sized up by astronomers BBC News 22 June 2010

- Pillar of physics challenged

- Merrifield, Michael; Copeland, Ed; Bowley, Roger (2010). "Neutrinos". *Sixty Symbols*. Brady Haran for the University of Nottingham.

- The Neutrino with Dr. Clyde L. Cowan (Lecture on Project Poltergeist by Clyde Cowan)

- Nuclear Reactor as the Source of Antineutrinos

3.3 Electron

For other uses, see Electron (disambiguation).

The **electron** is a subatomic particle, symbol e− or β−, with a negative elementary electric charge.[7] Electrons belong to the first generation of the lepton particle family,[8] and are generally thought to be elementary particles because they have no known components or substructure.[1] The electron has a mass that is approximately 1/1836 that of the proton.[9] Quantum mechanical properties of the electron include an intrinsic angular momentum (spin) of a half-integer value in units of $\hbar$, which means that it is a fermion. Being fermions, no two electrons can occupy the same quantum state, in accordance with the Pauli exclusion principle.[8] Like all matter, electrons have properties of both particles and waves, and so can collide with other particles and can be diffracted like light. The wave properties of electrons are easier to observe with experiments than those of other particles like neutrons and protons because electrons have a lower mass and hence a higher De Broglie wavelength for typical energies.

Many physical phenomena involve electrons in an essential role, such as electricity, magnetism, and thermal conductivity, and they also participate in gravitational, electromagnetic and weak interactions.[10] An electron generates an electric field surrounding it. An electron moving relative to an observer generates a magnetic field. External magnetic fields deflect an electron. Electrons radiate or absorb energy in the form of photons when accelerated. Laboratory instruments are capable of containing and observing individual electrons as well as electron plasma using electromagnetic fields, whereas dedicated telescopes can detect electron plasma in outer space. Electrons have many applications, including electronics, welding, cathode ray tubes, electron microscopes, radiation therapy, lasers, gaseous ionization detectors and particle accelerators.

Interactions involving electrons and other subatomic particles are of interest in fields such as chemistry and nuclear physics. The Coulomb force interaction between positive protons inside atomic nuclei and negative electrons composes atoms. Ionization or changes in the proportions of particles changes the binding energy of the system. The exchange or sharing of the electrons between two or more atoms is the main cause of chemical bonding.[11] British natural philosopher Richard Laming first hypothesized the concept of an indivisible quantity of electric charge to explain the chemical properties of atoms in 1838;[3] Irish physicist George Johnstone Stoney named this charge 'electron' in 1891, and J. J. Thomson and his team of British physicists identified it as a particle in 1897.[5][12][13] Electrons can also participate in nuclear reactions, such as nucleosynthesis in stars, where they are known as beta particles. Electrons may be created through beta decay of radioactive isotopes and in high-energy collisions, for instance when cosmic rays enter the atmosphere. The antiparticle of the electron is called the positron; it is identical to the electron except that it carries electrical and other charges of the opposite sign. When an electron collides with a positron, both particles may be totally annihilated, producing gamma ray photons.

3.3.1 History

See also: History of electromagnetism

The ancient Greeks noticed that amber attracted small objects when rubbed with fur. Along with lightning, this phenomenon is one of humanity's earliest recorded experiences with electricity. [14] In his 1600 treatise *De Magnete*, the English scientist William Gilbert coined the New Latin term *electricus*, to refer to this property of attracting small objects after being rubbed. [15] Both *electric* and *electricity* are derived from the Latin *ēlectrum* (also the root of the alloy of the same name), which came from the Greek word for amber, ἤλεκτρον (*ēlektron*).

In the early 1700s, Francis Hauksbee and French chemist Charles François de Fay independently discovered what they believed were two kinds of frictional electricity—one generated from rubbing glass, the other from rubbing resin. From this, Du Fay theorized that electricity consists of two electrical fluids, *vitreous* and *resinous*, that are separated by friction, and that neutralize each other when combined.[16] A decade later Benjamin Franklin proposed that electricity was not from different types of electrical fluid, but the same electrical fluid under different pressures. He gave them the modern charge nomenclature of positive and negative respectively.[17] Franklin thought of the charge carrier as being positive, but he did not correctly identify which situation was a surplus of the charge carrier, and which situation was a deficit.[18]

Between 1838 and 1851, British natural philosopher Richard Laming developed the idea that an atom is composed of a core of matter surrounded by subatomic particles that had unit electric charges.[2] Beginning in 1846, German physicist William Weber theorized that electricity was composed of positively and negatively charged fluids, and their interaction was governed by the inverse square law. After studying the phenomenon of electrolysis in 1874, Irish physicist George Johnstone Stoney suggested that there existed a "single definite quantity of electricity", the charge of a monovalent ion. He was able to estimate the value of this elementary charge *e* by means of Faraday's laws of electrolysis.[19] However, Stoney believed these charges were permanently attached to atoms and could not be removed. In 1881, German physicist Hermann von Helmholtz argued that both positive and negative charges were divided into elementary parts, each of which "behaves like atoms of electricity".[3]

Stoney initially coined the term *electrolion* in 1881. Ten years later, he switched to *electron* to describe these elementary charges, writing in 1894: "... an estimate was made of the actual amount of this most remarkable fundamental unit of electricity, for which I have since ventured to suggest the name *electron*". A 1906 proposal to change to *electrion* failed because Hendrik Lorentz preferred to keep *electron*.[20][21] The word *electron* is a combination of the words *electric* and *ion*.[22] The suffix *-on* which is now used to designate other subatomic particles, such as a proton or neutron, is in turn derived from electron.[23][24]

Discovery

A beam of electrons deflected in a circle by a magnetic field[25]

The German physicist Johann Wilhelm Hittorf studied electrical conductivity in rarefied gases: in 1869, he discovered a glow emitted from the cathode that increased in size with decrease in gas pressure. In 1876, the German physicist Eugen Goldstein showed that the rays from this glow cast a shadow, and he dubbed the rays cathode rays.[26] During the 1870s, the English chemist and physicist Sir William Crookes developed the first cathode ray tube to have a high vacuum inside.[27] He then showed that the luminescence rays appearing within the tube carried energy and moved from the cathode to the anode. Furthermore, by applying a magnetic field, he was able to deflect the rays, thereby demonstrating that the beam behaved as though it were negatively charged.[28][29] In 1879, he proposed that these properties could be explained by what he termed 'radiant matter'. He suggested that this was a fourth state of matter, consisting of negatively charged molecules that were being projected with high velocity from the cathode.[30]

The German-born British physicist Arthur Schuster expanded upon Crookes' experiments by placing metal plates parallel to the cathode rays and applying an electric potential between the plates. The field deflected the rays toward the positively charged plate, providing further evidence that the rays carried negative charge. By measuring the amount of deflection for a given level of current, in 1890 Schuster was able to estimate the charge-to-mass ratio of the ray components. However, this produced a value that was more than a thousand times greater than what was expected, so little credence was given to his calculations at the time.[28][31]

In 1892 Hendrik Lorentz suggested that the mass of these particles (electrons) could be a consequence of their electric charge.[32]

In 1896, the British physicist J. J. Thomson, with his colleagues John S. Townsend and H. A. Wilson,[12] performed experiments indicating that cathode rays really were unique particles, rather than waves, atoms or molecules as was believed earlier.[5] Thomson made good estimates of both the charge e and the mass m, finding that cathode ray particles, which he called "corpuscles," had perhaps one thousandth of the mass of the least massive ion known: hydrogen.[5][13] He showed that their charge to mass ratio, e/m, was independent of cathode material. He further showed that the negatively charged particles produced by radioactive materials, by heated materials and by illuminated materials were universal.[5][33] The name electron was again proposed for these particles by the Irish physicist George F. Fitzgerald, and the name has since gained universal acceptance.[28]

Robert Millikan

While studying naturally fluorescing minerals in 1896, the French physicist Henri Becquerel discovered that they emitted radiation without any exposure to an external energy source. These radioactive materials became the subject of much interest by scientists, including the New Zealand physicist Ernest Rutherford who discovered they emitted particles. He designated these particles alpha and beta, on the basis of their ability to penetrate matter.[34] In 1900, Becquerel showed that the beta rays emitted by radium could be deflected by an electric field, and that their mass-to-charge ratio was the same as for cathode rays.[35] This ev-

idence strengthened the view that electrons existed as components of atoms.[36][37]

The electron's charge was more carefully measured by the American physicists Robert Millikan and Harvey Fletcher in their oil-drop experiment of 1909, the results of which were published in 1911. This experiment used an electric field to prevent a charged droplet of oil from falling as a result of gravity. This device could measure the electric charge from as few as 1–150 ions with an error margin of less than 0.3%. Comparable experiments had been done earlier by Thomson's team,[5] using clouds of charged water droplets generated by electrolysis,[12] and in 1911 by Abram Ioffe, who independently obtained the same result as Millikan using charged microparticles of metals, then published his results in 1913.[38] However, oil drops were more stable than water drops because of their slower evaporation rate, and thus more suited to precise experimentation over longer periods of time.[39]

Around the beginning of the twentieth century, it was found that under certain conditions a fast-moving charged particle caused a condensation of supersaturated water vapor along its path. In 1911, Charles Wilson used this principle to devise his cloud chamber so he could photograph the tracks of charged particles, such as fast-moving electrons.[40]

Atomic theory

See also: The proton–electron model of the nucleus

By 1914, experiments by physicists Ernest Rutherford,

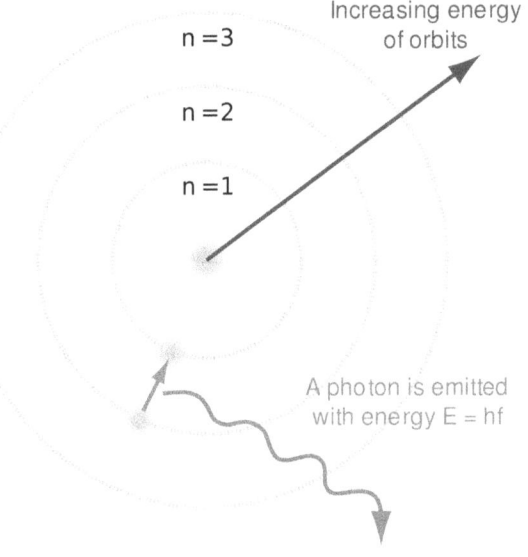

The Bohr model of the atom, showing states of electron with energy quantized by the number n. An electron dropping to a lower orbit emits a photon equal to the energy difference between the orbits.

Henry Moseley, James Franck and Gustav Hertz had largely

established the structure of an atom as a dense nucleus of positive charge surrounded by lower-mass electrons.[41] In 1913, Danish physicist Niels Bohr postulated that electrons resided in quantized energy states, with the energy determined by the angular momentum of the electron's orbits about the nucleus. The electrons could move between these states, or orbits, by the emission or absorption of photons at specific frequencies. By means of these quantized orbits, he accurately explained the spectral lines of the hydrogen atom.[42] However, Bohr's model failed to account for the relative intensities of the spectral lines and it was unsuccessful in explaining the spectra of more complex atoms.[41]

Chemical bonds between atoms were explained by Gilbert Newton Lewis, who in 1916 proposed that a covalent bond between two atoms is maintained by a pair of electrons shared between them.[43] Later, in 1927, Walter Heitler and Fritz London gave the full explanation of the electron-pair formation and chemical bonding in terms of quantum mechanics.[44] In 1919, the American chemist Irving Langmuir elaborated on the Lewis' static model of the atom and suggested that all electrons were distributed in successive "concentric (nearly) spherical shells, all of equal thickness".[45] The shells were, in turn, divided by him in a number of cells each containing one pair of electrons. With this model Langmuir was able to qualitatively explain the chemical properties of all elements in the periodic table,[44] which were known to largely repeat themselves according to the periodic law.[46]

In 1924, Austrian physicist Wolfgang Pauli observed that the shell-like structure of the atom could be explained by a set of four parameters that defined every quantum energy state, as long as each state was inhabited by no more than a single electron. (This prohibition against more than one electron occupying the same quantum energy state became known as the Pauli exclusion principle.)[47] The physical mechanism to explain the fourth parameter, which had two distinct possible values, was provided by the Dutch physicists Samuel Goudsmit and George Uhlenbeck. In 1925, Goudsmit and Uhlenbeck suggested that an electron, in addition to the angular momentum of its orbit, possesses an intrinsic angular momentum and magnetic dipole moment.[41][48] The intrinsic angular momentum became known as spin, and explained the previously mysterious splitting of spectral lines observed with a high-resolution spectrograph; this phenomenon is known as fine structure splitting.[49]

Quantum mechanics

See also: History of quantum mechanics

In his 1924 dissertation *Recherches sur la théorie des quanta* (Research on Quantum Theory), French physicist Louis de Broglie hypothesized that all matter possesses a de Broglie wave similar to light.[50] That is, under the appropriate conditions, electrons and other matter would show properties of either particles or waves. The corpuscular properties of a particle are demonstrated when it is shown to have a localized position in space along its trajectory at any given moment.[51] Wave-like nature is observed, for example, when a beam of light is passed through parallel slits and creates interference patterns. In 1927, the interference effect was found in a beam of electrons by English physicist George Paget Thomson with a thin metal film and by American physicists Clinton Davisson and Lester Germer using a crystal of nickel.[52]

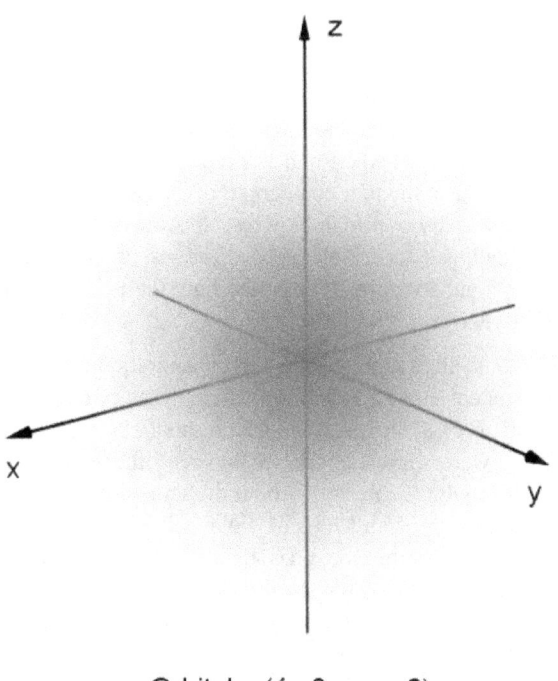

Orbital s ($\ell = 0$, $m_\ell = 0$)

In quantum mechanics, the behavior of an electron in an atom is described by an orbital, which is a probability distribution rather than an orbit. In the figure, the shading indicates the relative probability to "find" the electron, having the energy corresponding to the given quantum numbers, at that point.

De Broglie's prediction of a wave nature for electrons led Erwin Schrödinger to postulate a wave equation for electrons moving under the influence of the nucleus in the atom. In 1926, this equation, the Schrödinger equation, successfully described how electron waves propagated.[53] Rather than yielding a solution that determined the location of an electron over time, this wave equation also could be used to predict the probability of finding an electron near a position, especially a position near where the electron was bound in space, for which the electron wave equations did not change in time. This approach led to a second formulation of quantum mechanics (the first being by Heisenberg in 1925), and solutions of Schrödinger's equation, like Heisenberg's, provided derivations of the energy states of an electron in a hydrogen atom that were equivalent to those that had been derived first by Bohr in 1913, and that were known to reproduce the hydrogen spectrum.[54] Once spin and the interaction between multiple electrons were considered, quantum mechanics later made it possible to predict the configuration of electrons in atoms with higher atomic numbers than hydrogen.[55]

In 1928, building on Wolfgang Pauli's work, Paul Dirac produced a model of the electron – the Dirac equation, consistent with relativity theory, by applying relativistic and symmetry considerations to the hamiltonian formulation of the quantum mechanics of the electro-magnetic field.[56] To resolve some problems within his relativistic equation, in 1930 Dirac developed a model of the vacuum as an infinite sea of particles having negative energy, which was dubbed the Dirac sea. This led him to predict the existence of a positron, the antimatter counterpart of the electron.[57] This particle was discovered in 1932 by Carl Anderson, who proposed calling standard electrons *negatrons*, and using *electron* as a generic term to describe both the positively and negatively charged variants.

In 1947 Willis Lamb, working in collaboration with graduate student Robert Retherford, found that certain quantum states of hydrogen atom, which should have the same energy, were shifted in relation to each other, the difference being the Lamb shift. About the same time, Polykarp Kusch, working with Henry M. Foley, discovered the magnetic moment of the electron is slightly larger than predicted by Dirac's theory. This small difference was later called anomalous magnetic dipole moment of the electron. This difference was later explained by the theory of quantum electrodynamics, developed by Sin-Itiro Tomonaga, Julian Schwinger and Richard Feynman in the late 1940s.[58]

Particle accelerators

With the development of the particle accelerator during the first half of the twentieth century, physicists began to delve deeper into the properties of subatomic particles.[59] The first successful attempt to accelerate electrons using electromagnetic induction was made in 1942 by Donald Kerst. His initial betatron reached energies of 2.3 MeV, while subsequent betatrons achieved 300 MeV. In 1947, synchrotron radiation was discovered with a 70 MeV electron synchrotron at General Electric. This radiation was caused by the acceleration of electrons, moving near the speed of light, through a magnetic field.[60]

With a beam energy of 1.5 GeV, the first high-energy

particle collider was ADONE, which began operations in 1968.[61] This device accelerated electrons and positrons in opposite directions, effectively doubling the energy of their collision when compared to striking a static target with an electron.[62] The Large Electron–Positron Collider (LEP) at CERN, which was operational from 1989 to 2000, achieved collision energies of 209 GeV and made important measurements for the Standard Model of particle physics.[63][64]

Confinement of individual electrons

Individual electrons can now be easily confined in ultra small ($L = 20$ nm, $W = 20$ nm) CMOS transistors operated at cryogenic temperature over a range of −269 °C (4 K) to about −258 °C (15 K).[65] The electron wavefunction spreads in a semiconductor lattice and negligibly interacts with the valence band electrons, so it can be treated in the single particle formalism, by replacing its mass with the effective mass tensor.

3.3.2   Characteristics

Classification

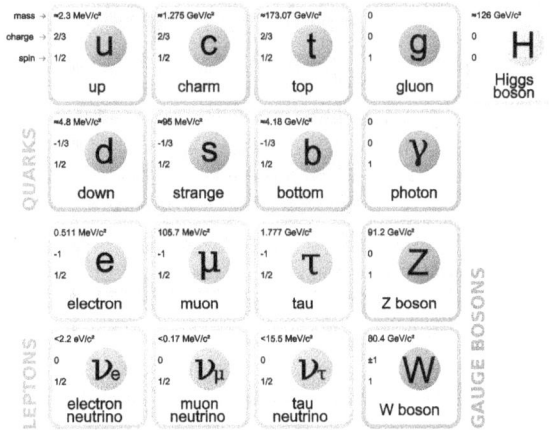

Standard Model of elementary particles. The electron (symbol e) is on the left.

In the Standard Model of particle physics, electrons belong to the group of subatomic particles called leptons, which are believed to be fundamental or elementary particles. Electrons have the lowest mass of any charged lepton (or electrically charged particle of any type) and belong to the first-generation of fundamental particles.[66] The second and third generation contain charged leptons, the muon and the tau, which are identical to the electron in charge, spin and interactions, but are more massive. Leptons differ from the other basic constituent of matter, the quarks, by their lack of strong interaction. All members of the lepton group are fermions, because they all have half-odd integer spin; the electron has spin $1/2$.[67]

Fundamental properties

The invariant mass of an electron is approximately 9.109×10^{-31} kilograms,[68] or 5.489×10^{-4} atomic mass units. On the basis of Einstein's principle of mass–energy equivalence, this mass corresponds to a rest energy of 0.511 MeV. The ratio between the mass of a proton and that of an electron is about 1836.[9][69] Astronomical measurements show that the proton-to-electron mass ratio has held the same value for at least half the age of the universe, as is predicted by the Standard Model.[70]

Electrons have an electric charge of -1.602×10^{-19} coulomb,[68] which is used as a standard unit of charge for subatomic particles, and is also called the elementary charge. This elementary charge has a relative standard uncertainty of 2.2×10^{-8}.[68] Within the limits of experimental accuracy, the electron charge is identical to the charge of a proton, but with the opposite sign.[71] As the symbol e is used for the elementary charge, the electron is commonly symbolized by e−, where the minus sign indicates the negative charge. The positron is symbolized by e+ because it has the same properties as the electron but with a positive rather than negative charge.[67][68]

The electron has an intrinsic angular momentum or spin of $1/2$.[68] This property is usually stated by referring to the electron as a spin-$1/2$ particle.[67] For such particles the spin magnitude is $\frac{\sqrt{3}}{2}\hbar$.[note 3] while the result of the measurement of a projection of the spin on any axis can only be $\pm\frac{\hbar}{2}$. In addition to spin, the electron has an intrinsic magnetic moment along its spin axis.[68] It is approximately equal to one Bohr magneton,[72][note 4] which is a physical constant equal to $9.27400915(23) \times 10^{-24}$ joules per tesla.[68] The orientation of the spin with respect to the momentum of the electron defines the property of elementary particles known as helicity.[73]

The electron has no known substructure.[1][74] and it is assumed to be a point particle with a point charge and no spatial extent.[8] In classical physics, the angular momentum and magnetic moment of an object depend upon its physical dimensions. Hence, the concept of a dimensionless electron possessing these properties might seem paradoxical and inconsistent to experimental observations in Penning traps which point to finite non-zero radius of the electron. A possible explanation of this paradoxical situation is given below in the "Virtual particles" subsection by taking into consideration the Foldy-Wouthuysen transformation. The issue of the radius of the electron is a challenging problem of the modern theoretical physics. The admission of the hypothesis of a finite radius of the electron is incompatible to the premises of the theory of relativity. On the other hand, a point-like electron (zero radius) generates serious mathematical difficulties due to the self-energy of the electron tending to infinity.[75] These aspects have been analyzed in detail by Dmitri Ivanenko and Arseny Sokolov.

Observation of a single electron in a Penning trap shows the upper limit of the particle's radius is 10^{-22} meters.[76] There *is* a physical constant called the "classical electron radius", with the much larger value of 2.8179×10^{-15} m, greater than the radius of the proton. However, the terminology comes from a simplistic calculation that ignores the effects of quantum mechanics; in reality, the so-called classical electron radius has little to do with the true fundamental structure of the electron.[77][note 5]

There are elementary particles that spontaneously decay into less massive particles. An example is the muon, which decays into an electron, a neutrino and an antineutrino, with a mean lifetime of 2.2×10^{-6} seconds. However, the electron is thought to be stable on theoretical grounds: the electron is the least massive particle with non-zero electric charge, so its decay would violate charge conservation.[78] The experimental lower bound for the electron's mean lifetime is 4.6×10^{26} years, at a 90% confidence level.[79][80]

Quantum properties

As with all particles, electrons can act as waves. This is called the wave–particle duality and can be demonstrated using the double-slit experiment.

The wave-like nature of the electron allows it to pass through two parallel slits simultaneously, rather than just one slit as would be the case for a classical particle. In quantum mechanics, the wave-like property of one particle can be described mathematically as a complex-valued function, the wave function, commonly denoted by the Greek letter psi (ψ). When the absolute value of this function is squared, it gives the probability that a particle will be observed near a location—a probability density.[81]:162–218

Electrons are identical particles because they cannot be distinguished from each other by their intrinsic physical properties. In quantum mechanics, this means that a pair of interacting electrons must be able to swap positions without an observable change to the state of the system. The wave function of fermions, including electrons, is antisymmetric, meaning that it changes sign when two electrons are swapped; that is, $\psi(r_1, r_2) = -\psi(r_2, r_1)$, where the variables r_1 and r_2 correspond to the first and second electrons, respectively. Since the absolute value is not changed by a sign swap, this corresponds to equal probabilities. Bosons, such as the photon, have symmetric wave functions instead.[81]:162–218

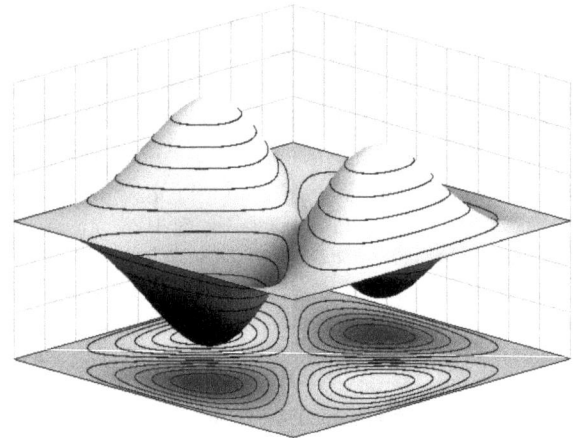

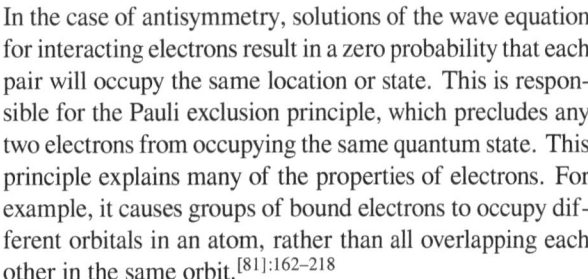

Example of an antisymmetric wave function for a quantum state of two identical fermions in a 1-dimensional box. If the particles swap position, the wave function inverts its sign.

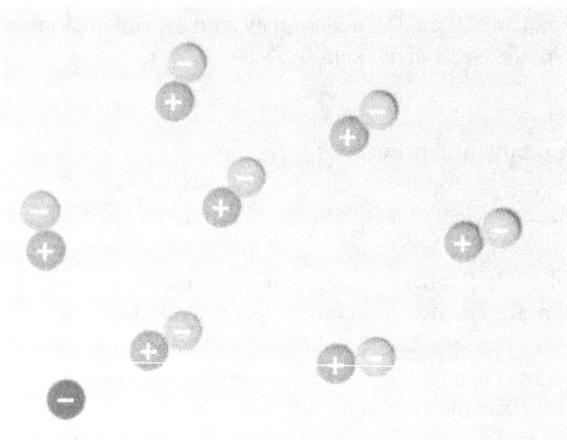

A schematic depiction of virtual electron–positron pairs appearing at random near an electron (at lower left)

In the case of antisymmetry, solutions of the wave equation for interacting electrons result in a zero probability that each pair will occupy the same location or state. This is responsible for the Pauli exclusion principle, which precludes any two electrons from occupying the same quantum state. This principle explains many of the properties of electrons. For example, it causes groups of bound electrons to occupy different orbitals in an atom, rather than all overlapping each other in the same orbit.[81]:162–218

Virtual particles

Main article: Virtual particle

In a simplified picture, every photon spends some time as a combination of a virtual electron plus its antiparticle, the virtual positron, which rapidly annihilate each other shortly thereafter.[82] The combination of the energy variation needed to create these particles, and the time during which they exist, fall under the threshold of detectability expressed by the Heisenberg uncertainty relation, $\Delta E \cdot \Delta t \geq \hbar$. In effect, the energy needed to create these virtual particles, ΔE, can be "borrowed" from the vacuum for a period of time, Δt, so that their product is no more than the reduced Planck constant, $\hbar \approx 6.6 \times 10^{-16}$ eV·s. Thus, for a virtual electron, Δt is at most 1.3×10^{-21} s.[83]

While an electron–positron virtual pair is in existence, the coulomb force from the ambient electric field surrounding an electron causes a created positron to be attracted to the original electron, while a created electron experiences a repulsion. This causes what is called vacuum polarization. In effect, the vacuum behaves like a medium having a dielectric permittivity more than unity. Thus the effec-

tive charge of an electron is actually smaller than its true value, and the charge decreases with increasing distance from the electron.[84][85] This polarization was confirmed experimentally in 1997 using the Japanese TRISTAN particle accelerator.[86] Virtual particles cause a comparable shielding effect for the mass of the electron.[87]

The interaction with virtual particles also explains the small (about 0.1%) deviation of the intrinsic magnetic moment of the electron from the Bohr magneton (the anomalous magnetic moment).[72][88] The extraordinarily precise agreement of this predicted difference with the experimentally determined value is viewed as one of the great achievements of quantum electrodynamics.[89]

The apparent paradox (mentioned above in the properties subsection) of a point particle electron having intrinsic angular momentum and magnetic moment can be explained by the formation of virtual photons in the electric field generated by the electron. These photons cause the electron to shift about in a jittery fashion (known as zitterbewegung),[90] which results in a net circular motion with precession. This motion produces both the spin and the magnetic moment of the electron.[8][91] In atoms, this creation of virtual photons explains the Lamb shift observed in spectral lines.[84]

Interaction

An electron generates an electric field that exerts an attractive force on a particle with a positive charge, such as the proton, and a repulsive force on a particle with a negative charge. The strength of this force is determined by Coulomb's inverse square law.[92] When an electron is in motion, it generates a magnetic field.[81]:140 The Ampère-Maxwell law relates the magnetic field to the mass motion of electrons (the current) with respect to an observer. This property of induction supplies the magnetic field that drives an electric motor.[93] The electromagnetic field of an arbitrary moving charged particle is expressed by the Liénard–Wiechert potentials, which are valid even when the particle's speed is close to that of light (relativistic).

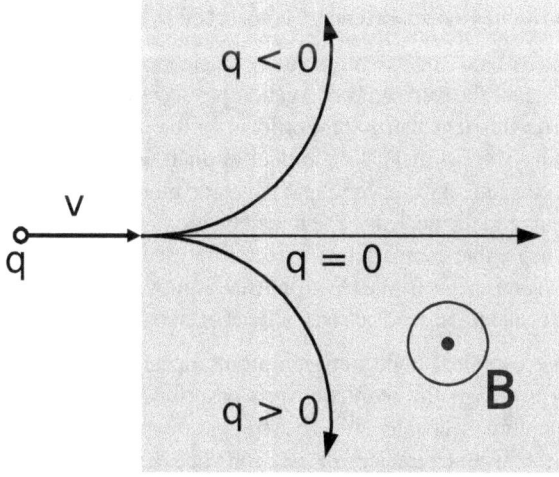

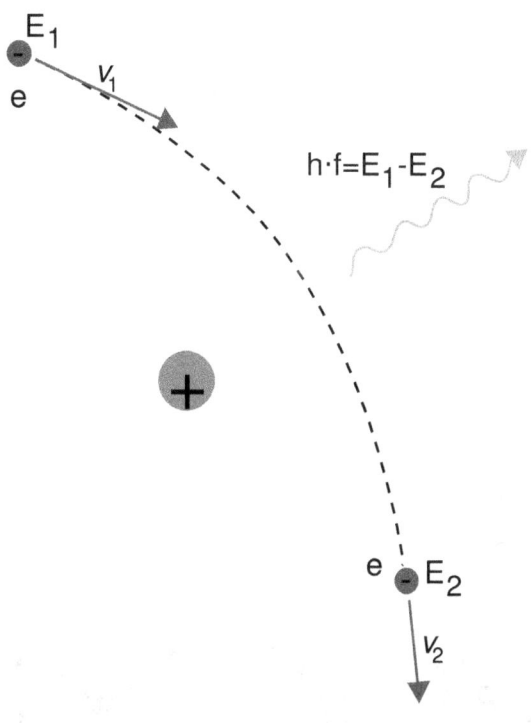

A particle with charge q *(at left) is moving with velocity* v *through a magnetic field* B *that is oriented toward the viewer. For an electron,* q *is negative so it follows a curved trajectory toward the top.*

Here, Bremsstrahlung is produced by an electron e *deflected by the electric field of an atomic nucleus. The energy change* $E_2 - E_1$ *determines the frequency* f *of the emitted photon.*

When an electron is moving through a magnetic field, it is subject to the Lorentz force that acts perpendicularly to the plane defined by the magnetic field and the electron velocity. This centripetal force causes the electron to follow a helical trajectory through the field at a radius called the gyroradius. The acceleration from this curving motion induces the electron to radiate energy in the form of synchrotron radiation.[81]:160[94][note 6] The energy emission in turn causes a recoil of the electron, known as the Abraham–Lorentz–Dirac Force, which creates a friction that slows the electron. This force is caused by a back-reaction of the electron's own field upon itself.[95]

Photons mediate electromagnetic interactions between particles in quantum electrodynamics. An isolated electron at a constant velocity cannot emit or absorb a real photon; doing so would violate conservation of energy and momentum. Instead, virtual photons can transfer momentum between two charged particles. This exchange of virtual photons, for example, generates the Coulomb force.[96] Energy emission can occur when a moving electron is deflected by a charged particle, such as a proton. The acceleration of the electron results in the emission of Bremsstrahlung radiation.[97]

An inelastic collision between a photon (light) and a solitary (free) electron is called Compton scattering. This collision results in a transfer of momentum and energy between the particles, which modifies the wavelength of the photon by an amount called the Compton shift.[note 7] The maximum

magnitude of this wavelength shift is $h/m_e c$, which is known as the Compton wavelength.[98] For an electron, it has a value of 2.43×10^{-12} m.[68] When the wavelength of the light is long (for instance, the wavelength of the visible light is 0.4–0.7 µm) the wavelength shift becomes negligible. Such interaction between the light and free electrons is called Thomson scattering or Linear Thomson scattering.[99]

The relative strength of the electromagnetic interaction between two charged particles, such as an electron and a proton, is given by the fine-structure constant. This value is a dimensionless quantity formed by the ratio of two energies: the electrostatic energy of attraction (or repulsion) at a separation of one Compton wavelength, and the rest energy of the charge. It is given by $\alpha \approx 7.297353 \times 10^{-3}$, which is approximately equal to $1/137$.[68]

When electrons and positrons collide, they annihilate each other, giving rise to two or more gamma ray photons. If the electron and positron have negligible momentum, a positronium atom can form before annihilation results in two or three gamma ray photons totalling 1.022 MeV.[100][101] On the other hand, high-energy photons may transform into an electron and a positron by a process called pair production, but only in the presence of a nearby charged particle, such as a nucleus.[102][103]

In the theory of electroweak interaction, the left-handed

component of electron's wavefunction forms a weak isospin doublet with the electron neutrino. This means that during weak interactions, electron neutrinos behave like electrons. Either member of this doublet can undergo a charged current interaction by emitting or absorbing a W and be converted into the other member. Charge is conserved during this reaction because the W boson also carries a charge, canceling out any net change during the transmutation. Charged current interactions are responsible for the phenomenon of beta decay in a radioactive atom. Both the electron and electron neutrino can undergo a neutral current interaction via a Z0 exchange, and this is responsible for neutrino-electron elastic scattering.[104]

Atoms and molecules

Main article: Atom
An electron can be *bound* to the nucleus of an atom by the

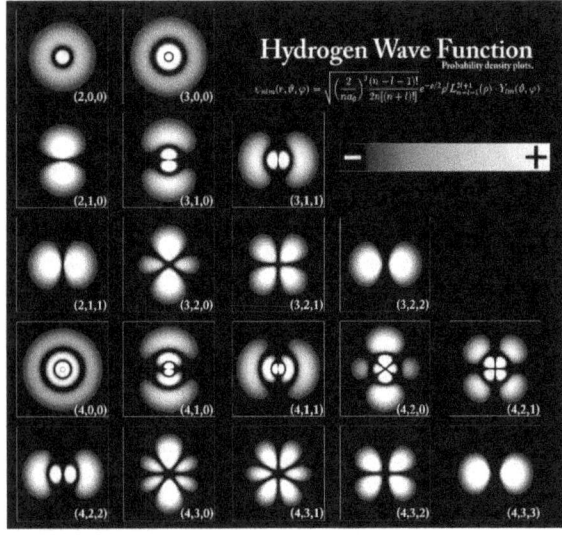

Probability densities for the first few hydrogen atom orbitals, seen in cross-section. The energy level of a bound electron determines the orbital it occupies, and the color reflects the probability of finding the electron at a given position.

attractive Coulomb force. A system of one or more electrons bound to a nucleus is called an atom. If the number of electrons is different from the nucleus' electrical charge, such an atom is called an ion. The wave-like behavior of a bound electron is described by a function called an atomic orbital. Each orbital has its own set of quantum numbers such as energy, angular momentum and projection of angular momentum, and only a discrete set of these orbitals exist around the nucleus. According to the Pauli exclusion principle each orbital can be occupied by up to two electrons, which must differ in their spin quantum number.

Electrons can transfer between different orbitals by the emission or absorption of photons with an energy that matches the difference in potential.[105] Other methods of orbital transfer include collisions with particles, such as electrons, and the Auger effect.[106] To escape the atom, the energy of the electron must be increased above its binding energy to the atom. This occurs, for example, with the photoelectric effect, where an incident photon exceeding the atom's ionization energy is absorbed by the electron.[107]

The orbital angular momentum of electrons is quantized. Because the electron is charged, it produces an orbital magnetic moment that is proportional to the angular momentum. The net magnetic moment of an atom is equal to the vector sum of orbital and spin magnetic moments of all electrons and the nucleus. The magnetic moment of the nucleus is negligible compared with that of the electrons. The magnetic moments of the electrons that occupy the same orbital (so called, paired electrons) cancel each other out.[108]

The chemical bond between atoms occurs as a result of electromagnetic interactions, as described by the laws of quantum mechanics.[109] The strongest bonds are formed by the sharing or transfer of electrons between atoms, allowing the formation of molecules.[11] Within a molecule, electrons move under the influence of several nuclei, and occupy molecular orbitals; much as they can occupy atomic orbitals in isolated atoms.[110] A fundamental factor in these molecular structures is the existence of electron pairs. These are electrons with opposed spins, allowing them to occupy the same molecular orbital without violating the Pauli exclusion principle (much like in atoms). Different molecular orbitals have different spatial distribution of the electron density. For instance, in bonded pairs (i.e. in the pairs that actually bind atoms together) electrons can be found with the maximal probability in a relatively small volume between the nuclei. On the contrary, in non-bonded pairs electrons are distributed in a large volume around nuclei.[111]

Conductivity

If a body has more or fewer electrons than are required to balance the positive charge of the nuclei, then that object has a net electric charge. When there is an excess of electrons, the object is said to be negatively charged. When there are fewer electrons than the number of protons in nuclei, the object is said to be positively charged. When the number of electrons and the number of protons are equal, their charges cancel each other and the object is said to be electrically neutral. A macroscopic body can develop an electric charge through rubbing, by the triboelectric effect.[115]

Independent electrons moving in vacuum are termed *free* electrons. Electrons in metals also behave as if they were free. In reality the particles that are commonly termed

A lightning discharge consists primarily of a flow of electrons.[112] *The electric potential needed for lightning may be generated by a triboelectric effect.*[113][114]

electrons in metals and other solids are quasi-electrons—quasiparticles, which have the same electrical charge, spin and magnetic moment as real electrons but may have a different mass.[116] When free electrons—both in vacuum and metals—move, they produce a net flow of charge called an electric current, which generates a magnetic field. Likewise a current can be created by a changing magnetic field. These interactions are described mathematically by Maxwell's equations.[117]

At a given temperature, each material has an electrical conductivity that determines the value of electric current when an electric potential is applied. Examples of good conductors include metals such as copper and gold, whereas glass and Teflon are poor conductors. In any dielectric material, the electrons remain bound to their respective atoms and the material behaves as an insulator. Most semiconductors have a variable level of conductivity that lies between the extremes of conduction and insulation.[118] On the other hand, metals have an electronic band structure containing partially filled electronic bands. The presence of such bands allows electrons in metals to behave as if they were free or delocalized electrons. These electrons are not associated with specific atoms, so when an electric field is applied, they are free to move like a gas (called Fermi gas)[119] through the material much like free electrons.

Because of collisions between electrons and atoms, the drift velocity of electrons in a conductor is on the order of millimeters per second. However, the speed at which a change

of current at one point in the material causes changes in currents in other parts of the material, the velocity of propagation, is typically about 75% of light speed.[120] This occurs because electrical signals propagate as a wave, with the velocity dependent on the dielectric constant of the material.[121]

Metals make relatively good conductors of heat, primarily because the delocalized electrons are free to transport thermal energy between atoms. However, unlike electrical conductivity, the thermal conductivity of a metal is nearly independent of temperature. This is expressed mathematically by the Wiedemann–Franz law,[119] which states that the ratio of thermal conductivity to the electrical conductivity is proportional to the temperature. The thermal disorder in the metallic lattice increases the electrical resistivity of the material, producing a temperature dependence for electric current.[122]

When cooled below a point called the critical temperature, materials can undergo a phase transition in which they lose all resistivity to electric current, in a process known as superconductivity. In BCS theory, this behavior is modeled by pairs of electrons entering a quantum state known as a Bose–Einstein condensate. These Cooper pairs have their motion coupled to nearby matter via lattice vibrations called phonons, thereby avoiding the collisions with atoms that normally create electrical resistance.[123] (Cooper pairs have a radius of roughly 100 nm, so they can overlap each other.)[124] However, the mechanism by which higher temperature superconductors operate remains uncertain.

Electrons inside conducting solids, which are quasiparticles themselves, when tightly confined at temperatures close to absolute zero, behave as though they had split into three other quasiparticles: spinons, Orbitons and holons.[125][126] The former carries spin and magnetic moment, the next carries its orbital location while the latter electrical charge.

Motion and energy

According to Einstein's theory of special relativity, as an electron's speed approaches the speed of light, from an observer's point of view its relativistic mass increases, thereby making it more and more difficult to accelerate it from within the observer's frame of reference. The speed of an electron can approach, but never reach, the speed of light in a vacuum, c. However, when relativistic electrons—that is, electrons moving at a speed close to c—are injected into a dielectric medium such as water, where the local speed of light is significantly less than c, the electrons temporarily travel faster than light in the medium. As they interact with the medium, they generate a faint light called Cherenkov radiation.[127]

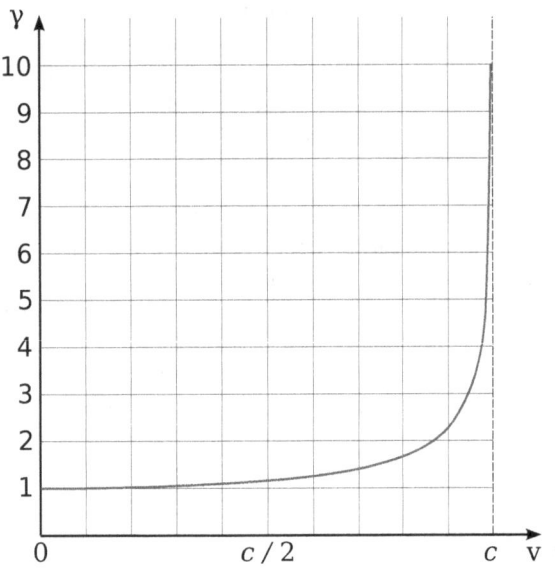

Lorentz factor as a function of velocity. It starts at value 1 and goes to infinity as v approaches c.

The effects of special relativity are based on a quantity known as the Lorentz factor, defined as $\gamma = 1/\sqrt{1-v^2/c^2}$ where v is the speed of the particle. The kinetic energy K_e of an electron moving with velocity v is:

$$K_e = (\gamma - 1)m_e c^2,$$

where m_e is the mass of electron. For example, the Stanford linear accelerator can accelerate an electron to roughly 51 GeV.[128] Since an electron behaves as a wave, at a given velocity it has a characteristic de Broglie wavelength. This is given by $\lambda_e = h/p$ where h is the Planck constant and p is the momentum.[50] For the 51 GeV electron above, the wavelength is about 2.4×10^{-17} m, small enough to explore structures well below the size of an atomic nucleus.[129]

3.3.3   Formation

The Big Bang theory is the most widely accepted scientific theory to explain the early stages in the evolution of the Universe.[130] For the first millisecond of the Big Bang, the temperatures were over 10 billion Kelvin and photons had mean energies over a million electronvolts. These photons were sufficiently energetic that they could react with each other to form pairs of electrons and positrons. Likewise, positron-electron pairs annihilated each other and emitted energetic photons:

$$\gamma + \gamma \leftrightarrow e+ + e-$$

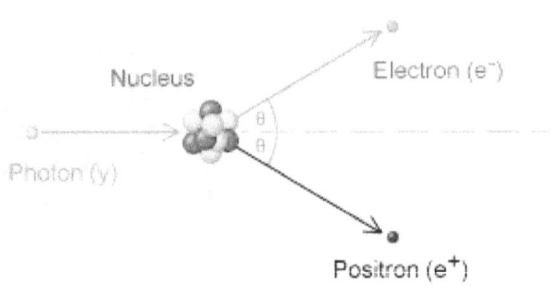

Pair production caused by the collision of a photon with an atomic nucleus

An equilibrium between electrons, positrons and photons was maintained during this phase of the evolution of the Universe. After 15 seconds had passed, however, the temperature of the universe dropped below the threshold where electron-positron formation could occur. Most of the surviving electrons and positrons annihilated each other, releasing gamma radiation that briefly reheated the universe.[131]

For reasons that remain uncertain, during the process of leptogenesis there was an excess in the number of electrons over positrons.[132] Hence, about one electron in every billion survived the annihilation process. This excess matched the excess of protons over antiprotons, in a condition known as baryon asymmetry, resulting in a net charge of zero for the universe.[133][134] The surviving protons and neutrons began to participate in reactions with each other—in the process known as nucleosynthesis, forming isotopes of hydrogen and helium, with trace amounts of lithium. This process peaked after about five minutes.[135] Any leftover neutrons underwent negative beta decay with a half-life of about a thousand seconds, releasing a proton and electron in the process,

$$n \rightarrow p + e- + \nu_e$$

For about the next 300000–400000 years, the excess electrons remained too energetic to bind with atomic nuclei.[136] What followed is a period known as recombination, when neutral atoms were formed and the expanding universe became transparent to radiation.[137]

Roughly one million years after the big bang, the first generation of stars began to form.[137] Within a star, stellar nucleosynthesis results in the production of positrons from the fusion of atomic nuclei. These antimatter particles immediately annihilate with electrons, releasing gamma rays. The net result is a steady reduction in the number of electrons, and a matching increase in the number of neutrons. However, the process of stellar evolution can result in the synthesis of radioactive isotopes. Selected isotopes can subsequently undergo negative beta decay, emitting an electron and antineutrino from the nucleus.[138] An example is the cobalt-60 (^{60}Co) isotope, which decays to form nickel-60 (60Ni).[139]

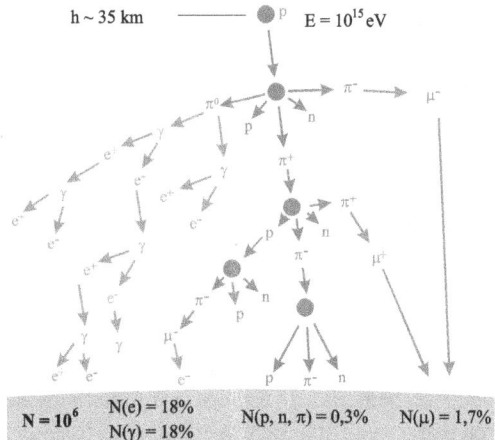

An extended air shower generated by an energetic cosmic ray striking the Earth's atmosphere

At the end of its lifetime, a star with more than about 20 solar masses can undergo gravitational collapse to form a black hole.[140] According to classical physics, these massive stellar objects exert a gravitational attraction that is strong enough to prevent anything, even electromagnetic radiation, from escaping past the Schwarzschild radius. However, quantum mechanical effects are believed to potentially allow the emission of Hawking radiation at this distance. Electrons (and positrons) are thought to be created at the event horizon of these stellar remnants.

When pairs of virtual particles (such as an electron and positron) are created in the vicinity of the event horizon, the random spatial distribution of these particles may permit one of them to appear on the exterior; this process is called quantum tunnelling. The gravitational potential of the black hole can then supply the energy that transforms this virtual particle into a real particle, allowing it to radiate away into space.[141] In exchange, the other member of the pair is given negative energy, which results in a net loss of mass-energy by the black hole. The rate of Hawking radiation increases with decreasing mass, eventually causing the black hole to evaporate away until, finally, it explodes.[142]

Cosmic rays are particles traveling through space with high energies. Energy events as high as 3.0×10^{20} eV have been recorded.[143] When these particles collide with nucleons in the Earth's atmosphere, a shower of particles is generated, including pions.[144] More than half of the cosmic radiation observed from the Earth's surface consists of muons. The particle called a muon is a lepton produced in the upper atmosphere by the decay of a pion.

$$\pi- \rightarrow \mu- + \nu$$
$$\mu$$

A muon, in turn, can decay to form an electron or positron.[145]

$$\mu- \rightarrow e- + \nu$$
$$e + \nu$$
$$\mu$$

3.3.4 Observation

Aurorae are mostly caused by energetic electrons precipitating into the atmosphere.[146]

Remote observation of electrons requires detection of their radiated energy. For example, in high-energy environments such as the corona of a star, free electrons form a plasma that radiates energy due to Bremsstrahlung radiation. Electron gas can undergo plasma oscillation, which is waves caused by synchronized variations in electron density, and these produce energy emissions that can be detected by using radio telescopes.[147]

The frequency of a photon is proportional to its energy. As a bound electron transitions between different energy levels of an atom, it absorbs or emits photons at characteristic frequencies. For instance, when atoms are irradiated by a source with a broad spectrum, distinct absorption lines appear in the spectrum of transmitted radiation. Each element or molecule displays a characteristic set of spectral lines, such as the hydrogen spectral series. Spectroscopic measurements of the strength and width of these lines allow the composition and physical properties of a substance to be determined.[148][149]

In laboratory conditions, the interactions of individual electrons can be observed by means of particle detectors, which allow measurement of specific properties such as energy, spin and charge.[107] The development of the Paul trap and Penning trap allows charged particles to be contained within a small region for long durations. This enables precise measurements of the particle properties. For example, in one instance a Penning trap was used to contain a single electron for a period of 10 months.[150] The magnetic moment of the electron was measured to a precision of eleven digits, which, in 1980, was a greater accuracy than for any other physical constant.[151]

The first video images of an electron's energy distribution were captured by a team at Lund University in Sweden, February 2008. The scientists used extremely short flashes of light, called attosecond pulses, which allowed an electron's motion to be observed for the first time.[152][153]

The distribution of the electrons in solid materials can be visualized by angle-resolved photoemission spectroscopy (ARPES). This technique employs the photoelectric effect to measure the reciprocal space—a mathematical representation of periodic structures that is used to infer the original structure. ARPES can be used to determine the direction, speed and scattering of electrons within the material.[154]

3.3.5   Plasma applications

Particle beams

During a NASA wind tunnel test, a model of the Space Shuttle is targeted by a beam of electrons, simulating the effect of ionizing gases during re-entry.[155]

Electron beams are used in welding.[156] They allow energy densities up to 10^7 W·cm^{-2} across a narrow focus diameter of 0.1–1.3 mm and usually require no filler material. This welding technique must be performed in a vacuum to prevent the electrons from interacting with the gas before reaching their target, and it can be used to join conductive materials that would otherwise be considered unsuitable for welding.[157][158]

Electron-beam lithography (EBL) is a method of etching semiconductors at resolutions smaller than a micrometer.[159] This technique is limited by high costs, slow performance, the need to operate the beam in the vacuum and the tendency of the electrons to scatter in solids. The last problem limits the resolution to about 10 nm. For this reason, EBL is primarily used for the production of small numbers of specialized integrated circuits.[160]

Electron beam processing is used to irradiate materials in order to change their physical properties or sterilize medical and food products.[161] Electron beams fluidise or quasi-melt glasses without significant increase of temperature on intensive irradiation: e.g. intensive electron radiation causes a many orders of magnitude decrease of viscosity and stepwise decrease of its activation energy.[162]

Linear particle accelerators generate electron beams for treatment of superficial tumors in radiation therapy. Electron therapy can treat such skin lesions as basal-cell carcinomas because an electron beam only penetrates to a limited depth before being absorbed, typically up to 5 cm for electron energies in the range 5–20 MeV. An electron beam can be used to supplement the treatment of areas that have been irradiated by X-rays.[163][164]

Particle accelerators use electric fields to propel electrons and their antiparticles to high energies. These particles emit synchrotron radiation as they pass through magnetic fields. The dependency of the intensity of this radiation upon spin polarizes the electron beam—a process known as the Sokolov–Ternov effect.[note 8] Polarized electron beams can be useful for various experiments. Synchrotron radiation can also cool the electron beams to reduce the momentum spread of the particles. Electron and positron beams are collided upon the particles' accelerating to the required energies; particle detectors observe the resulting energy emissions, which particle physics studies .[165]

Imaging

Low-energy electron diffraction (LEED) is a method of bombarding a crystalline material with a collimated beam of electrons and then observing the resulting diffraction patterns to determine the structure of the material. The required energy of the electrons is typically in the range 20–200 eV.[166] The reflection high-energy electron diffraction (RHEED) technique uses the reflection of a beam of electrons fired at various low angles to characterize the surface of crystalline materials. The beam energy is typically in the range 8–20 keV and the angle of incidence is 1–4°.[167][168]

The electron microscope directs a focused beam of electrons at a specimen. Some electrons change their properties, such as movement direction, angle, and relative phase and energy as the beam interacts with the material. Microscopists can record these changes in the electron beam to produce atomically resolved images of the material.[169] In blue light, conventional optical microscopes have a diffraction-limited resolution of about 200 nm.[170] By comparison, electron microscopes are limited by the de Broglie wavelength of the electron. This wavelength, for example, is equal to 0.0037 nm for electrons accelerated across a 100,000-volt potential.[171] The Transmission Electron Aberration-Corrected Microscope is capable of sub-0.05 nm resolution, which is more than enough to resolve individual atoms.[172] This capability makes the electron microscope a useful laboratory instrument for high resolution imaging. However, electron microscopes are expensive instruments that are costly to maintain.

Two main types of electron microscopes exist: transmission and scanning. Transmission electron microscopes function like overhead projectors, with a beam of electrons passing through a slice of material then being projected by lenses on a photographic slide or a charge-coupled device. Scanning electron microscopes rasteri a finely focused electron beam, as in a TV set, across the studied sample to produce the image. Magnifications range from 100× to 1,000,000× or higher for both microscope types. The scanning tunneling microscope uses quantum tunneling of electrons from a sharp metal tip into the studied material and can produce atomically resolved images of its surface.[173][174][175]

Other applications

In the free-electron laser (FEL), a relativistic electron beam passes through a pair of undulators that contain arrays of dipole magnets whose fields point in alternating directions. The electrons emit synchrotron radiation that coherently interacts with the same electrons to strongly amplify the radiation field at the resonance frequency. FEL can emit a coherent high-brilliance electromagnetic radiation with a wide range of frequencies, from microwaves to soft X-rays. These devices may find manufacturing, communication and various medical applications, such as soft tissue surgery.[176]

Electrons are important in cathode ray tubes, which have been extensively used as display devices in laboratory instruments, computer monitors and television sets.[177] In a photomultiplier tube, every photon striking the photocathode initiates an avalanche of electrons that produces a detectable current pulse.[178] Vacuum tubes use the flow of electrons to manipulate electrical signals, and they played a critical role in the development of electronics technology. However, they have been largely supplanted by solid-state devices such as the transistor.[179]

3.3.6 See also

- Anyon
- Electride
- Electron bubble
- Exoelectron emission
- *g*-factor
- Periodic systems of small molecules
- Spintronics
- Stern–Gerlach experiment
- Townsend discharge
- Zeeman effect
- List of particles
- Lepton

3.3.7 Notes

[1] The fractional version's denominator is the inverse of the decimal value (along with its relative standard uncertainty of 4.2×10^{-13} u).

[2] The electron's charge is the negative of elementary charge, which has a positive value for the proton.

[3] This magnitude is obtained from the spin quantum number as

$$S = \sqrt{s(s+1)} \cdot \frac{h}{2\pi}$$
$$= \frac{\sqrt{3}}{2}\hbar$$

for quantum number $s = \frac{1}{2}$.
See: Gupta, M.C. (2001). *Atomic and Molecular Spectroscopy*. New Age Publishers. p. 81. ISBN 81-224-1300-5.

[4] Bohr magneton:

$$\mu_B = \frac{e\hbar}{2m_e}.$$

[5] The classical electron radius is derived as follows. Assume that the electron's charge is spread uniformly throughout a spherical volume. Since one part of the sphere would repel the other parts, the sphere contains electrostatic potential energy. This energy is assumed to equal the electron's rest energy, defined by special relativity ($E = mc^2$).

From electrostatics theory, the potential energy of a sphere with radius r and charge e is given by:

$$E_\mathrm{p} = \frac{e^2}{8\pi\varepsilon_0 r},$$

where ε_0 is the vacuum permittivity. For an electron with rest mass m_0, the rest energy is equal to:

$$E_\mathrm{p} = m_0 c^2,$$

where c is the speed of light in a vacuum. Setting them equal and solving for r gives the classical electron radius.
See: Haken, H.; Wolf, H.C.; Brewer, W.D. (2005). *The Physics of Atoms and Quanta: Introduction to Experiments and Theory*. Springer. p. 70. ISBN 3-540-67274-5.

[6] Radiation from non-relativistic electrons is sometimes termed cyclotron radiation.

[7] The change in wavelength, $\Delta\lambda$, depends on the angle of the recoil, θ, as follows,

$$\Delta\lambda = \frac{h}{m_\mathrm{e} c}(1 - \cos\theta),$$

where c is the speed of light in a vacuum and m_e is the electron mass. See Zombeck (2007: 393, 396).

[8] The polarization of an electron beam means that the spins of all electrons point into one direction. In other words, the projections of the spins of all electrons onto their momentum vector have the same sign.

3.3.8 References

[1] Eichten, E.J.; Peskin, M.E.; Peskin, M. (1983). "New Tests for Quark and Lepton Substructure". *Physical Review Letters* **50** (11): 811–814. Bibcode:1983PhRvL..50..811E. doi:10.1103/PhysRevLett.50.811.

[2] Farrar, W.V. (1969). "Richard Laming and the Coal-Gas Industry, with His Views on the Structure of Matter". *Annals of Science* **25** (3): 243–254. doi:10.1080/00033796900200141.

[3] Arabatzis, T. (2006). *Representing Electrons: A Biographical Approach to Theoretical Entities*. University of Chicago Press. pp. 70–74. ISBN 0-226-02421-0.

[4] Buchwald, J.Z.; Warwick, A. (2001). *Histories of the Electron: The Birth of Microphysics*. MIT Press. pp. 195–203. ISBN 0-262-52424-4.

[5] Thomson, J.J. (1897). "Cathode Rays". *Philosophical Magazine* **44** (269): 293. doi:10.1080/14786449708621070.

[6] P.J. Mohr, B.N. Taylor, and D.B. Newell (2011), "The 2010 CODATA Recommended Values of the Fundamental Physical Constants" (Web Version 6.0). This database was developed by J. Baker, M. Douma, and S. Kotochigova. Available: http://physics.nist.gov/constants [Thursday, 02-Jun-2011 21:00:12 EDT]. National Institute of Standards and Technology, Gaithersburg, MD 20899.

[7] "JERRY COFF". Retrieved 10 September 2010.

[8] Curtis, L.J. (2003). *Atomic Structure and Lifetimes: A Conceptual Approach*. Cambridge University Press. p. 74. ISBN 0-521-53635-9.

[9] "CODATA value: proton-electron mass ratio". *2006 CODATA recommended values*. National Institute of Standards and Technology. Retrieved 2009-07-18.

[10] Anastopoulos, C. (2008). *Particle Or Wave: The Evolution of the Concept of Matter in Modern Physics*. Princeton University Press. pp. 236–237. ISBN 0-691-13512-6.

[11] Pauling, L.C. (1960). *The Nature of the Chemical Bond and the Structure of Molecules and Crystals: an introduction to modern structural chemistry* (3rd ed.). Cornell University Press. pp. 4–10. ISBN 0-8014-0333-2.

[12] Dahl (1997:122–185).

[13] Wilson, R. (1997). *Astronomy Through the Ages: The Story of the Human Attempt to Understand the Universe*. CRC Press. p. 138. ISBN 0-7484-0748-0.

[14] Shipley, J.T. (1945). *Dictionary of Word Origins*. The Philosophical Library. p. 133. ISBN 0-88029-751-4.

[15] Baigrie, B. (2006). *Electricity and Magnetism: A Historical Perspective*. Greenwood Press. pp. 7–8. ISBN 0-313-33358-0.

[16] Keithley, J.F. (1999). *The Story of Electrical and Magnetic Measurements: From 500 B.C. to the 1940s*. IEEE Press. pp. 15, 20. ISBN 0-7803-1193-0.

[17] "Benjamin Franklin (1706–1790)". *Eric Weisstein's World of Biography*. Wolfram Research. Retrieved 2010-12-16.

[18] Myers, R.L. (2006). *The Basics of Physics*. Greenwood Publishing Group. p. 242. ISBN 0-313-32857-9.

[19] Barrow, J.D. (1983). "Natural Units Before Planck". *Quarterly Journal of the Royal Astronomical Society* **24**: 24–26. Bibcode:1983QJRAS..24...24B.

[20] Sōgo Okamura (1994). *History of Electron Tubes*. IOS Press. p. 11. ISBN 978-90-5199-145-1. Retrieved 29 May 2015. In 1881, Stoney named this electromagnetic 'electrolion'. It came to be called 'electron' from 1891. [...] In 1906, the suggestion to call cathode ray particles 'electrions' was brought up but through the opinion of Lorentz of Holland 'electrons' came to be widely used.

[21] Stoney, G.J. (1894). "Of the "Electron," or Atom of Electricity". *Philosophical Magazine* **38** (5): 418–420. doi:10.1080/14786449408620653.

[22] "electron, n.2". OED Online. March 2013. Oxford University Press. Accessed 12 April 2013

[23] Soukhanov, A.H. ed. (1986). *Word Mysteries & Histories*. Houghton Mifflin Company. p. 73. ISBN 0-395-40265-4.

[24] Guralnik, D.B. ed. (1970). *Webster's New World Dictionary*. Prentice Hall. p. 450.

[25] Born, M.; Blin-Stoyle, R.J.; Radcliffe, J.M. (1989). *Atomic Physics*. Courier Dover. p. 26. ISBN 0-486-65984-4.

[26] Dahl (1997:55–58).

[27] DeKosky, R.K. (1983). "William Crookes and the quest for absolute vacuum in the 1870s". *Annals of Science* **40** (1): 1–18. doi:10.1080/00033798300200101.

[28] Leicester, H.M. (1971). *The Historical Background of Chemistry*. Courier Dover. pp. 221–222. ISBN 0-486-61053-5.

[29] Dahl (1997:64–78).

[30] Zeeman, P.; Zeeman, P. (1907). "Sir William Crookes, F.R.S". *Nature* **77** (1984): 1–3. Bibcode:1907Natur..77....1C. doi:10.1038/077001a0.

[31] Dahl (1997:99).

[32] Frank Wilczek: "Happy Birthday, Electron" *Scientific American*, June 2012.

[33] Thomson, J.J. (1906). "Nobel Lecture: Carriers of Negative Electricity" (PDF). The Nobel Foundation. Retrieved 2008-08-25.

[34] Trenn, T.J. (1976). "Rutherford on the Alpha-Beta-Gamma Classification of Radioactive Rays". *Isis* **67** (1): 61–75. doi:10.1086/351545. JSTOR 231134.

[35] Becquerel, H. (1900). "Déviation du Rayonnement du Radium dans un Champ Électrique". *Comptes rendus de l'Académie des sciences* (in French) **130**: 809–815.

[36] Buchwald and Warwick (2001:90–91).

[37] Myers, W.G. (1976). "Becquerel's Discovery of Radioactivity in 1896". *Journal of Nuclear Medicine* **17** (7): 579–582. PMID 775027.

[38] Kikoin, I.K.; Sominskiĭ, I.S. (1961). "Abram Fedorovich Ioffe (on his eightieth birthday)". *Soviet Physics Uspekhi* **3** (5): 798–809. Bibcode:1961SvPhU...3..798K. doi:10.1070/PU1961v003n05ABEH005812. Original publication in Russian: Кикоин, И.К.; Соминский, М.С. (1960). "Академик А.Ф. Иоффе" (PDF). *Успехи Физических Наук* **72** (10): 303–321.

[39] Millikan, R.A. (1911). "The Isolation of an Ion, a Precision Measurement of its Charge, and the Correction of Stokes' Law". *Physical Review* **32** (2): 349–397. Bibcode:1911PhRvI..32..349M. doi:10.1103/PhysRevSeriesI.32.349.

[40] Das Gupta, N.N.; Ghosh, S.K. (1999). "A Report on the Wilson Cloud Chamber and Its Applications in Physics". *Reviews of Modern Physics* **18** (2): 225–290. Bibcode:1946RvMP...18..225G. doi:10.1103/RevModPhys.18.225.

[41] Smirnov, B.M. (2003). *Physics of Atoms and Ions*. Springer. pp. 14–21. ISBN 0-387-95550-X.

[42] Bohr, N. (1922). "Nobel Lecture: The Structure of the Atom" (PDF). The Nobel Foundation. Retrieved 2008-12-03.

[43] Lewis, G.N. (1916). "The Atom and the Molecule". *Journal of the American Chemical Society* **38** (4): 762–786. doi:10.1021/ja02261a002.

[44] Arabatzis, T.; Gavroglu, K. (1997). "The chemists' electron". *European Journal of Physics* **18** (3): 150–163. Bibcode:1997EJPh...18..150A. doi:10.1088/0143-0807/18/3/005.

[45] Langmuir, I. (1919). "The Arrangement of Electrons in Atoms and Molecules". *Journal of the American Chemical Society* **41** (6): 868–934. doi:10.1021/ja02227a002.

[46] Scerri, E.R. (2007). *The Periodic Table*. Oxford University Press. pp. 205–226. ISBN 0-19-530573-6.

[47] Massimi, M. (2005). *Pauli's Exclusion Principle, The Origin and Validation of a Scientific Principle*. Cambridge University Press. pp. 7–8. ISBN 0-521-83911-4.

[48] Uhlenbeck, G.E.; Goudsmith, S. (1925). "Ersetzung der Hypothese vom unmechanischen Zwang durch eine Forderung bezüglich des inneren Verhaltens jedes einzelnen Elektrons". *Die Naturwissenschaften* (in German) **13** (47): 953. Bibcode:1925NW.....13..953E. doi:10.1007/BF01558878.

[49] Pauli, W. (1923). "Über die Gesetzmäßigkeiten des anomalen Zeemaneffektes". *Zeitschrift für Physik* (in German) **16** (1): 155–164. Bibcode:1923ZPhy...16..155P. doi:10.1007/BF01327386.

[50] de Broglie, L. (1929). "Nobel Lecture: The Wave Nature of the Electron" (PDF). The Nobel Foundation. Retrieved 2008-08-30.

[51] Falkenburg, B. (2007). *Particle Metaphysics: A Critical Account of Subatomic Reality*. Springer. p. 85. ISBN 3-540-33731-8.

[52] Davisson, C. (1937). "Nobel Lecture: The Discovery of Electron Waves" (PDF). The Nobel Foundation. Retrieved 2008-08-30.

[53] Schrödinger, E. (1926). "Quantisierung als Eigenwertproblem". *Annalen der Physik* (in German) **385** (13): 437–490. Bibcode:1926AnP...385..437S. doi:10.1002/andp.19263851302.

[54] Rigden, J.S. (2003). *Hydrogen*. Harvard University Press. pp. 59–86. ISBN 0-674-01252-6.

[55] Reed, B.C. (2007). *Quantum Mechanics*. Jones & Bartlett Publishers. pp. 275–350. ISBN 0-7637-4451-4.

[56] Dirac, P.A.M. (1928). "The Quantum Theory of the Electron". *Proceedings of the Royal Society A* **117** (778): 610–624. Bibcode:1928RSPSA.117..610D. doi:10.1098/rspa.1928.0023.

[57] Dirac, P.A.M. (1933). "Nobel Lecture: Theory of Electrons and Positrons" (PDF). The Nobel Foundation. Retrieved 2008-11-01.

[58] "The Nobel Prize in Physics 1965". The Nobel Foundation. Retrieved 2008-11-04.

[59] Panofsky, W.K.H. (1997). "The Evolution of Particle Accelerators & Colliders" (PDF). *Beam Line* (Stanford University) **27** (1): 36–44. Retrieved 2008-09-15.

[60] Elder, F.R. et al. (1947). "Radiation from Electrons in a Synchrotron". *Physical Review* **71** (11): 829–830. Bibcode:1947PhRv...71..829E. doi:10.1103/PhysRev.71.829.5.

[61] Hoddeson, L. et al. (1997). *The Rise of the Standard Model: Particle Physics in the 1960s and 1970s.* Cambridge University Press. pp. 25–26. ISBN 0-521-57816-7.

[62] Bernardini, C. (2004). "AdA: The First Electron–Positron Collider". *Physics in Perspective* **6** (2): 156–183. Bibcode:2004PhP.....6..156B. doi:10.1007/s00016-003-0202-y.

[63] "Testing the Standard Model: The LEP experiments". CERN. 2008. Retrieved 2008-09-15.

[64] "LEP reaps a final harvest". *CERN Courier* **40** (10). 2000.

[65] Prati, E.; De Michielis, M.; Belli, M.; Cocco, S.; Fanciulli, M.; Kotekar-Patil, D.; Ruoff, M.; Kern, D. P.; Wharam, D. A.; Verduijn, J.; Tettamanzi, G. C.; Rogge, S.; Roche, B.; Wacquez, R.; Jehl, X.; Vinet, M.; Sanquer, M. (2012). "Few electron limit of n-type metal oxide semiconductor single electron transistors". *Nanotechnology* **23** (21): 215204. doi:10.1088/0957-4484/23/21/215204. PMID 22552118.

[66] Frampton, P.H.; Hung, P.Q.; Sher, Marc (2000). "Quarks and Leptons Beyond the Third Generation". *Physics Reports* **330** (5–6): 263–348. arXiv:hep-ph/9903387. Bibcode:2000PhR...330..263F. doi:10.1016/S0370-1573(99)00095-2.

[67] Raith, W.; Mulvey, T. (2001). *Constituents of Matter: Atoms, Molecules, Nuclei and Particles.* CRC Press. pp. 777–781. ISBN 0-8493-1202-7.

[68] The original source for CODATA is Mohr, P.J.; Taylor, B.N.; Newell, D.B. (2006). "CODATA recommended values of the fundamental physical constants". *Reviews of Modern Physics* **80** (2): 633–730. arXiv:0801.0028. Bibcode:2008RvMP...80..633M. doi:10.1103/RevModPhys.80.633.

Individual physical constants from the CODATA are available at: "The NIST Reference on Constants, Units and Uncertainty". National Institute of Standards and Technology. Retrieved 2009-01-15.

[69] Zombeck, M.V. (2007). *Handbook of Space Astronomy and Astrophysics* (3rd ed.). Cambridge University Press. p. 14. ISBN 0-521-78242-2.

[70] Murphy, M.T. et al. (2008). "Strong Limit on a Variable Proton-to-Electron Mass Ratio from Molecules in the Distant Universe". *Science* **320** (5883): 1611–1613. arXiv:0806.3081. Bibcode:2008Sci...320.1611M. doi:10.1126/science.1156352. PMID 18566280.

[71] Zorn, J.C.; Chamberlain, G.E.; Hughes, V.W. (1963). "Experimental Limits for the Electron-Proton Charge Difference and for the Charge of the Neutron". *Physical Review* **129** (6): 2566–2576. Bibcode:1963PhRv..129.2566Z. doi:10.1103/PhysRev.129.2566.

[72] Odom, B. et al. (2006). "New Measurement of the Electron Magnetic Moment Using a One-Electron Quantum Cyclotron". *Physical Review Letters* **97** (3): 030801. Bibcode:2006PhRvL..97c0801O. doi:10.1103/PhysRevLett.97.030801. PMID 16907490.

[73] Anastopoulos, C. (2008). *Particle Or Wave: The Evolution of the Concept of Matter in Modern Physics.* Princeton University Press. pp. 261–262. ISBN 0-691-13512-6.

[74] Gabrielse, G. et al. (2006). "New Determination of the Fine Structure Constant from the Electron *g* Value and QED". *Physical Review Letters* **97** (3): 030802(1–4). Bibcode:2006PhRvL..97c0802G. doi:10.1103/PhysRevLett.97.030802.

[75] Eduard Shpolsky, Atomic physics (Atomnaia fizika),second edition, 1951

[76] Dehmelt, H. (1988). "A Single Atomic Particle Forever Floating at Rest in Free Space: New Value for Electron Radius". *Physica Scripta* **T22**: 102–10. Bibcode:1988PhST...22..102D. doi:10.1088/0031-8949/1988/T22/016.

[77] Meschede, D. (2004). *Optics, light and lasers: The Practical Approach to Modern Aspects of Photonics and Laser Physics.* Wiley-VCH. p. 168. ISBN 3-527-40364-7.

[78] Steinberg, R.I. et al. (1999). "Experimental test of charge conservation and the stability of the electron". *Physical Review D* **61** (2): 2582–2586. Bibcode:1975PhRvD..12.2582S. doi:10.1103/PhysRevD.12.2582.

[79] J. Beringer (Particle Data Group) et al. (2012). "Review of Particle Physics: [electron properties]" (PDF). *Physical Review D* **86** (1): 010001. Bibcode:2012PhRvD..86a0001B. doi:10.1103/PhysRevD.86.010001.

[80] Back, H. O. et al. (2002). "Search for electron decay mode e → γ + ν with prototype of Borexino detector". *Physics Letters B* **525**: 29–40. Bibcode:2002PhLB..525...29B. doi:10.1016/S0370-2693(01)01440-X.

[81] Munowitz, M. (2005). *Knowing, The Nature of Physical Law*. Oxford University Press. ISBN 0-19-516737-6.

[82] Kane, G. (October 9, 2006). "Are virtual particles really constantly popping in and out of existence? Or are they merely a mathematical bookkeeping device for quantum mechanics?". Scientific American. Retrieved 2008-09-19.

[83] Taylor, J. (1989). "Gauge Theories in Particle Physics". In Davies, Paul. *The New Physics*. Cambridge University Press. p. 464. ISBN 0-521-43831-4.

[84] Genz, H. (2001). *Nothingness: The Science of Empty Space*. Da Capo Press. pp. 241–243, 245–247. ISBN 0-7382-0610-5.

[85] Gribbin, J. (January 25, 1997). "More to electrons than meets the eye". *New Scientist*. Retrieved 2008-09-17.

[86] Levine, I. et al. (1997). "Measurement of the Electromagnetic Coupling at Large Momentum Transfer". *Physical Review Letters* **78** (3): 424–427. Bibcode:1997PhRvL..78..424L. doi:10.1103/PhysRevLett.78.424.

[87] Murayama, H. (March 10–17, 2006). *Supersymmetry Breaking Made Easy, Viable and Generic. Proceedings of the XLIInd Rencontres de Moriond on Electroweak Interactions and Unified Theories* (La Thuile, Italy). arXiv:0709.3041.—lists a 9% mass difference for an electron that is the size of the Planck distance.

[88] Schwinger, J. (1948). "On Quantum-Electrodynamics and the Magnetic Moment of the Electron". *Physical Review* **73** (4): 416–417. Bibcode:1948PhRv...73..416S. doi:10.1103/PhysRev.73.416.

[89] Huang, K. (2007). *Fundamental Forces of Nature: The Story of Gauge Fields*. World Scientific. pp. 123–125. ISBN 981-270-645-3.

[90] Foldy, L.L.; Wouthuysen, S. (1950). "On the Dirac Theory of Spin 1/2 Particles and Its Non-Relativistic Limit". *Physical Review* **78**: 29–36. Bibcode:1950PhRv...78...29F. doi:10.1103/PhysRev.78.29.

[91] Sidharth, B.G. (2008). "Revisiting Zitterbewegung". *International Journal of Theoretical Physics* **48** (2): 497–506. arXiv:0806.0985. Bibcode:2009IJTP...48..497S. doi:10.1007/s10773-008-9825-8.

[92] Elliott, R.S. (1978). "The History of Electromagnetics as Hertz Would Have Known It". *IEEE Transactions on Microwave Theory and Techniques* **36** (5): 806–823. Bibcode:1988ITMTT..36..806E. doi:10.1109/22.3600.

[93] Crowell, B. (2000). *Electricity and Magnetism*. Light and Matter. pp. 129–152. ISBN 0-9704670-4-4.

[94] Mahadevan, R.; Narayan, R.; Yi, I. (1996). "Harmony in Electrons: Cyclotron and Synchrotron Emission by Thermal Electrons in a Magnetic Field". *The Astrophysical Journal* **465**: 327–337. arXiv:astro-ph/9601073. Bibcode:1996ApJ...465..327M. doi:10.1086/177422.

[95] Rohrlich, F. (1999). "The Self-Force and Radiation Reaction". *American Journal of Physics* **68** (12): 1109–1112. Bibcode:2000AmJPh..68.1109R. doi:10.1119/1.1286430.

[96] Georgi, H. (1989). "Grand Unified Theories". In Davies, Paul. *The New Physics*. Cambridge University Press. p. 427. ISBN 0-521-43831-4.

[97] Blumenthal, G.J.; Gould, R. (1970). "Bremsstrahlung, Synchrotron Radiation, and Compton Scattering of High-Energy Electrons Traversing Dilute Gases". *Reviews of Modern Physics* **42** (2): 237–270. Bibcode:1970RvMP...42..237B. doi:10.1103/RevModPhys.42.237.

[98] Staff (2008). "The Nobel Prize in Physics 1927". The Nobel Foundation. Retrieved 2008-09-28.

[99] Chen, S.-Y.; Maksimchuk, A.; Umstadter, D. (1998). "Experimental observation of relativistic nonlinear Thomson scattering". *Nature* **396** (6712): 653–655. arXiv:physics/9810036. Bibcode:1998Natur.396..653C. doi:10.1038/25303.

[100] Beringer, R.; Montgomery, C.G. (1942). "The Angular Distribution of Positron Annihilation Radiation". *Physical Review* **61** (5–6): 222–224. Bibcode:1942PhRv...61..222B. doi:10.1103/PhysRev.61.222.

[101] Buffa, A. (2000). *College Physics* (4th ed.). Prentice Hall. p. 888. ISBN 0-13-082444-5.

[102] Eichler, J. (2005). "Electron–positron pair production in relativistic ion–atom collisions". *Physics Letters A* **347** (1–3): 67–72. Bibcode:2005PhLA..347...67E. doi:10.1016/j.physleta.2005.06.105.

[103] Hubbell, J.H. (2006). "Electron positron pair production by photons: A historical overview". *Radiation Physics and Chemistry* **75** (6): 614–623. Bibcode:2006RaPC...75..614H. doi:10.1016/j.radphyschem.2005.10.008.

[104] Quigg, C. (June 4–30, 2000). *The Electroweak Theory. TASI 2000: Flavor Physics for the Millennium* (Boulder, Colorado): 80. arXiv:hep-ph/0204104.

[105] Mulliken, R.S. (1967). "Spectroscopy, Molecular Orbitals, and Chemical Bonding". *Science* **157** (3784): 13–24. Bibcode:1967Sci...157...13M. doi:10.1126/science.157.3784.13. PMID 5338306.

[106] Burhop, E.H.S. (1952). *The Auger Effect and Other Radiationless Transitions*. Cambridge University Press. pp. 2–3. ISBN 0-88275-966-3.

[107] Grupen, C. (2000). "Physics of Particle Detection". *AIP Conference Proceedings* **536**: 3–34. arXiv:physics/9906063. doi:10.1063/1.1361756.

[108] Jiles, D. (1998). *Introduction to Magnetism and Magnetic Materials*. CRC Press. pp. 280–287. ISBN 0-412-79860-3.

[109] Löwdin, P.O.; Erkki Brändas, E.; Kryachko, E.S. (2003). *Fundamental World of Quantum Chemistry: A Tribute to the Memory of Per- Olov Löwdin.* Springer. pp. 393–394. ISBN 1-4020-1290-X.

[110] McQuarrie, D.A.; Simon, J.D. (1997). *Physical Chemistry: A Molecular Approach.* University Science Books. pp. 325–361. ISBN 0-935702-99-7.

[111] Daudel, R. et al. (1973). "The Electron Pair in Chemistry". *Canadian Journal of Chemistry* **52** (8): 1310–1320. doi:10.1139/v74-201.

[112] Rakov, V.A.; Uman, M.A. (2007). *Lightning: Physics and Effects.* Cambridge University Press. p. 4. ISBN 0-521-03541-4.

[113] Freeman, G.R.; March, N.H. (1999). "Triboelectricity and some associated phenomena". *Materials Science and Technology* **15** (12): 1454–1458. doi:10.1179/026708399101505464.

[114] Forward, K.M.; Lacks, D.J.; Sankaran, R.M. (2009). "Methodology for studying particle–particle triboelectrification in granular materials". *Journal of Electrostatics* **67** (2–3): 178–183. doi:10.1016/j.elstat.2008.12.002.

[115] Weinberg, S. (2003). *The Discovery of Subatomic Particles.* Cambridge University Press. pp. 15–16. ISBN 0-521-82351-X.

[116] Lou, L.-F. (2003). *Introduction to phonons and electrons.* World Scientific. pp. 162, 164. ISBN 978-981-238-461-4.

[117] Guru, B.S.; Hızıroğlu, H.R. (2004). *Electromagnetic Field Theory.* Cambridge University Press. pp. 138, 276. ISBN 0-521-83016-8.

[118] Achuthan, M.K.; Bhat, K.N. (2007). *Fundamentals of Semiconductor Devices.* Tata McGraw-Hill. pp. 49–67. ISBN 0-07-061220-X.

[119] Ziman, J.M. (2001). *Electrons and Phonons: The Theory of Transport Phenomena in Solids.* Oxford University Press. p. 260. ISBN 0-19-850779-8.

[120] Main, P. (June 12, 1993). "When electrons go with the flow: Remove the obstacles that create electrical resistance, and you get ballistic electrons and a quantum surprise". *New Scientist* **1887**: 30. Retrieved 2008-10-09.

[121] Blackwell, G.R. (2000). *The Electronic Packaging Handbook.* CRC Press. pp. 6.39–6.40. ISBN 0-8493-8591-1.

[122] Durrant, A. (2000). *Quantum Physics of Matter: The Physical World.* CRC Press. pp. 43, 71–78. ISBN 0-7503-0721-8.

[123] Staff (2008). "The Nobel Prize in Physics 1972". The Nobel Foundation. Retrieved 2008-10-13.

[124] Kadin, A.M. (2007). "Spatial Structure of the Cooper Pair". *Journal of Superconductivity and Novel Magnetism* **20** (4): 285–292. arXiv:cond-mat/0510279. doi:10.1007/s10948-006-0198-z.

[125] "Discovery About Behavior Of Building Block Of Nature Could Lead To Computer Revolution". *ScienceDaily.* July 31, 2009. Retrieved 2009-08-01.

[126] Jompol, Y. et al. (2009). "Probing Spin-Charge Separation in a Tomonaga-Luttinger Liquid". *Science* **325** (5940): 597–601. arXiv:1002.2782. Bibcode:2009Sci...325..597J. doi:10.1126/science.1171769. PMID 19644117.

[127] Staff (2008). "The Nobel Prize in Physics 1958, for the discovery and the interpretation of the Cherenkov effect". The Nobel Foundation. Retrieved 2008-09-25.

[128] Staff (August 26, 2008). "Special Relativity". Stanford Linear Accelerator Center. Retrieved 2008-09-25.

[129] Adams, S. (2000). *Frontiers: Twentieth Century Physics.* CRC Press. p. 215. ISBN 0-7484-0840-1.

[130] Lurquin, P.F. (2003). *The Origins of Life and the Universe.* Columbia University Press. p. 2. ISBN 0-231-12655-7.

[131] Silk, J. (2000). *The Big Bang: The Creation and Evolution of the Universe* (3rd ed.). Macmillan. pp. 110–112, 134–137. ISBN 0-8050-7256-X.

[132] Christianto, V. (2007). "Thirty Unsolved Problems in the Physics of Elementary Particles" (PDF). *Progress in Physics* **4**: 112–114.

[133] Kolb, E.W.; Wolfram, Stephen (1980). "The Development of Baryon Asymmetry in the Early Universe". *Physics Letters B* **91** (2): 217–221. Bibcode:1980PhLB...91..217K. doi:10.1016/0370-2693(80)90435-9.

[134] Sather, E. (Spring–Summer 1996). "The Mystery of Matter Asymmetry" (PDF). *Beam Line.* University of Stanford. Retrieved 2008-11-01.

[135] Burles, S.; Nollett, K.M.; Turner, M.S. (1999). "Big-Bang Nucleosynthesis: Linking Inner Space and Outer Space". arXiv:astro-ph/9903300 [astro-ph].

[136] Boesgaard, A.M.; Steigman, G. (1985). "Big bang nucleosynthesis – Theories and observations". *Annual Review of Astronomy and Astrophysics* **23** (2): 319–378. Bibcode:1985ARA&A..23..319B. doi:10.1146/annurev.aa.23.090185.001535.

[137] Barkana, R. (2006). "The First Stars in the Universe and Cosmic Reionization". *Science* **313** (5789): 931–934. arXiv:astro-ph/0608450. Bibcode:2006Sci...313..931B. doi:10.1126/science.1125644. PMID 16917052.

[138] Burbidge, E.M. et al. (1957). "Synthesis of Elements in Stars". *Reviews of Modern Physics* **29** (4): 548–647. Bibcode:1957RvMP...29..547B. doi:10.1103/RevModPhys.29.547.

[139] Rodberg, L.S.; Weisskopf, V. (1957). "Fall of Parity: Recent Discoveries Related to Symmetry of Laws of Nature". *Science* **125** (3249): 627–633. Bibcode:1957Sci...125..627R. doi:10.1126/science.125.3249.627. PMID 17810563.

[140] Fryer, C.L. (1999). "Mass Limits For Black Hole Formation". *The Astrophysical Journal* **522** (1): 413–418. arXiv:astro-ph/9902315. Bibcode:1999ApJ...522..413F. doi:10.1086/307647.

[141] Parikh, M.K.; Wilczek, F. (2000). "Hawking Radiation As Tunneling". *Physical Review Letters* **85** (24): 5042–5045. arXiv:hep-th/9907001. Bibcode:2000PhRvL..85.5042P. doi:10.1103/PhysRevLett.85.5042. PMID 11102182.

[142] Hawking, S.W. (1974). "Black hole explosions?". *Nature* **248** (5443): 30–31. Bibcode:1974Natur.248...30H. doi:10.1038/248030a0.

[143] Halzen, F.; Hooper, D. (2002). "High-energy neutrino astronomy: the cosmic ray connection". *Reports on Progress in Physics* **66** (7): 1025–1078. arXiv:astro-ph/0204527. Bibcode:2002astro.ph..4527H. doi:10.1088/0034-4885/65/7/201.

[144] Ziegler, J.F. (1998). "Terrestrial cosmic ray intensities". *IBM Journal of Research and Development* **42** (1): 117–139. doi:10.1147/rd.421.0117.

[145] Sutton, C. (August 4, 1990). "Muons, pions and other strange particles". *New Scientist*. Retrieved 2008-08-28.

[146] Wolpert, S. (July 24, 2008). "Scientists solve 30-year-old aurora borealis mystery". University of California. Retrieved 2008-10-11.

[147] Gurnett, D.A.; Anderson, R. (1976). "Electron Plasma Oscillations Associated with Type III Radio Bursts". *Science* **194** (4270): 1159–1162. Bibcode:1976Sci...194.1159G. doi:10.1126/science.194.4270.1159. PMID 17790910.

[148] Martin, W.C.; Wiese, W.L. (2007). "Atomic Spectroscopy: A Compendium of Basic Ideas, Notation, Data, and Formulas". National Institute of Standards and Technology. Retrieved 2007-01-08.

[149] Fowles, G.R. (1989). *Introduction to Modern Optics*. Courier Dover. pp. 227–233. ISBN 0-486-65957-7.

[150] Staff (2008). "The Nobel Prize in Physics 1989". The Nobel Foundation. Retrieved 2008-09-24.

[151] Ekstrom, P.; Wineland, David (1980). "The isolated Electron" (PDF). *Scientific American* **243** (2): 91–101. doi:10.1038/scientificamerican0880-104. Retrieved 2008-09-24.

[152] Mauritsson, J. "Electron filmed for the first time ever" (PDF). Lund University. Archived from the original (PDF) on March 25, 2009. Retrieved 2008-09-17.

[153] Mauritsson, J. et al. (2008). "Coherent Electron Scattering Captured by an Attosecond Quantum Stroboscope". *Physical Review Letters* **100** (7): 073003. arXiv:0708.1060. Bibcode:2008PhRvL.100g3003M. doi:10.1103/PhysRevLett.100.073003. PMID 18352546.

[154] Damascelli, A. (2004). "Probing the Electronic Structure of Complex Systems by ARPES". *Physica Scripta* **T109**: 61–74. arXiv:cond-mat/0307085. Bibcode:2004PhST..109...61D. doi:10.1238/Physica.Topical.109a00061.

[155] Staff (April 4, 1975). "Image # L-1975-02972". Langley Research Center, NASA. Retrieved 2008-09-20.

[156] Elmer, J. (March 3, 2008). "Standardizing the Art of Electron-Beam Welding". Lawrence Livermore National Laboratory. Retrieved 2008-10-16.

[157] Schultz, H. (1993). *Electron Beam Welding*. Woodhead Publishing. pp. 2–3. ISBN 1-85573-050-2.

[158] Benedict, G.F. (1987). *Nontraditional Manufacturing Processes*. Manufacturing engineering and materials processing **19**. CRC Press. p. 273. ISBN 0-8247-7352-7.

[159] Ozdemir, F.S. (June 25–27, 1979). *Electron beam lithography*. Proceedings of the 16th Conference on Design automation (San Diego, CA, USA: IEEE Press): 383–391. Retrieved 2008-10-16.

[160] Madou, M.J. (2002). *Fundamentals of Microfabrication: the Science of Miniaturization* (2nd ed.). CRC Press. pp. 53–54. ISBN 0-8493-0826-7.

[161] Jongen, Y.; Herer, A. (May 2–5, 1996). *Electron Beam Scanning in Industrial Applications*. APS/AAPT Joint Meeting (American Physical Society). Bibcode:1996APS..MAY.H9902J.

[162] Mobus G. et al. (2010). Journal of Nuclear Materials, v. 396, 264–271, doi:10.1016/j.jnucmat.2009.11.020

[163] Beddar, A.S.; Domanovic, Mary Ann; Kubu, Mary Lou; Ellis, Rod J.; Sibata, Claudio H.; Kinsella, Timothy J. (2001). "Mobile linear accelerators for intraoperative radiation therapy". *AORN Journal* **74** (5): 700. doi:10.1016/S0001-2092(06)61769-9.

[164] Gazda, M.J.; Coia, L.R. (June 1, 2007). "Principles of Radiation Therapy" (PDF). Retrieved 2013-10-31.

[165] Chao, A.W.; Tigner, M. (1999). *Handbook of Accelerator Physics and Engineering*. World Scientific. pp. 155, 188. ISBN 981-02-3500-3.

[166] Oura, K. et al. (2003). *Surface Science: An Introduction*. Springer. pp. 1–45. ISBN 3-540-00545-5.

[167] Ichimiya, A.; Cohen, P.I. (2004). *Reflection High-energy Electron Diffraction*. Cambridge University Press. p. 1. ISBN 0-521-45373-9.

[168] Heppell, T.A. (1967). "A combined low energy and reflection high energy electron diffraction apparatus". *Journal of Scientific Instruments* **44** (9): 686–688. Bibcode:1967JScI...44..686H. doi:10.1088/0950-7671/44/9/311.

[169] McMullan, D. (1993). "Scanning Electron Microscopy: 1928–1965". University of Cambridge. Retrieved 2009-03-23.

[170] Slayter, H.S. (1992). *Light and electron microscopy.* Cambridge University Press. p. 1. ISBN 0-521-33948-0.

[171] Cember, H. (1996). *Introduction to Health Physics.* McGraw-Hill Professional. pp. 42–43. ISBN 0-07-105461-8.

[172] Erni, R. et al. (2009). "Atomic-Resolution Imaging with a Sub-50-pm Electron Probe". *Physical Review Letters* **102** (9): 096101. Bibcode:2009PhRvL.102i6101E. doi:10.1103/PhysRevLett.102.096101. PMID 19392535.

[173] Bozzola, J.J.; Russell, L.D. (1999). *Electron Microscopy: Principles and Techniques for Biologists.* Jones & Bartlett Publishers. pp. 12, 197–199. ISBN 0-7637-0192-0.

[174] Flegler, S.L.; Heckman Jr., J.W.; Klomparens, K.L. (1995). *Scanning and Transmission Electron Microscopy: An Introduction* (Reprint ed.). Oxford University Press. pp. 43–45. ISBN 0-19-510751-9.

[175] Bozzola, J.J.; Russell, L.D. (1999). *Electron Microscopy: Principles and Techniques for Biologists* (2nd ed.). Jones & Bartlett Publishers. p. 9. ISBN 0-7637-0192-0.

[176] Freund, H.P.; Antonsen, T. (1996). *Principles of Free-Electron Lasers.* Springer. pp. 1–30. ISBN 0-412-72540-1.

[177] Kitzmiller, J.W. (1995). *Television Picture Tubes and Other Cathode-Ray Tubes: Industry and Trade Summary.* DIANE Publishing. pp. 3–5. ISBN 0-7881-2100-6.

[178] Sclater, N. (1999). *Electronic Technology Handbook.* McGraw-Hill Professional. pp. 227–228. ISBN 0-07-058048-0.

[179] Staff (2008). "The History of the Integrated Circuit". The Nobel Foundation. Retrieved 2008-10-18.

3.3.9 External links

- "The Discovery of the Electron". American Institute of Physics, Center for History of Physics.

- "Particle Data Group". University of California.

- Bock, R.K.; Vasilescu, A. (1998). *The Particle Detector BriefBook* (14th ed.). Springer. ISBN 3-540-64120-3.

- Copeland, Ed. "Spherical Electron". *Sixty Symbols*. Brady Haran for the University of Nottingham.

3.4 Electron neutrino

The **electron neutrino** (ν e) is a subatomic lepton elementary particle which has no net electric charge. Together with the electron it forms the first generation of leptons, hence its name *electron neutrino*. It was first hypothesized by Wolfgang Pauli in 1930, to account for missing momentum and missing energy in beta decay, and was discovered in 1956 by a team led by Clyde Cowan and Frederick Reines (see Cowan–Reines neutrino experiment).[1]

3.4.1 Proposal

In the early 1900s, theories predicted that the electrons resulting from beta decay should have been emitted at a specific energy. However, in 1914, James Chadwick showed that electrons were instead emitted in a continuous spectrum.[1]

$$n0 \rightarrow p+ + e-$$

The early understanding of beta decay

In 1930, Wolfgang Pauli theorized that an undetected particle was carrying away the observed difference between the energy, momentum, and angular momentum of the initial and final particles.[nb 1][2]

$$n0 \rightarrow p+ + e- + \nu0$$
e

Pauli's version of beta decay

Pauli's letter

On 4 December 1930, Pauli wrote a letter to the Physical Institute of the Federal Institute of Technology, Zürich, in which he proposed the electron neutrino as a potential solution to solve the problem of the continuous beta decay spectrum. An excerpt of the letter reads:[1]

> Dear radioactive ladies and gentlemen,
> As the bearer of these lines [...] will explain more exactly, considering the 'false' statistics of N-14 and Li-6 nuclei, as well as the continuous β-spectrum, I have hit upon a desperate remedy to save the "exchange theorem" of statistics and the energy theorem. Namely [there is] the possibility that there could exist in the nuclei electrically neutral particles that I wish to call neutrons,[nb 2]

which have spin 1/2 and obey the exclusion principle, and additionally differ from light quanta in that they do not travel with the velocity of light: The mass of the neutron must be of the same order of magnitude as the electron mass and, in any case, not larger than 0.01 proton mass. The continuous β-spectrum would then become understandable by the assumption that in β decay a neutron is emitted together with the electron, in such a way that the sum of the energies of neutron and electron is constant.

[...]

But I don't feel secure enough to publish anything about this idea, so I first turn confidently to you, dear radioactives, with a question as to the situation concerning experimental proof of such a neutron, if it has something like about 10 times the penetrating capacity of a γ ray.

I admit that my remedy may appear to have a small *a priori* probability because neutrons, if they exist, would probably have long ago been seen. However, only those who wager can win, and the seriousness of the situation of the continuous β-spectrum can be made clear by the saying of my honored predecessor in office, Mr. Debye, [...] "One does best not to think about that at all, like the new taxes." [...] So, dear radioactives, put it to test and set it right. [...]

With many greetings to you, also to Mr. Back, your devoted servant,

W. Pauli

A translated reprint of the full letter can be found in the September 1978 issue of *Physics Today*.[3]

3.4.2 Discovery

Main article: Cowan–Reines neutrino experiment

The electron neutrino was discovered by Clyde Cowan and Frederick Reines in 1956.[1][4]

3.4.3 Name

Pauli originally named his proposed light particle a *neutron*. When James Chadwick discovered a much more massive nuclear particle in 1932 and also named it a neutron, this left the two particles with the same name. Enrico Fermi, who developed the theory of beta decay, coined the term *neutrino* in 1934 to resolve the confusion. It was a pun on *neutrone*, the Italian equivalent of *neutron*: the *-one* ending

can be an augmentative in Italian, so *neutrone* could be read as the "large neutral thing"; *-ino* replaces the augmentative suffix with a diminutive one. [5]

Upon the prediction and discovery of a second neutrino, it became important to distinguish between different types of neutrinos. Pauli's neutrino is now identified as the *electron neutrino*, while the second neutrino is identified as the *muon neutrino*.

3.4.4 Electron antineutrino

Like all fermions, the electron neutrino has a corresponding antiparticle, the electron antineutrino (ν
e), which differs only in that some of its properties have equal magnitude but opposite sign. The process of beta decay produces both beta particles and electron antineutrinos. Wolfgang Pauli proposed the existence of these particles, in 1930, to ensure that beta decay conserved energy (the electrons in beta decay have a continuum of energies and momentum (the momentum of the electron and recoil nucleus – in beta decay – do not add up to zero).

3.4.5 Notes

[1] Niels Bohr was notably opposed to this interpretation of beta decay and was ready to accept that energy, momentum and angular momentum were not conserved quantities.

[2] See *Name*.

3.4.6 See also

- Muon neutrino

- PMNS matrix

- Tau neutrino

3.4.7 References

[1] "The Reines-Cowan Experiments: Detecting the Poltergeist" (PDF). *Los Alamos Science* **25**: 3. 1997. Retrieved 2010-02-10.

[2] K. Riesselmann (2007). "Logbook: Neutrino Invention". *Symmetry Magazine* **4** (2).

[3] L.M. Brown (1978). "The idea of the neutrino". *Physics Today* **31** (9): 23. Bibcode:1978PhT....31i..23B. doi:10.1063/1.2995181.

[4] F. Reines, C.L. Cowan, Jr. (1956). "The Neutrino". *Nature* **178** (4531): 446. Bibcode:1956Natur.178..446R. doi:10.1038/178446a0.

[5] M.F. L'Annunziata (2007). *Radioactivity*. Elsevier. p. 100. ISBN 978-0-444-52715-8.

3.4.8 Further reading

- F. Reines, C.L. Cowan, Jr. (1956). "The Neutrino". *Nature* **178** (4531): 446. Bibcode:1956Natur.178..446R. doi:10.1038/178446a0.

- C.L. Cowan, Jr., F. Reines, F.B. Harrison, H.W. Kruse, A.D. McGuire (1956). "Detection of the Free Neutrino: A Confirmation". *Science* **124** (3212): 103–4. Bibcode:1956Sci...124..103C. doi:10.1126/science.124.3212.103. PMID 17796274.

3.5   Muon

The **muon** (/ˈmjuːɒn/; from the Greek letter mu (μ) used to represent it) is an elementary particle similar to the electron, with electric charge of −1 e and a spin of $1/2$, but with a much greater mass (105.7 MeV/c^2). It is classified as a lepton, together with the electron (mass 0.511 MeV/c^2), the tau (mass 1776.82 MeV/c^2), and the three neutrinos (electron neutrino ν
e, muon neutrino ν
μ and tau neutrino ν
τ). As is the case with other leptons, the muon is not believed to have any sub-structure—that is, it is not thought to be composed of any simpler particles.

The muon is an unstable subatomic particle with a mean lifetime of 2.2 μs. Among all known unstable subatomic particles, only the neutron (lasting around 15 minutes) and some atomic nuclei have a longer decay lifetime; others decay significantly faster. The decay of the muon (as well as of the neutron, the longest-lived unstable baryon), is mediated by the weak interaction exclusively. Muon decay always produces at least three particles, which must include an electron of the same charge as the muon and two neutrinos of different types.

Like all elementary particles, the muon has a corresponding antiparticle of opposite charge (+1 e) but equal mass and spin: the **antimuon** (also called a *positive muon*). Muons are denoted by μ− and antimuons by μ+. Muons were previously called **mu mesons**, but are not classified as mesons by modern particle physicists (see § History), and that name is no longer used by the physics community.

Muons have a mass of 105.7 MeV/c^2, which is about 207 times that of the electron. Due to their greater mass, muons are not as sharply accelerated when they encounter electromagnetic fields, and do not emit as much bremsstrahlung (deceleration radiation). This allows muons of a given energy to penetrate far more deeply into matter than electrons, since the deceleration of electrons and muons is primarily due to energy loss by the bremsstrahlung mechanism. As an example, so-called "secondary muons", generated by cosmic rays hitting the atmosphere, can penetrate to the Earth's surface, and even into deep mines.

Because muons have a very large mass and energy compared with the decay energy of radioactivity, they are never produced by radioactive decay. They are, however, produced in copious amounts in high-energy interactions in normal matter, in certain particle accelerator experiments with hadrons, or naturally in cosmic ray interactions with matter. These interactions usually produce pi mesons initially, which most often decay to muons.

As with the case of the other charged leptons, the muon has an associated muon neutrino, denoted by ν
μ, which is not the same particle as the electron neutrino, and does not participate in the same nuclear reactions.

3.5.1   History

Muons were discovered by Carl D. Anderson and Seth Neddermeyer at Caltech in 1936, while studying cosmic radiation. Anderson had noticed particles that curved differently from electrons and other known particles when passed through a magnetic field. They were negatively charged but curved less sharply than electrons, but more sharply than protons, for particles of the same velocity. It was assumed that the magnitude of their negative electric charge was equal to that of the electron, and so to account for the difference in curvature, it was supposed that their mass was greater than an electron but smaller than a proton. Thus Anderson initially called the new particle a *mesotron*, adopting the prefix *meso-* from the Greek word for "mid-". The existence of the muon was confirmed in 1937 by J. C. Street and E. C. Stevenson's cloud chamber experiment.[2]

A particle with a mass in the meson range had been predicted before the discovery of any mesons, by theorist Hideki Yukawa:[3]

> "It seems natural to modify the theory of Heisenberg and Fermi in the following way. The transition of a heavy particle from neutron state to proton state is not always accompanied by the emission of light particles. The transition is sometimes taken up by another heavy particle."

Because of its mass, the mu meson was initially thought to be Yukawa's particle, but it later proved to have the wrong properties. Yukawa's predicted particle, the pi meson, was finally identified in 1947 (again from cosmic ray interactions), and shown to differ from the earlier-discovered mu meson by having the correct properties to be a particle which mediated the nuclear force.

With two particles now known with the intermediate mass, the more general term *meson* was adopted to refer to any such particle within the correct mass range between electrons and nucleons. Further, in order to differentiate between the two different types of mesons after the second meson was discovered, the initial mesotron particle was renamed the *mu meson* (the Greek letter *μ* (*mu*) corresponds to *m*), and the new 1947 meson (Yukawa's particle) was named the pi meson.

As more types of mesons were discovered in accelerator experiments later, it was eventually found that the mu meson significantly differed not only from the pi meson (of about the same mass), but also from all other types of mesons. The difference, in part, was that mu mesons did not interact with the nuclear force, as pi mesons did (and were required to do, in Yukawa's theory). Newer mesons also showed evidence of behaving like the pi meson in nuclear interactions, but not like the mu meson. Also, the mu meson's decay products included both a neutrino and an antineutrino, rather than just one or the other, as was observed in the decay of other charged mesons.

In the eventual Standard Model of particle physics codified in the 1970s, all mesons other than the mu meson were understood to be hadrons—that is, particles made of quarks—and thus subject to the nuclear force. In the quark model, a *meson* was no longer defined by mass (for some had been discovered that were very massive—more than nucleons), but instead were particles composed of exactly two quarks (a quark and antiquark), unlike the baryons, which are defined as particles composed of three quarks (protons and neutrons were the lightest baryons). Mu mesons, however, had shown themselves to be fundamental particles (leptons) like electrons, with no quark structure. Thus, mu mesons were not mesons at all, in the new sense and use of the term *meson* used with the quark model of particle structure.

With this change in definition, the term *mu meson* was abandoned, and replaced whenever possible with the modern term *muon*, making the term mu meson only historical. In the new quark model, other types of mesons sometimes continued to be referred to in shorter terminology (e.g., *pion* for pi meson), but in the case of the muon, it retained the shorter name and was never again properly referred to by older "mu meson" terminology.

The eventual recognition of the "mu meson" muon as a simple "heavy electron" with no role at all in the nuclear interaction, seemed so incongruous and surprising at the time, that Nobel laureate I. I. Rabi famously quipped, "Who ordered that?"

In the Rossi–Hall experiment (1941), muons were used to observe the time dilation (or alternately, length contraction) predicted by special relativity, for the first time.

3.5.2 Muon sources

On Earth, most naturally occurring muons are created by quasars and supernovas, which consist mostly of protons, many arriving from deep space at very high energy[4]

> About 10,000 muons reach every square meter of the earth's surface a minute; these charged particles form as by-products of cosmic rays colliding with molecules in the upper atmosphere. Traveling at relativistic speeds, muons can penetrate tens of meters into rocks and other matter before attenuating as a result of absorption or deflection by other atoms.[5]

When a cosmic ray proton impacts atomic nuclei in the upper atmosphere, pions are created. These decay within a relatively short distance (meters) into muons (their preferred decay product), and muon neutrinos. The muons from these high energy cosmic rays generally continue in about the same direction as the original proton, at a velocity near the speed of light. Although their lifetime *without* relativistic effects would allow a half-survival distance of only about 456 m (2,197 μs×ln(2) × 0,9997×c) at most (as seen from Earth) the time dilation effect of special relativity (from the viewpoint of the Earth) allows cosmic ray secondary muons to survive the flight to the Earth's surface, since in the Earth frame, the muons have a longer half life due to their velocity. From the viewpoint (inertial frame) of the muon, on the other hand, it is the length contraction effect of special relativity which allows this penetration, since in the muon frame, its lifetime is unaffected, but the length contraction causes distances through the atmosphere and Earth to be far shorter than these distances in the Earth rest-frame. Both effects are equally valid ways of explaining the fast muon's unusual survival over distances.

Since muons are unusually penetrative of ordinary matter, like neutrinos, they are also detectable deep underground (700 meters at the Soudan 2 detector) and underwater, where they form a major part of the natural background ionizing radiation. Like cosmic rays, as noted, this secondary muon radiation is also directional.

The same nuclear reaction described above (i.e. hadron-hadron impacts to produce pion beams, which then quickly decay to muon beams over short distances) is used by particle physicists to produce muon beams, such as the beam used for the muon $g - 2$ experiment.[6]

3.5.3   Muon decay

See also: Michel parameters

Muons are unstable elementary particles and are heavier

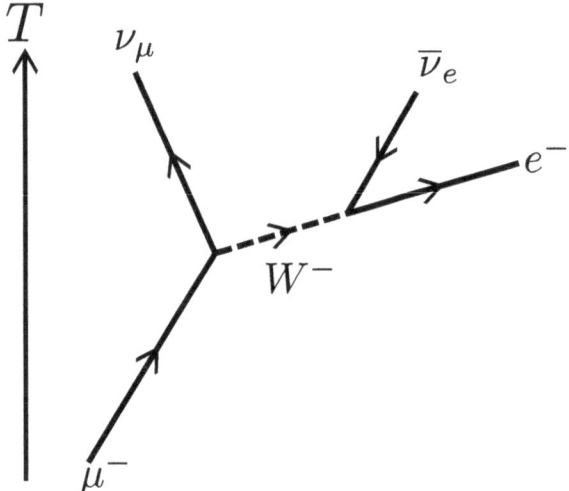

The most common decay of the muon

than electrons and neutrinos but lighter than all other matter particles. They decay via the weak interaction. Because lepton numbers must be conserved, one of the product neutrinos of muon decay must be a muon-type neutrino and the other an electron-type antineutrino (antimuon decay produces the corresponding antiparticles, as detailed below). Because charge must be conserved, one of the products of muon decay is always an electron of the same charge as the muon (a positron if it is a positive muon). Thus all muons decay to at least an electron, and two neutrinos. Sometimes, besides these necessary products, additional other particles that have no net charge and spin of zero (e.g., a pair of photons, or an electron-positron pair), are produced.

The dominant muon decay mode (sometimes called the Michel decay after Louis Michel) is the simplest possible: the muon decays to an electron, an electron antineutrino, and a muon neutrino. Antimuons, in mirror fashion, most often decay to the corresponding antiparticles: a positron, an electron neutrino, and a muon antineutrino. In formulaic terms, these two decays are:

μ− → e− + ν
e + ν
μ

μ+ → e+ + ν
e + ν
μ

The mean lifetime, τ = 1/Γ, of the (positive) muon is (2.1969811±0.0000022) μs.[1] The equality of the muon

and antimuon lifetimes has been established to better than one part in 10^4.

The muon decay width which follows from Fermi's golden rule follows Sargent's law of fifth-power dependence on m_μ,

$$\Gamma = \frac{G_F^2 m_\mu^5}{192\pi^3} I\left(\frac{m_e^2}{m_\mu^2}\right),$$

where $I(x) = 1 - 8x - 12x^2 \ln x + 8x^3 - x^4$, G_F is the Fermi coupling constant and $x = 2E_e/m_\mu c^2$ is the fraction of the maximum energy transmitted to the electron.

The decay distributions of the electron in muon decays have been parameterised using the so-called Michel parameters. The values of these four parameters are predicted unambiguously in the Standard Model of particle physics, thus muon decays represent a good test of the space-time structure of the weak interaction. No deviation from the Standard Model predictions has yet been found.

For the decay of the muon, the expected decay distribution for the Standard Model values of Michel parameters is

$$\frac{d^2\Gamma}{dx\,d\cos\theta} \sim x^2[(3 - 2x) + P_\mu \cos\theta(1 - 2x)]$$

where θ is the angle between the muon's polarization vector $\mathbf{P}_\mu$ and the decay-electron momentum vector, and $P_\mu = |\mathbf{P}_\mu|$ is the fraction of muons that are forward-polarized. Integrating this expression over electron energy gives the angular distribution of the daughter electrons:

$$\frac{d\Gamma}{d\cos\theta} \sim 1 - \frac{1}{3}P_\mu \cos\theta.$$

The electron energy distribution integrated over the polar angle (valid for $x < 1$) is

$$\frac{d\Gamma}{dx} \sim (3x^2 - 2x^3).$$

Due to the muons decaying by the weak interaction, parity conservation is violated. Replacing the $\cos\theta$ term in the expected decay values of the Michel Parameters with a $\cos\omega t$ term, where ω is the Larmor frequency from Larmor precession of the muon in a uniform magnetic field, given by:

$$\omega = \frac{egB}{2m}$$

where m is mass of the muon, e is charge, g is the muon g-factor and B is applied field.

A change in the electron distribution computed using the standard, unprecessional, Michel Parameters can be seen displaying a periodicity of π radians. This can be shown to physically correspond to a phase change of π, introduced in the electron distribution as the angular momentum is changed by the action of the charge conjugation operator, which is conserved by the weak interaction.

The observation of Parity violation in muon decay can be compared to the concept of violation of parity in weak interactions in general as an extension of The Wu Experiment, as well as the change of angular momentum introduced by a phase change of π corresponding to the charge-parity operator being invariant in this interaction. This fact is true for all lepton interactions in The Standard Model.

Certain neutrino-less decay modes are kinematically allowed but forbidden in the Standard Model. Examples forbidden by lepton flavour conservation are:

$$\mu- \rightarrow e- + \gamma \text{ and}$$

$$\mu- \rightarrow e- + e+ + e- \,.$$

Observation of such decay modes would constitute clear evidence for theories beyond the Standard Model. Upper limits for the branching fractions of such decay modes were measured in many experiments starting more than 50 years ago. The current upper limit for the $\mu+ \rightarrow e+ + \gamma$ branching fraction was measured 2013 in the MEG experiment and is 5.7×10^{-13}.[7]

3.5.4 Muonic atoms

The muon was the first elementary particle discovered that does not appear in ordinary atoms. Negative muons can, however, form muonic atoms (also called mu-mesic atoms), by replacing an electron in ordinary atoms. Muonic hydrogen atoms are much smaller than typical hydrogen atoms because the much larger mass of the muon gives it a much more localized ground-state wavefunction than is observed for the electron. In multi-electron atoms, when only one of the electrons is replaced by a muon, the size of the atom continues to be determined by the other electrons, and the atomic size is nearly unchanged. However, in such cases the orbital of the muon continues to be smaller and far closer to the nucleus than the atomic orbitals of the electrons.

Muonic helium is created by substituting a muon for one of the electrons in helium-4. The muon orbits much closer to the nucleus, so muonic helium can therefore be regarded like an isotope of helium whose nucleus consists of two neutrons, two protons and a muon, with a single electron outside. Colloquially, it could be called "helium 4.1", since the mass of the muon is roughly 0.1 amu. Chemically, muonic helium, possessing an unpaired valence electron, can bond with other atoms, and behaves more like a hydrogen atom than an inert helium atom.[8][9][10]

A positive muon, when stopped in ordinary matter, can also bind an electron and form an exotic atom known as muonium (Mu) atom, in which the muon acts as the nucleus. The positive muon, in this context, can be considered a pseudo-isotope of hydrogen with one ninth of the mass of the proton. Because the reduced mass of muonium, and hence its Bohr radius, is very close to that of hydrogen, this short-lived "atom" behaves chemically — to a first approximation — like hydrogen, deuterium and tritium.

3.5.5 Use in measurement of the proton charge radius

The recent culmination of a twelve year experiment at investigating the proton's charge radius involved the use of muonic hydrogen. This form of hydrogen is composed of a muon orbiting a proton.[11] The Lamb shift in muonic hydrogen was measured by driving the muon from its 2s state up to an excited 2p state using a laser. The frequency of the photon required to induce this transition was revealed to be 50 terahertz which, according to present theories of quantum electrodynamics, yields a value of 0.84184 ± 0.00067 femtometres for the charge radius of the proton.[12]

3.5.6 Anomalous magnetic dipole moment

The anomalous magnetic dipole moment is the difference between the experimentally observed value of the magnetic dipole moment and the theoretical value predicted by the Dirac equation. The measurement and prediction of this value is very important in the precision tests of QED (quantum electrodynamics). The E821 experiment[13] at Brookhaven National Laboratory (BNL) studied the precession of muon and anti-muon in a constant external magnetic field as they circulated in a confining storage ring. E821 reported the following average value[14] in 2006:

$$a = \frac{g - 2}{2} = 0.00116592080(54)(33)$$

where the first errors are statistical and the second systematic.

The prediction for the value of the muon anomalous magnetic moment includes three parts:

$$a\mu^{SM} = a\mu^{QED} + a\mu^{EW} + a\mu^{had} \,.$$

The difference between the g-factors of the muon and the electron is due to their difference in mass. Because of the muon's larger mass, contributions to the theoretical calculation of its anomalous magnetic dipole moment from

Standard Model weak interactions and from contributions involving hadrons are important at the current level of precision, whereas these effects are not important for the electron. The muon's anomalous magnetic dipole moment is also sensitive to contributions from new physics beyond the Standard Model, such as supersymmetry. For this reason, the muon's anomalous magnetic moment is normally used as a probe for new physics beyond the Standard Model rather than as a test of QED.[15] A new experiment at Fermilab using the E821 magnet will improve the precision of this measurement.[16]

3.5.7   Muon radiography and tomography

Main article: Muon tomography

Since muons are much more deeply penetrating than X-rays or gamma rays, muon imaging can be used with much thicker material or, with cosmic ray sources, larger objects. An important advantage of muon non-ionizing radiation is that it is safe for humans, plants, and animals. One example is commercial muon tomography used to image entire cargo containers to detect shielded nuclear material, as well as explosives or other contraband.[17]

The technique of muon transmission radiography based on cosmic ray sources was first used in the 1950s to measure the depth of the overburden of a tunnel in Australia[18] and in the 1960s to search for possible hidden chambers in the Pyramid of Chephren in Giza.[19]

In 2003, the scientists at Los Alamos National Laboratory developed a new imaging technique: **muon scattering tomography**. With muon scattering tomography, both incoming and outgoing trajectories for each particle are reconstructed, such as with sealed aluminum drift tubes.[20] Since the development of this technique, several companies have started to use it.

In August 2014, Decision Sciences International Corporation announced it had been awarded a contract by Toshiba for use of its muon tracking detectors in reclaiming the Fukushima nuclear complex.[21] The Fukushima Daiichi Tracker (FDT) was proposed to make a few months of muon measurements to show the distribution of the reactor cores.

In December 2014, Tepco reported that they would be using two different muon imaging techniques at Fukushima, "Muon Scanning Method" on Unit 1 (the most badly damaged, where the fuel may have left the reactor vessel) and "Muon Scattering Method" on Unit 2.[22]

The International Research Institute for Nuclear Decommissioning IRID in Japan and the High Energy Accelerator Research Organization KEK call the method they developed for Unit 1 the **muon permeation method**; 1,200 optical fibers for wavelength conversion light up when muons come into contact with them.[23] After a month of data collection, it is hoped to reveal the location and amount of fuel debris still inside the reactor. The measurements began in February 2015.[24]

3.5.8   See also

- Muonic atoms

- Muon spin spectroscopy

- Muon-catalyzed fusion

- Muon Tomography

- Mu2e, an experiment to detect neutrinoless conversion of muons to electrons

- List of particles

3.5.9   References

[1] J. Beringer et al. (Particle Data Group) (2012). "PDGLive Particle Summary 'Leptons (e, mu, tau, ... neutrinos ...)'" (PDF). Particle Data Group. Retrieved 2013-01-12.

[2] New Evidence for the Existence of a Particle Intermediate Between the Proton and Electron", Phys. Rev. 52, 1003 (1937).

[3] Yukaya Hideka, On the Interaction of Elementary Particles 1, Proceedings of the Physico-Mathematical Society of Japan (3) 17, 48, pp 139–148 (1935). (Read 17 November 1934)

[4] S. Carroll (2004). *Spacetime and Geometry: An Introduction to General Relativity*. Addison Wesley. p. 204

[5] Mark Wolverton (September 2007). "Muons for Peace: New Way to Spot Hidden Nukes Gets Ready to Debut". *Scientific American* **297** (3): 26–28. doi:10.1038/scientificamerican0907-26.

[6] "Physicists Announce Latest Muon g-2 Measurement" (Press release). Brookhaven National Laboratory. 30 July 2002. Retrieved 2009-11-14.

[7] J. Adam (MEG Collaboration) et al. (2013). "New Constraint on the Existence of the mu+ -> e+ gamma Decay". *Physical Review Letters* **110** (20): 201801. arXiv:1303.0754. Bibcode:2013PhRvL.110t1801A. doi:10.1103/PhysRevLett.110.201801.

[8] Fleming, D. G.; Arseneau, D. J.; Sukhorukov, O.; Brewer, J. H.; Mielke, S. L.; Schatz, G. C.; Garrett, B. C.; Peterson, K. A.; Truhlar, D. G. (28 Jan 2011). "Kinetic Isotope Effects for the Reactions

of Muonic Helium and Muonium with H2". *Science* **331** (6016): 448–450. Bibcode:2011Sci...331..448F. doi:10.1126/science.1199421. PMID 21273484.

[9] Moncada, F.; Cruz, D.; Reyes, A. "Muonic alchemy: Transmuting elements with the inclusion of negative muons". *Chemical Physics Letters* **539**: 209–213. Bibcode:2012CPL...539..209M. doi:10.1016/j.cplett.2012.04.062.

[10] Moncada, F.; Cruz, D.; Reyes, A (10 May 2013). "Electronic properties of atoms and molecules containing one and two negative muons". *Chemical Physics Letters* **570**: 16–21. Bibcode:2013CPL...570...16M. doi:10.1016/j.cplett.2013.03.004.

[11] TRIUMF Muonic Hydrogen collaboration. "A brief description of Muonic Hydrogen research". Retrieved 2010-11-7

[12] Pohl, Randolf et al. *"The Size of the Proton"* Nature 466, 213–216 (8 July 2010)

[13] "The Muon g-2 Experiment Home Page". G-2.bnl.gov. 2004-01-08. Retrieved 2012-01-06.

[14] "(from the July 2007 review by Particle Data Group)" (PDF). Retrieved 2012-01-06.

[15] Hagiwara, K; Martin, A; Nomura, D; Teubner, T (2007). "Improved predictions for g−2g−2 of the muon and αQED(MZ2)". *Physics Letters B* **649** (2–3): 173. arXiv:hep-ph/0611102. Bibcode:2007PhLB..649..173H. doi:10.1016/j.physletb.2007.04.012.

[16] "Revolutionary muon experiment to begin with 3,200-mile move of 50-foot-wide particle storage ring". May 8, 2013. Retrieved Mar 16, 2015.

[17] "Decision Sciences Corp".

[18] George, E.P. (July 1, 1955). "Cosmic rays measure overburden of tunnel". *Commonwealth Engineer*: 455.

[19] Alvarez, L.W. (1970). "Search for hidden chambers in the pyramids using cosmic rays". *Science* **167**: 832. Bibcode:1970Sci...167..832A. doi:10.1126/science.167.3919.832.

[20] Konstantin N. Borozdin, Gary E. Hogan, Christopher Morris, William C. Priedhorsky, Alexander Saunders, Larry J. Schultz & Margaret E. Teasdale. "Radiographic imaging with cosmic-ray muons". Nature.

[21] http://www.decisionsciencescorp.com/ds-awarded-toshiba-contract-fukushima-daiichi-nuclear-project

[22] Tepco to start "scanning" inside of Reactor 1 in early February by using muon Fukushima Diary

[23] "Muon measuring instrument production for "muon permeation method" and its review by international experts". IRID.or.jp.

[24] Muon Scans Begin At Fukushima Daiichi - SimplyInfo

- S.H. Neddermeyer, C.D. Anderson; Anderson (1937). "Note on the Nature of Cosmic-Ray Particles". *Physical Review* **51** (10): 884–886. Bibcode:1937PhRv...51..884N. doi:10.1103/PhysRev.51.884.

- J.C. Street, E.C. Stevenson; Stevenson (1937). "New Evidence for the Existence of a Particle of Mass Intermediate Between the Proton and Electron". *Physical Review* **52** (9): 1003–1004. Bibcode:1937PhRv...52.1003S. doi:10.1103/PhysRev.52.1003.

- G. Feinberg, S. Weinberg; Weinberg (1961). "Law of Conservation of Muons". *Physical Review Letters* **6** (7): 381–383. Bibcode:1961PhRvL...6..381F. doi:10.1103/PhysRevLett.6.381.

- Serway & Faughn (1995). *College Physics* (4th ed.). Saunders. p. 841.

- M. Knecht (2003). "The Anomalous Magnetic Moments of the Electron and the Muon". In B. Duplantier, V. Rivasseau. *Poincaré Seminar 2002: Vacuum Energy – Renormalization*. Progress in Mathematical Physics **30**. Birkhäuser Verlag. p. 265. ISBN 3-7643-0579-7.

- E. Derman (2004). *My Life As A Quant*. Wiley. pp. 58–62.

3.5.10 External links

- Muon anomalous magnetic moment and supersymmetry

- g-2 (muon anomalous magnetic moment) experiment

- muLan (Measurement of the Positive Muon Lifetime) experiment

- The Review of Particle Physics

- The TRIUMF Weak Interaction Symmetry Test

- The MEG Experiment (Search for the decay Muon → Positron + Gamma)

- King, Philip. "Making Muons". *Backstage Science*. Brady Haran.

3.6 Muon neutrino

The **muon neutrino** is a subatomic lepton elementary particle which has the symbol ν
μ and no net electric charge. Together with the muon it forms the second generation of leptons, hence its name *muon neutrino*. It was first hypothesized in the early 1940s by several people, and was discovered in 1962 by Leon Lederman, Melvin Schwartz and Jack Steinberger. The discovery was rewarded with the 1988 Nobel Prize in Physics.

3.6.1 Discovery

In 1962 Leon M. Lederman, Melvin Schwartz and Jack Steinberger established by performing an experiment at the Brookhaven National Laboratory[1] that more than one type of neutrino exists by first detecting interactions of the muon neutrino (already hypothesised with the name *neutretto*[2]), which earned them the 1988 Nobel Prize.[3]

3.6.2 Speed

Main article: Faster-than-light neutrino anomaly

In September 2011, OPERA researchers reported that muon neutrinos were apparently traveling at faster than light speed. This result was confirmed again in a second experiment in November 2011. These results have been viewed skeptically by the scientific community at large, and more experiments have/are investigating the phenomenon. In March 2012, the ICARUS team published results directly contradicting the results of OPERA.[4]

Later in July 2012 the apparent anomalous super-luminous propagation of neutrinos was traced to a faulty element of the fibre optic timing system in Gran-Sasso. After it was corrected the neutrinos appeared to travel with the speed of light within the errors of the experiment.[5]

3.6.3 See also

- Electron neutrino
- Neutrino oscillation
- PMNS matrix
- Tau neutrino

3.6.4 References

[1] G. Danby, J.-M. Gaillard, K. Goulianos, L. M. Lederman, N. B. Mistry, M. Schwartz, J. Steinberger (1962). "Observation of high-energy neutrino reactions and the existence of two kinds of neutrinos". *Physical Review Letters* **9**: 36. Bibcode:1962PhRvL...9...36D. doi:10.1103/PhysRevLett.9.36.

[2] I.V. Anicin (2005). "The Neutrino – Its Past, Present and Future". *SFIN (Institute of Physics, Belgrade) year XV, Series A: Conferences, No. A* **2**: 3–59. arXiv:physics/0503172. Bibcode:2005physics...3172A.

[3] "The Nobel Prize in Physics 1988". The Nobel Foundation. Retrieved 2010-02-11.

[4] M. Antonello et at. (2012). "Measurement of the neutrino velocity with the ICARUS detector at the CNGS beam". http://arxiv.org/abs/1203.3433v3

[5] "OPERA experiment reports anomaly in flight time of neutrinos from CERN to Gran Sasso (UPDATE 8 June 2012)". *CERN press office.* 8 June 2012. Retrieved 19 April 2013.

3.6.5 Further reading

- Leon M. Lederman (1988). "Observations in Particle Physics from Two Neutrinos to the Standard Model" (PDF). *Nobel Lectures.* The Nobel Foundation. Retrieved 2010-02-11.

- Melvin Schwartz (1988). "The First High Energy Neutrino Experiment" (PDF). *Nobel Lectures.* The Nobel Foundation. Retrieved 2010-02-11.

- Jack Steinberger (1988). "Experiments with High-Energy Neutrino Beams" (PDF). *Nobel Lectures.* The Nobel Foundation. Retrieved 2010-02-11.

3.7 Tau

Not to be confused with the τ^+ of the τ–θ puzzle, which is now identified as a kaon.

The **tau** (τ), also called the **tau lepton**, **tau particle** or **tauon**, is an elementary particle similar to the electron, with negative electric charge and a spin of $1/2$. Together with the electron, the muon, and the three neutrinos, it is a lepton. Like all elementary particles with half-integral spin, the tau has a corresponding antiparticle of opposite charge but equal mass and spin, which in the tau's case is the **anti-tau** (also called the *positive tau*). Tau particles are denoted by $\tau-$ and the antitau by $\tau+$.

Tau leptons have a lifetime of 2.9×10^{-13} s and a mass of 1776.82 MeV/c^2 (compared to 105.7 MeV/c^2 for muons and 0.511 MeV/c^2 for electrons). Since their interactions are very similar to those of the electron, a tau can be thought of as a much heavier version of the electron. Because of their greater mass, tau particles do not emit as much bremsstrahlung radiation as electrons; consequently they are potentially highly penetrating, much more so than electrons. However, because of their short lifetime, the range of the tau is mainly set by their decay length, which is too small for bremsstrahlung to be noticeable: their penetrating power appears only at ultra high energy (above PeV energies).[4]

As with the case of the other charged leptons, the tau has an associated tau neutrino, denoted by ν_τ.

3.7.1 History

The tau was detected in a series of experiments between 1974 and 1977 by Martin Lewis Perl with his colleagues at the SLAC-LBL group.[2] Their equipment consisted of SLAC's then-new e+–e– colliding ring, called SPEAR, and the LBL magnetic detector. They could detect and distinguish between leptons, hadrons and photons. They did not detect the tau directly, but rather discovered anomalous events:

"*We have discovered 64 events of the form*

e+ + e− → e± + μ∓ + at least two undetected particles

for which we have no conventional explanation."

The need for at least two undetected particles was shown by the inability to conserve energy and momentum with only one. However, no other muons, electrons, photons, or hadrons were detected. It was proposed that this event was the production and subsequent decay of a new particle pair:

e+ + e− → τ+ + τ− → e± + μ∓ + 4ν

This was difficult to verify, because the energy to produce the τ+τ− pair is similar to the threshold for D meson production. Work done at DESY-Hamburg, and with the Direct Electron Counter (DELCO) at SPEAR, subsequently established the mass and spin of the tau.

The symbol τ was derived from the Greek τρίτον (*triton*, meaning "third" in English), since it was the third charged lepton discovered.[5]

Martin Perl shared the 1995 Nobel Prize in Physics with Frederick Reines. The latter was awarded his share of the prize for experimental discovery of the neutrino.

3.7.2 Tau decay

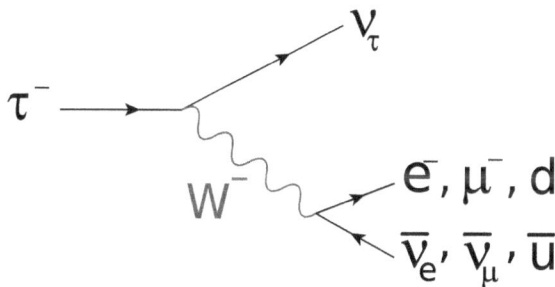

Feynman diagram of the common decays of the tau by emission of a W boson.

The tau is the only lepton that can decay into hadrons – the other leptons do not have the necessary mass. Like the other decay modes of the tau, the hadronic decay is through the weak interaction.[6]

The branching ratio of the dominant hadronic tau decays are:[3]

- 25.52% for decay into a charged pion, a neutral pion, and a tau neutrino;
- 10.83% for decay into a charged pion and a tau neutrino;
- 9.30% for decay into a charged pion, two neutral pions, and a tau neutrino;
- 8.99% for decay into three charged pions (of which two have the same electrical charge) and a tau neutrino;
- 2.70% for decay into three charged pions (of which two have the same electrical charge), a neutral pion, and a tau neutrino;
- 1.05% for decay into three neutral pions, a charged pion, and a tau neutrino.

In total, the tau lepton will decay hadronically approximately 64.79% of the time.

Since the tauonic lepton number is conserved in weak decays, a tau neutrino is always created when a tau decays.[6]

The branching ratio of the common purely leptonic tau decays are:[3]

- 17.82% for decay into a tau neutrino, electron and electron antineutrino;
- 17.39% for decay into a tau neutrino, muon and muon antineutrino.

The similarity of values of the two branching ratios is a consequence of lepton universality.

3.7.3 Exotic atoms

The tau lepton is predicted to form exotic atoms like other charged subatomic particles. One of such, called **tauonium** by the analogy to muonium, consists in antitauon and an electron: τ+e−.[7]

Another one is an onium atom τ+τ− called *true tauonium* and is difficult to detect due to tau's extremely short lifetime at low (non-relativistic) energies needed to form this atom. Its detection is important for quantum electrodynamics.[7]

3.7.4 See also

- Koide formula

3.7.5 References

[1] L. B. Okun (1980). *Leptons and Quarks*. V.I. Kisin (trans.). North-Holland Publishing. p. 103. ISBN 978-0444869241.

[2] Perl, M. L.; Abrams, G.; Boyarski, A.; Breidenbach, M.; Briggs, D.; Bulos, F.; Chinowsky, W.; Dakin, J. et al. (1975). "Evidence for Anomalous Lepton Production in e+e− Annihilation". *Physical Review Letters* **35** (22): 1489. Bibcode:1975PhRvL..35.1489P. doi:10.1103/PhysRevLett.35.1489.

[3] J. Beringer *et al.* (Particle Data Group) (2012). "Review of Particle Physics". *Journal of Physics G* **86** (1): 581–651. Bibcode:2012PhRvD..86a0001B. doi:10.1103/PhysRevD.86.010001. |chapter= ignored (help)

[4] D. Fargion, P.G. De Sanctis Lucentini, M. De Santis, M. Grossi (2004). "Tau Air Showers from Earth". *The Astrophysical Journal* **613** (2): 1285. arXiv:hep-ph/0305128. Bibcode:2004ApJ...613.1285F. doi:10.1086/423124.

[5] M.L. Perl (1977). "Evidence for, and properties of, the new charged heavy lepton" (PDF). In T. Thanh Van (ed.). *Proceedings of the XII Rencontre de Moriond*. SLAC-PUB-1923.

[6] Riazuddin (2009). "Non-standard interactions" (PDF). *NCP 5th Particle Physics Sypnoisis* (Islamabad,: Riazuddin, Head of High-Energy Theory Group at National Center for Physics) **1** (1): 1–25.

[7] Brodsky, Stanley J.; Lebed, Richard F. (2009). "Production of the Smallest QED Atom: True Muonium (μ+μ−)". *Physical Review Letters* **102** (21): 213401. arXiv:0904.2225. Bibcode:2009PhRvL.102u3401B. doi:10.1103/PhysRevLett.102.213401.

3.7.6 External links

- Nobel Prize in Physics 1995
- Perl's logbook showing tau discovery
- A Tale of Three Papers gives the covers of the three original papers announcing the discovery.

3.8 Tau neutrino

The **tau neutrino** or **tauon neutrino** is a subatomic elementary particle which has the symbol ν τ and no net electric charge. Together with the tau, it forms the third generation of leptons, hence its name *tau neutrino*. Its existence was immediately implied after the tau particle was detected in a series of experiments between 1974 and 1977 by Martin Lewis Perl with his colleagues at the SLAC–LBL group.[1] The discovery of the tau neutrino was announced in July 2000 by the DONUT collaboration.[2][3]

3.8.1 Discovery

Main article: DONUT

The tau neutrino is last of the leptons, and is the second most recent particle of the Standard Model to be discovered. The DONUT experiment (which stands for *Direct Observation of the Nu Tau*) from Fermilab was built during the 1990s to specifically detect the tau neutrino. These efforts came to fruition in July 2000, when the DONUT collaboration reported its detection.[2][3]

3.8.2 See also

- Electron neutrino
- Muon neutrino
- PMNS matrix

3.8.3 References

[1] M. L. Perl; Abrams, G.; Boyarski, A.; Breidenbach, M.; Briggs, D.; Bulos, F.; Chinowsky, W.; Dakin, J.; Feldman, G.; Friedberg, C.; Fryberger, D.; Goldhaber, G.; Hanson, G.; Heile, F.; Jean-Marie, B.; Kadyk, J.; Larsen, R.; Litke, A.; Lüke, D.; Lulu, B.; Lüth, V.; Lyon, D.; Morehouse, C.; Paterson, J.; Pierre, F.; Pun, T.; Rapidis, P.; Richter, B.; Sadoulet, B. et al. (1975). "Evidence for Anomalous Lepton Production in e+e− Annihilation". *Physical Review Letters* **35** (22): 1489. Bibcode:1975PhRvL..35.1489P. doi:10.1103/PhysRevLett.35.1489.

[2] "Physicists Find First Direct Evidence for Tau Neutrino at Fermilab" (Press release). Fermilab. 20 July 2000.

[3] K. Kodama *et al.* (DONUT Collaboration; Kodama; Ushida; Andreopoulos; Saoulidou; Tzanakos; Yager; Baller; Boehnlein; Freeman; Lundberg; Morfin; Rameika; Yun; Song; Yoon; Chung; Berghaus; Kubantsev; Reay; Sidwell; Stanton; Yoshida; Aoki; Hara; Rhee; Ciampa; Erickson; Graham et al. (2001). "Observation of tau neutrino interactions". *Physics Letters B* **504** (3): 218. arXiv:hep-ex/0012035. Bibcode:2001PhLB..504..218D. doi:10.1016/S0370-2693(01)00307-0.

Chapter 4

Gauge bosons

4.1 Gauge boson

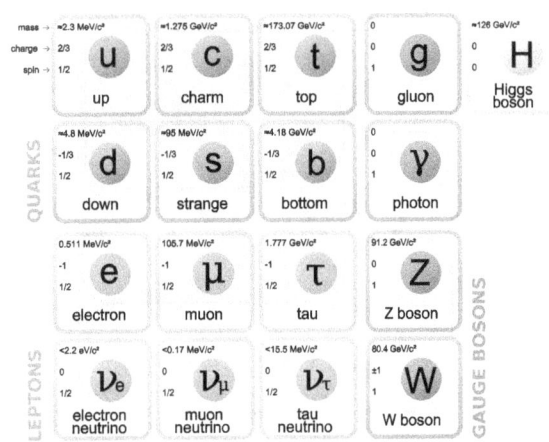

The Standard Model of elementary particles, with the gauge bosons in the fourth column in red

In particle physics, a **gauge boson** is a force carrier, a bosonic particle that carries any of the fundamental interactions of nature.[1][2] Elementary particles, whose interactions are described by a gauge theory, interact with each other by the exchange of gauge bosons—usually as virtual particles.

4.1.1 Gauge bosons in the Standard Model

The Standard Model of particle physics recognizes four kinds of gauge bosons: photons, which carry the electromagnetic interaction; W and Z bosons, which carry the weak interaction; and gluons, which carry the strong interaction.[3]

Isolated gluons do not occur at low energies because they are color-charged, and subject to color confinement.

Multiplicity of gauge bosons

In a quantized gauge theory, gauge bosons are quanta of the gauge fields. Consequently, there are as many gauge bosons as there are generators of the gauge field. In quantum electrodynamics, the gauge group is $U(1)$; in this simple case, there is only one gauge boson. In quantum chromodynamics, the more complicated group $SU(3)$ has eight generators, corresponding to the eight gluons. The three W and Z bosons correspond (roughly) to the three generators of $SU(2)$ in GWS theory.

Massive gauge bosons

For technical reasons involving gauge invariance, gauge bosons are described mathematically by field equations for massless particles. Therefore, at a naïve theoretical level all gauge bosons are required to be massless, and the forces that they describe are required to be long-ranged. The conflict between this idea and experimental evidence that the weak interaction has a very short range requires further theoretical insight.

According to the Standard Model, the W and Z bosons gain mass via the Higgs mechanism. In the Higgs mechanism, the four gauge bosons (of $SU(2) \times U(1)$ symmetry) of the unified electroweak interaction couple to a Higgs field. This field undergoes spontaneous symmetry breaking due to the shape of its interaction potential. As a result, the universe is permeated by a nonzero Higgs vacuum expectation value (VEV). This VEV couples to three of the electroweak gauge bosons (the Ws and Z), giving them mass; the remaining gauge boson remains massless (the photon). This theory also predicts the existence of a scalar Higgs boson, which has been observed in experiments that were reported on 4 July 2012.[4]

4.1.2 Beyond the Standard Model

Grand unification theories

A grand unified theory predicts additional gauge bosons named X and Y bosons. The hypothetical X and Y bosons direct interactions between quarks and leptons, hence violating conservation of baryon number and causing proton decay. Such bosons would be even more massive than W and Z bosons due to symmetry breaking. Analysis of data collected from such sources as the Super-Kamiokande neutrino detector has yielded no evidence of X and Y bosons.

Gravitons

The fourth fundamental interaction, gravity, may also be carried by a boson, called the graviton. In the absence of experimental evidence and a mathematically coherent theory of quantum gravity, it is unknown whether this would be a gauge boson or not. The role of gauge invariance in general relativity is played by a similar symmetry: diffeomorphism invariance.

W' and Z' bosons

Main article: W' and Z' bosons

W' and Z' bosons refer to hypothetical new gauge bosons (named in analogy with the Standard Model W and Z bosons).

4.1.3 See also

- 1964 PRL symmetry breaking papers

- Boson

- Glueball

- Quantum chromodynamics

- Quantum electrodynamics

4.1.4 References

[1] Gribbin, John (2000). *Q is for Quantum – An Encyclopedia of Particle Physics*. Simon & Schuster. ISBN 0-684-85578-X.

[2] Clark, John, E.O. (2004). *The Essential Dictionary of Science*. Barnes & Noble. ISBN 0-7607-4616-8.

[3] Veltman, Martinus (2003). *Facts and Mysteries in Elementary Particle Physics*. World Scientific. ISBN 981-238-149-X.

[4] "CERN experiments observe particle consistent with long-sought Higgs boson". CERN. Retrieved 4 July 2012.

4.1.5 External links

- Explanation of gauge boson and gauge fields by Christopher T. Hill

4.2 W and Z bosons

The **W and Z bosons** (together known as the **weak bosons** or, less specifically, the **intermediate vector bosons**) are the elementary particles that mediate the weak interaction; their symbols are W+, W−, and Z. The W bosons have a positive and negative electric charge of 1 elementary charge respectively and are each other's antiparticles. The Z boson is electrically neutral and is its own antiparticle. The three particles have a spin of 1, and the W bosons have a magnetic moment, while the Z has none. All three of these particles are very short-lived, with a half-life of about 3×10^{-25} s. Their discovery was a major success for what is now called the Standard Model of particle physics.

The W bosons are named after the *w*eak force. The physicist Steven Weinberg named the additional particle the "Z particle",[3] later giving the explanation that it was the last additional particle needed by the model – the W bosons had already been named – and that it has *z*ero electric charge.[4]

The two **W bosons** are best known as mediators of neutrino absorption and emission, where their charge is associated with electron or positron emission or absorption, always causing nuclear transmutation. The Z boson is not involved in the absorption or emission of electrons and positrons.

The **Z boson** mediates the transfer of momentum, spin, and energy when neutrinos scatter *elastically* from matter, something that must happen without the production or absorption of new, charged particles. Such behaviour (which is almost as common as inelastic neutrino interactions) is seen in bubble chambers irradiated with neutrino beams. Whenever an electron simply "appears" in such a chamber as a new free particle suddenly moving with kinetic energy, and moves in the direction of the neutrinos as the apparent result of a new impulse, and this behavior happens more often when the neutrino beam is present, it is inferred to be a result of a neutrino interacting directly with the electron. Here, the neutrino simply strikes the electron and scatters away from it, transferring some of the neutrino's momentum to the electron. Since (i) neither neutrinos nor electrons are affected by the strong force, (ii) neutrinos are electrically neutral (therefore don't interact electromagnetically), and (iii) the incredibly small masses of these particles make any gravitational force between them negligible, such an interaction can only happen via the weak force. Since such an electron is not created from a nucleon, and is unchanged except for the new force impulse imparted by the neutrino, this weak force interaction between the neutrino and the electron must be mediated by a weak-force boson particle with no charge. Thus, this interaction requires a Z boson.

4.2.1 Basic properties

These bosons are among the heavyweights of the elementary particles. With masses of 80.4 GeV/c^2 and 91.2 GeV/c^2, respectively, the W and Z bosons are almost 100 times as large as the proton – heavier, even, than entire atoms of iron. The masses of these bosons are significant because they act as the force carriers of a quite short-range fundamental force: their high masses thus limit the range of the weak nuclear force. By way of contrast, the electromagnetic force has an infinite range, because its force carrier, the photon, has zero mass, and the same is supposed of the hypothetical graviton.

All three bosons have particle spin $s = 1$. The emission of a W+ or W− boson either raises or lowers the electric charge of the emitting particle by one unit, and also alters the spin by one unit. At the same time, the emission or absorption of a W boson can change the type of the particle – for example changing a strange quark into an up quark. The neutral Z boson cannot change the electric charge of any particle, nor can it change any other of the so-called "charges" (such as strangeness, baryon number, charm, etc.). The emission or absorption of a Z boson can only change the spin, momentum, and energy of the other particle. (See also *weak neutral current*.)

4.2.2 Weak nuclear force

The W and Z bosons are carrier particles that mediate the weak nuclear force, much as the photon is the carrier particle for the electromagnetic force.

W bosons

The W bosons are best known for their role in nuclear decay. Consider, for example, the beta decay of cobalt-60.

    60
    27Co → 60
    28Ni⁺ + e− + ν
    e

This reaction does not involve the whole cobalt-60 nucleus, but affects only one of its 33 neutrons. The neutron is con-

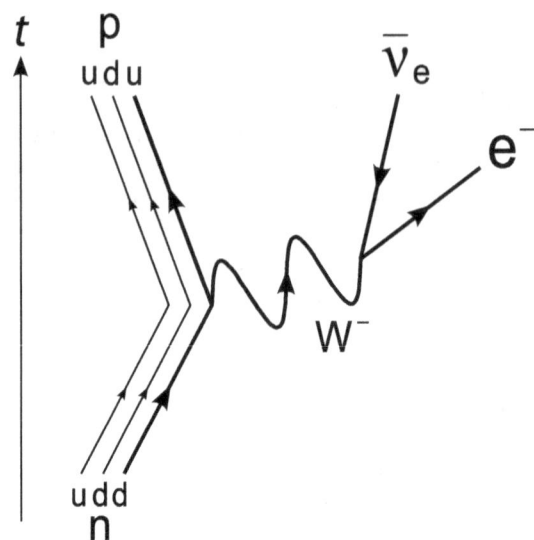

The Feynman diagram for beta decay of a neutron into a proton, electron, and electron antineutrino via an intermediate heavy W boson

verted into a proton while also emitting an electron (called a beta particle in this context) and an electron antineutrino:

    n0 → p+ + e− + ν
    e

Again, the neutron is not an elementary particle but a composite of an up quark and two down quarks (udd). It is in fact one of the down quarks that interacts in beta decay, turning into an up quark to form a proton (uud). At the most fundamental level, then, the weak force changes the flavour of a single quark:

    d → u + W−

which is immediately followed by decay of the W− itself:

    W− → e− + ν
    e

Z boson

The Z boson is its own antiparticle. Thus, all of its flavour quantum numbers and charges are zero. The exchange of a Z boson between particles, called a neutral current interaction, therefore leaves the interacting particles unaffected, except for a transfer of momentum. Z boson interactions involving neutrinos have distinctive signatures: They provide the only known mechanism for elastic scattering of neutrinos in matter; neutrinos are almost as likely to scatter elastically (via Z boson exchange) as inelastically (via W boson exchange). The first prediction of Z bosons was made by Brazilian physicist José Leite Lopes in 1958,[5] by devising an equation which showed the analogy of the weak nuclear interactions with electromagnetism. Steve Weinberg, Sheldon Glashow and Abdus Salam used later these results to develop the electroweak unification,[6] in 1973. Weak neutral currents via Z boson exchange were confirmed shortly thereafter in 1974, in a neutrino experiment in the Gargamelle bubble chamber at CERN.

4.2.3 Predicting the W and Z

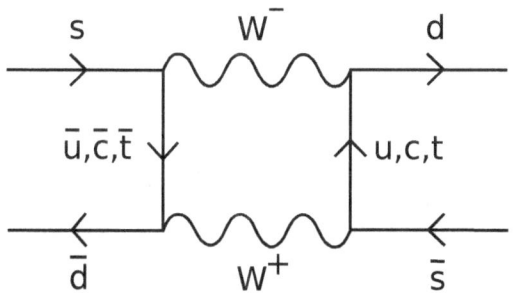

A Feynman diagram showing the exchange of a pair of W bosons. This is one of the leading terms contributing to neutral Kaon oscillation.

Following the spectacular success of quantum electrodynamics in the 1950s, attempts were undertaken to formulate a similar theory of the weak nuclear force. This culminated around 1968 in a unified theory of electromagnetism and weak interactions by Sheldon Glashow, Steven Weinberg, and Abdus Salam, for which they shared the 1979 Nobel Prize in Physics.[7] Their electroweak theory postulated not only the W bosons necessary to explain beta decay, but also a new Z boson that had never been observed.

The fact that the W and Z bosons have mass while photons are massless was a major obstacle in developing electroweak theory. These particles are accurately described by an SU(2) gauge theory, but the bosons in a gauge theory must be massless. As a case in point, the photon is massless because electromagnetism is described by a U(1) gauge theory. Some mechanism is required to break the SU(2) symmetry, giving mass to the W and Z in the process. One explanation, the Higgs mechanism, was forwarded by the 1964 PRL symmetry breaking papers. It predicts the existence of yet another new particle; the Higgs boson. Of the four components of a Goldstone boson created by the Higgs field, three are "eaten" by the W^+, Z^0, and W^- bosons to form their longitudinal components and the remainder appears as the spin 0 Higgs boson.

The combination of the SU(2) gauge theory of the weak interaction, the electromagnetic interaction, and the Higgs mechanism is known as the Glashow-Weinberg-Salam model. These days it is widely accepted as one of the pillars of the Standard Model of particle physics. As of 13 December 2011, intensive search for the Higgs boson carried out at CERN has indicated that if the particle is to be found, it seems likely to be found around 125 GeV. On 4 July 2012, the CMS and the ATLAS experimental collaborations at CERN announced the discovery of a new particle with a mass of 125.3 ± 0.6 GeV that appears consistent with a Higgs boson.

4.2.4 Discovery

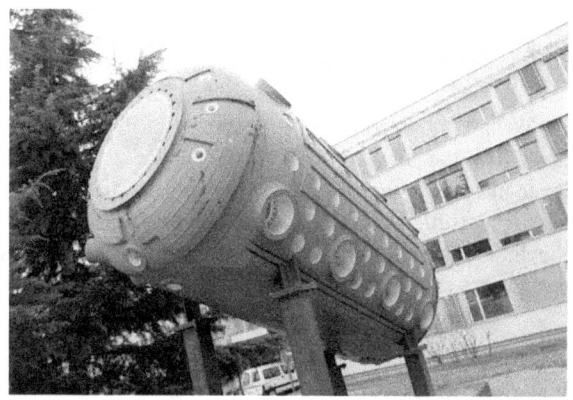

The Gargamelle bubble chamber, now exhibited at CERN

Unlike beta decay, the observation of neutral current interactions that involve particles *other than neutrinos* requires huge investments in particle accelerators and detectors, such as are available in only a few high-energy physics laboratories in the world (and then only after 1983). This is because Z-bosons behave in somewhat the same manner as photons, but do not become important until the energy of the interaction is comparable with the relatively huge mass of the Z boson.

The discovery of the W and Z bosons was considered a major success for CERN. First, in 1973, came the observation of neutral current interactions as predicted by electroweak theory. The huge Gargamelle bubble chamber photographed the tracks of a few electrons suddenly starting to move, seemingly of their own accord. This is interpreted as a neutrino interacting with the electron by the exchange of an unseen Z boson. The neutrino is otherwise undetectable, so the only observable effect is the momentum imparted to the electron by the interaction.

The discovery of the W and Z bosons themselves had to wait for the construction of a particle accelerator powerful enough to produce them. The first such machine that became available was the Super Proton Synchrotron, where unambiguous signals of W bosons were seen in January 1983 during a series of experiments made possible by Carlo Rubbia and Simon van der Meer. The actual experiments were called UA1 (led by Rubbia) and UA2 (led by Pierre Darriulat),[8] and were the collaborative effort of many people. Van der Meer was the driving force on the accelerator end (stochastic cooling). UA1 and UA2 found the Z boson a few months later, in May 1983. Rubbia and van der Meer were promptly awarded the 1984 Nobel Prize in Physics, a most unusual step for the conservative Nobel Foundation.[9]

The W+, W−, and Z0 bosons, together with the photon (γ), comprise the four gauge bosons of the electroweak interaction.

4.2.5   Decay

The W and Z bosons decay to fermion–antifermion pairs but neither the W nor the Z bosons can decay into the higher-mass top quark. Neglecting phase space effects and higher order corrections, simple estimates of their branching fractions can be calculated from the coupling constants.

W bosons

W bosons can decay to a lepton and neutrino or to an up-type quark and a down-type quark. The decay width of the W boson to a quark–antiquark pair is proportional to the corresponding squared CKM matrix element and the number of quark colours, $NC = 3$. The decay widths for the W bosons are then proportional to:

Here, e+, μ+, τ+ denote the three flavours of leptons (more exactly, the positive charged antileptons). ν

e, ν

μ, ν

τ denote the three flavours of neutrinos. The other particles, starting with u and d, all denote quarks and antiquarks (factor NC is applied). The various Vij denote the corresponding CKM matrix coefficients.

Unitarity of the CKM matrix implies that $|V_{ud}|^2 + |V_{us}|^2 + |V_{ub}|^2 = |V_{cd}|^2 + |V_{cs}|^2 + |V_{cb}|^2 = 1$. Therefore the leptonic branching ratios of the W boson are approximately B(e+ν

e) = B(μ+ν

μ) = B(τ+ν

τ) = $^1/_9$. The hadronic branching ratio is dominated by the CKM-favored ud and cs final states. The sum of the hadronic branching ratios has been measured experimentally to be 67.60±0.27%, with B(l+ν$_l$) = 10.80±0.09%.[1]

Z bosons

Z bosons decay into a fermion and its antiparticle. As the Z-boson is a mixture of the pre-symmetry-breaking W^0 and B^0 bosons (see weak mixing angle), each vertex factor includes a factor $T_3 - Q sin^2\theta W$, where T_3 is the third component of the weak isospin of the fermion, Q is the electric charge of the fermion (in units of the elementary charge), and θW is the weak mixing angle. Because the weak isospin is different for fermions of different chirality, either left-handed or right-handed, the coupling is different as well.

The **relative** strengths of each coupling can be estimated by considering that the decay rates include the square of these factors, and all possible diagrams (e.g. sum over quark families, and left and right contributions). This is just an estimate, as we are considering only tree-level diagrams in the Fermi theory.

Here, L and R denote the left- and right-handed chiralities of the fermions respectively. (The right-handed neutrinos do not exist in the standard model. However, in some extensions beyond the standard model they do.) The notation $x = sin^2\theta W$ is used.

4.2.6   See also

- Bose–Einstein statistics

- Boson

- List of particles

- Standard Model (mathematical formulation)

- W' and Z' bosons

- X and Y bosons: analogous pair of bosons predicted by the Grand Unified Theory

4.2.7   References

[1] J. Beringer et al. (2012). "2012 Review of Particle Physics - Gauge and Higgs Bosons" (PDF). *Physical Review D* **86**: 1. Bibcode:2012PhRvD..86a0001B. doi:10.1103/PhysRevD.86.010001.

[2] (PDF)   http://pdg.lbl.gov/2013/reviews/ rpp2013-rev-w-mass.pdf. Missing or empty |title= (help)

[3] Steven Weinberg, A Model of Leptons, Phys. Rev. Lett. 19, 1264–1266 (1967) – the electroweak unification paper.

[4] Weinberg, Steven (1993). *Dreams of a Final Theory: the search for the fundamental laws of nature*. Vintage Press. p. 94. ISBN 0-09-922391-0.

[5] "Forty years of the first attempt at the electroweak unification and of the prediction of the weak neutral boson".

[6] "The Nobel Prize in Physics 1979". Nobel Foundation. Retrieved 2008-09-10.

[7] Nobel Prize in Physics for 1979 (see also Nobel Prize in Physics on Wikipedia)

[8] The UA2 Collaboration collection

[9] 1984 Nobel Prize in physics

[10] C. Amsler et al. (Particle Data Group), PL B667, 1 (2008) and 2009 partial update for the 2010 edition

4.2.8 External links

- The Review of Particle Physics, the ultimate source of information on particle properties.

- The W and Z particles: a personal recollection by Pierre Darriulat

- When CERN saw the end of the alphabet by Daniel Denegri

- W and Z particles at Hyperphysics

4.3 Gluon

Gluons /ˈɡluːɒnz/ are elementary particles that act as the exchange particles (or gauge bosons) for the strong force between quarks, analogous to the exchange of photons in the electromagnetic force between two charged particles.[6]

In technical terms, gluons are vector gauge bosons that mediate strong interactions of quarks in quantum chromodynamics (QCD). Gluons themselves carry the color charge of the strong interaction. This is unlike the photon, which mediates the electromagnetic interaction but lacks an electric charge. Gluons therefore participate in the strong interaction in addition to mediating it, making QCD significantly harder to analyze than QED (quantum electrodynamics).

4.3.1 Properties

The gluon is a vector boson; like the photon, it has a spin of 1. While massive spin-1 particles have three polarization states, massless gauge bosons like the gluon have only two polarization states because gauge invariance requires the polarization to be transverse. In quantum field theory, unbroken gauge invariance requires that gauge bosons have zero mass (experiment limits the gluon's rest mass to less than a few meV/c^2). The gluon has negative intrinsic parity.

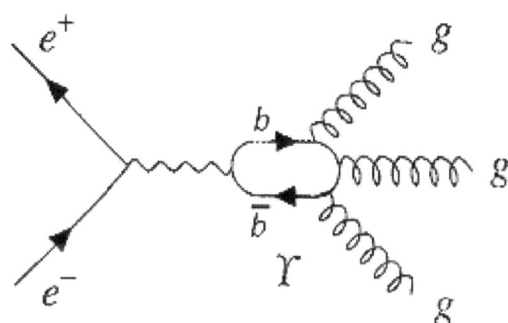

Diagram 2: $e^+ e^- \to \Upsilon(9.46) \to 3g$

4.3.2 Numerology of gluons

Unlike the single photon of QED or the three W and Z bosons of the weak interaction, there are eight independent types of gluon in QCD.

This may be difficult to understand intuitively. Quarks carry three types of color charge; antiquarks carry three types of anticolor. Gluons may be thought of as carrying both color and anticolor, but to correctly understand how they are combined, it is necessary to consider the mathematics of color charge in more detail.

Color charge and superposition

In quantum mechanics, the states of particles may be added according to the principle of superposition; that is, they may be in a "combined state" with a *probability*, if some particular quantity is measured, of giving several different outcomes. A relevant illustration in the case at hand would be a gluon with a color state described by:

$$(r\bar{b} + b\bar{r})/\sqrt{2}.$$

This is read as "red–antiblue plus blue–antired". (The factor of the square root of two is required for normalization, a detail that is not crucial to understand in this discussion.) If one were somehow able to make a direct measurement of the color of a gluon in this state, there would be a 50% chance of it having red-antiblue color charge and a 50% chance of blue-antired color charge.

Color singlet states

It is often said that the stable strongly interacting particles (such as the proton and the neutron, i.e. hadrons) observed in nature are "colorless", but more precisely they are in a "color singlet" state, which is mathematically analogous to a *spin* singlet state.[7] Such states allow interaction with other color singlets, but not with other color states; because long-range gluon interactions do not exist, this illustrates that gluons in the singlet state do not exist either.[7]

The color singlet state is:[7]

$$(r\bar{r} + b\bar{b} + g\bar{g})/\sqrt{3}.$$

In words, if one could measure the color of the state, there would be equal probabilities of it being red-antired, blue-antiblue, or green-antigreen.

Eight gluon colors

There are eight remaining independent color states, which correspond to the "eight types" or "eight colors" of gluons. Because states can be mixed together as discussed above, there are many ways of presenting these states, which are known as the "color octet". One commonly used list is:[7]

These are equivalent to the Gell-Mann matrices; the translation between the two is that red-antired is the upper-left matrix entry, red-antiblue is the upper middle entry, blue-antigreen is the middle right entry, and so on. The critical feature of these particular eight states is that they are linearly independent, and also independent of the singlet state; there is no way to add any combination of states to produce any other. (It is also impossible to add them to make rr, gg, or bb[8] otherwise the forbidden singlet state could also be made.) There are many other possible choices, but all are mathematically equivalent, at least equally complex, and give the same physical results.

Group theory details

Technically, QCD is a gauge theory with SU(3) gauge symmetry. Quarks are introduced as spinor fields in N_f flavors, each in the fundamental representation (triplet, denoted **3**) of the color gauge group, SU(3). The gluons are vector fields in the adjoint representation (octets, denoted **8**) of color SU(3). For a general gauge group, the number of force-carriers (like photons or gluons) is always equal to the dimension of the adjoint representation. For the simple case of SU(N), the dimension of this representation is $N^2 - 1$.

In terms of group theory, the assertion that there are no color singlet gluons is simply the statement that quantum chromodynamics has an SU(3) rather than a U(3) symmetry. There is no known *a priori* reason for one group to be preferred over the other, but as discussed above, the experimental evidence supports SU(3).[7] The U(1) group for

electromagnetic field combines with a slightly more complicated group known as SU(2),S stands for "special", which means the corresponding matrices have derterminant 1.

4.3.3   Confinement

Main article: Color confinement

Since gluons themselves carry color charge, they participate in strong interactions. These gluon-gluon interactions constrain color fields to string-like objects called "flux tubes", which exert constant force when stretched. Due to this force, quarks are confined within composite particles called hadrons. This effectively limits the range of the strong interaction to 1×10^{-15} meters, roughly the size of an atomic nucleus. Beyond a certain distance, the energy of the flux tube binding two quarks increases linearly. At a large enough distance, it becomes energetically more favorable to pull a quark-antiquark pair out of the vacuum rather than increase the length of the flux tube.

Gluons also share this property of being confined within hadrons. One consequence is that gluons are not directly involved in the nuclear forces between hadrons. The force mediators for these are other hadrons called mesons.

Although in the normal phase of QCD single gluons may not travel freely, it is predicted that there exist hadrons that are formed entirely of gluons — called glueballs. There are also conjectures about other exotic hadrons in which real gluons (as opposed to virtual ones found in ordinary hadrons) would be primary constituents. Beyond the normal phase of QCD (at extreme temperatures and pressures), quark–gluon plasma forms. In such a plasma there are no hadrons; quarks and gluons become free particles.

4.3.4   Experimental observations

Quarks and gluons (colored) manifest themselves by fragmenting into more quarks and gluons, which in turn hadronize into normal (colorless) particles, correlated in jets. As shown in 1978 summer conferences[2] the PLUTO detector at the electron-positron collider DORIS (DESY) produced the first evidence that the hadronic decays of the very narrow resonance $\Upsilon(9.46)$ could be interpreted as three-jet event topologies produced by three gluons. Later published analyses by the same experiment confirmed this interpretation and also the spin 1 nature of the gluon[9][10] (see also the recollection[2] and PLUTO experiments).

In summer 1979 at higher energies at the electron-positron collider PETRA (DESY) again three-jet topologies were observed, now interpreted as qq gluon bremsstrahlung, now clearly visible, by TASSO,[11] MARK-J[12] and PLUTO experiments[13] (later in 1980 also by JADE[14]). The spin 1 of the gluon was confirmed in 1980 by TASSO[15] and PLUTO experiments[16] (see also the review[3]). In 1991 a subsequent experiment at the LEP storage ring at CERN again confirmed this result.[17]

The gluons play an important role in the elementary strong interactions between quarks and gluons, described by QCD and studied particularly at the electron-proton collider HERA at DESY. The number and momentum distribution of the gluons in the proton (gluon density) have been measured by two experiments, H1 and ZEUS,[18] in the years 1996 till today (2012). The gluon contribution to the proton spin has been studied by the HERMES experiment at HERA.[19] The gluon density in the proton (when behaving hadronically) also has been measured.[20]

Color confinement is verified by the failure of free quark searches (searches of fractional charges). Quarks are normally produced in pairs (quark + antiquark) to compensate the quantum color and flavor numbers; however at Fermilab single production of top quarks has been shown (technically this still involves a pair production, but quark and antiquark are of different flavor).[21] No glueball has been demonstrated.

Deconfinement was claimed in 2000 at CERN SPS[22] in heavy-ion collisions, and it implies a new state of matter: quark–gluon plasma, less interacting than in the nucleus, almost as in a liquid. It was found at the Relativistic Heavy Ion Collider (RHIC) at Brookhaven in the years 2004–2010 by four contemporaneous experiments.[23] A quark–gluon plasma state has been confirmed at the CERN Large Hadron Collider (LHC) by the three experiments ALICE, ATLAS and CMS in 2010.[24]

4.3.5 See also

- Quark
- Hadron
- Meson
- Gauge boson
- Quark model
- Quantum chromodynamics
- Quark–gluon plasma
- Color confinement
- Glueball
- Gluon field
- Gluon field strength tensor

- Exotic hadrons
- Standard Model
- Three-jet events
- Deep inelastic scattering

4.3.6 References

[1] M. Gell-Mann (1962). "Symmetries of Baryons and Mesons". *Physical Review* **125** (3): 1067–1084. Bibcode:1962PhRv..125.1067G. doi:10.1103/PhysRev.125.1067.

[2] B.R. Stella and H.-J. Meyer (2011). "Υ(9.46 GeV) and the gluon discovery (a critical recollection of PLUTO results)". *European Physical Journal H* **36** (2): 203–243. arXiv:1008.1869v3. Bibcode:2011EPJH...36..203S. doi:10.1140/epjh/e2011-10029-3.

[3] P. Söding (2010). "On the discovery of the gluon". *European Physical Journal H* **35** (1): 3–28. Bibcode:2010EPJH...35....3S. doi:10.1140/epjh/e2010-00002-5.

[4] W.-M. Yao et al. (2006). "Review of Particle Physics" (PDF). *Journal of Physics G* **33**: 1. arXiv:astro-ph/0601168. Bibcode:2006JPhG...33....1Y. doi:10.1088/0954-3899/33/1/001.

[5] F. Yndurain (1995). "Limits on the mass of the gluon". *Physics Letters B* **345** (4): 524. Bibcode:1995PhLB..345..524Y. doi:10.1016/0370-2693(94)01677-5.

[6] C.R. Nave. "The Color Force". *HyperPhysics*. Georgia State University, Department of Physics. Retrieved 2012-04-02.

[7] David Griffiths (1987). *Introduction to Elementary Particles*. John Wiley & Sons. pp. 280–281. ISBN 0-471-60386-4.

[8] J. Baez. "Why are there eight gluons and not nine?". Retrieved 2009-09-13.

[9] Ch. Berger *et al.* (PLUTO Collaboration) (1979). "Jet analysis of the Υ(9.46) decay into charged hadrons". *Physics Letters B* **82** (3–4): 449. Bibcode:1979PhLB...82..449B. doi:10.1016/0370-2693(79)90265-X.

[10] Ch. Berger *et al.* (PLUTO Collaboration) (1981). "Topology of the Υ decay". *Zeitschrift für Physik C* **8** (2): 101. Bibcode:1981ZPhyC...8..101B. doi:10.1007/BF01547873.

[11] R. Brandelik *et al.* (TASSO collaboration) (1979). "Evidence for Planar Events in e⁺e⁻ Annihilation at High Energies". *Physics Letters B* **86** (2): 243–249. Bibcode:1979PhLB...86..243B. doi:10.1016/0370-2693(79)90830-X.

[12] D.P. Barber *et al.* (MARK-J collaboration) (1979). "Discovery of Three-Jet Events and a Test of Quantum Chromodynamics at PETRA". *Physical Review Letters* **43** (12): 830. Bibcode:1979PhRvL..43..830B. doi:10.1103/PhysRevLett.43.830.

[13] Ch. Berger *et al.* (PLUTO Collaboration) (1979). "Evidence for Gluon Bremsstrahlung in e⁺e⁻ Annihilations at High Energies". *Physics Letters B* **86** (3–4): 418. Bibcode:1979PhLB...86..418B. doi:10.1016/0370-2693(79)90869-4.

[14] W. Bartel *et al.* (JADE Collaboration) (1980). "Observation of planar three-jet events in e⁺e⁻ annihilation and evidence for gluon bremsstrahlung". *Physics Letters B* **91**: 142. Bibcode:1980PhLB...91..142B. doi:10.1016/0370-2693(80)90680-2.

[15] R. Brandelik *et al.* (TASSO Collaboration) (1980). "Evidence for a spin-1 gluon in three-jet events". *Physics Letters B* **97** (3–4): 453. Bibcode:1980PhLB...97..453B. doi:10.1016/0370-2693(80)90639-5.

[16] Ch. Berger *et al.* (PLUTO Collaboration) (1980). "A study of multi-jet events in e⁺e⁻ annihilation". *Physics Letters B* **97** (3–4): 459. Bibcode:1980PhLB...97..459B. doi:10.1016/0370-2693(80)90640-1.

[17] G. Alexander *et al.* (OPAL Collaboration) (1991). "Measurement of Three-Jet Distributions Sensitive to the Gluon Spin in e⁺e⁻ Annihilations at √s = 91 GeV". *Zeitschrift für Physik C* **52** (4): 543. Bibcode:1991ZPhyC..52..543A. doi:10.1007/BF01562326.

[18] L. Lindeman (H1 and ZEUS collaborations) (1997). "Proton structure functions and gluon density at HERA". *Nuclear Physics B Proceedings Supplements* **64**: 179–183. Bibcode:1998NuPhS..64..179L. doi:10.1016/S0920-5632(97)01057-8.

[19] http://www-hermes.desy.de

[20] C. Adloff *et al.* (H1 collaboration) (1999). "Charged particle cross sections in the photoproduction and extraction of the gluon density in the photon". *European Physical Journal C* **10**: 363–372. arXiv:hep-ex/9810020. Bibcode:1999EPJC...10..363H. doi:10.1007/s100520050761.

[21] M. Chalmers (6 March 2009). "Top result for Tevatron". *Physics World*. Retrieved 2012-04-02.

[22] M.C. Abreu et al. (2000). "Evidence for deconfinement of quark and antiquark from the J/Ψ suppression pattern measured in Pb-Pb collisions at the CERN SpS". *Physics Letters B* **477**: 28–36. Bibcode:2000PhLB..477...28A. doi:10.1016/S0370-2693(00)00237-9.

[23] D. Overbye (15 February 2010). "In Brookhaven Collider, Scientists Briefly Break a Law of Nature". *New York Times*. Retrieved 2012-04-02.

[24] "LHC experiments bring new insight into primordial universe" (Press release). CERN. 26 November 2010. Retrieved 2012-04-02.

4.3.7 Further reading

- A. Ali and G. Kramer (2011). "JETS and QCD: A historical review of the discovery of the quark and gluon jets and its impact on QCD". *European Physical Journal H* **36** (2): 245–326. arXiv:1012.2288. Bibcode:2011EPJH...36..245A. doi:10.1140/epjh/e2011-10047-1.

4.4 Photon

This article is about the elementary particle of light. For other uses, see Photon (disambiguation).

A **photon** is an elementary particle, the quantum of light and all other forms of electromagnetic radiation. It is the force carrier for the electromagnetic force, even when static via virtual photons. The effects of this force are easily observable at the microscopic and at the macroscopic level, because the photon has zero rest mass; this allows long distance interactions. Like all elementary particles, photons are currently best explained by quantum mechanics and exhibit wave–particle duality, exhibiting properties of waves and of particles. For example, a single photon may be refracted by a lens or exhibit wave interference with itself, but also act as a particle giving a definite result when its position is measured. Waves and quanta, being two observable aspects of a single phenomenon cannot have their true nature described in terms of any mechanical model. [2] A representation of this dual property of light, which assumes certain points on the wave front to be the seat of the energy is also impossible. Thus, the quanta in a light wave cannot be spatially localized. Some defined physical parameters of a photon are listed.

The modern photon concept was developed gradually by Albert Einstein in the first years of the 20th century to explain experimental observations that did not fit the classical wave model of light. In particular, the photon model accounted for the frequency dependence of light's energy, and explained the ability of matter and radiation to be in thermal equilibrium. It also accounted for anomalous observations, including the properties of black-body radiation, that other physicists, most notably Max Planck, had sought to explain using *semiclassical models*, in which light is still described by Maxwell's equations, but the material objects that emit and absorb light do so in amounts of energy that are *quantized* (i.e., they change energy only by certain particular discrete amounts and cannot change energy in any arbitrary way). Although these semiclassical models contributed to the development of quantum mechanics, many further experiments[3][4] starting with Compton scattering of single photons by electrons, first observed in 1923, validated Einstein's hypothesis that *light itself* is quantized. In 1926 the optical physicist Frithiof Wolfers and the chemist Gilbert N. Lewis coined the name *photon* for these particles, and after 1927, when Arthur H. Compton won the Nobel Prize for his scattering studies, most scientists accepted the validity that quanta of light have an independent existence, and the term *photon* for light quanta was accepted.

In the Standard Model of particle physics, photons and other elementary particles are described as a necessary consequence of physical laws having a certain symmetry at every point in spacetime. The intrinsic properties of particles, such as charge, mass and spin, are determined by the properties of this gauge symmetry. The photon concept has led to momentous advances in experimental and theoretical physics, such as lasers, Bose–Einstein condensation, quantum field theory, and the probabilistic interpretation of quantum mechanics. It has been applied to photochemistry, high-resolution microscopy, and measurements of molecular distances. Recently, photons have been studied as elements of quantum computers and for applications in optical imaging and optical communication such as quantum cryptography.

4.4.1 Nomenclature

In 1900, the German physicist Max Planck was working on black-body radiation and suggested that the energy in electromagnetic waves could only be released in "packets" of energy. In his 1901 article [5] in Annalen der Physik he called these packets "energy elements". The word *quanta* (singular *quantum*) was used even before 1900 to mean particles or amounts of different quantities, including electricity. Later, in 1905, Albert Einstein went further by suggesting that electromagnetic waves could only exist in these discrete wave-packets.[6] He called such a wave-packet *the light quantum* (German: *das Lichtquant*).[Note 1] The name *photon* derives from the Greek word for light, φῶς (transliterated *phôs*). Arthur Compton used *photon* in 1928, referring to Gilbert N. Lewis.[7] The same name was used earlier, by the American physicist and psychologist Leonard T. Troland, who coined the word in 1916, in 1921 by the Irish physicist John Joly, in 1924 by the French physiologist René Wurmser (1890-1993) and in 1926 by the French physicist Frithiof Wolfers (1891-1971).[8] The name was suggested initially as a unit related to the illumination of the eye and the resulting sensation of light and was used later on in a physiological context. Although Wolfers's and Lewis's theories were never accepted, as they were contradicted by many experiments, the new name was adopted very soon by most physicists after Compton used it.[8][Note 2]

In physics, a photon is usually denoted by the symbol γ (the Greek letter gamma). This symbol for the photon probably derives from gamma rays, which were discovered in 1900 by Paul Villard,[9][10] named by Ernest Rutherford in 1903, and shown to be a form of electromagnetic radiation in 1914 by Rutherford and Edward Andrade.[11] In chemistry and optical engineering, photons are usually symbolized by $h\nu$, the energy of a photon, where h is Planck's constant and the Greek letter ν (nu) is the photon's frequency. Much less commonly, the photon can be symbolized by hf, where its frequency is denoted by f.

4.4.2 Physical properties

See also: Special relativity and Photonic molecule

A photon is massless,[Note 3] has no electric charge,[12] and

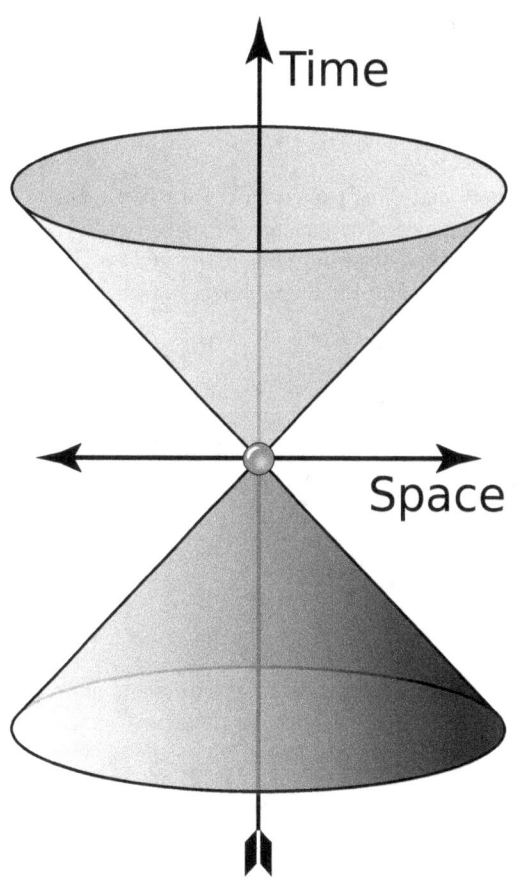

The cone shows possible values of wave 4-vector of a photon. The "time" axis gives the angular frequency (rad⋅s⁻¹) and the "space" axes represent the angular wavenumber (rad⋅m⁻¹). Green and indigo represent left and right polarization

is stable. A photon has two possible polarization states. In the momentum representation, which is preferred in quantum field theory, a photon is described by its wave vector, which determines its wavelength λ and its direction of propagation. A photon's wave vector may not be zero and can be represented either as a spatial 3-vector or as a (relativistic) four-vector; in the latter case it belongs to the light cone (pictured). Different signs of the four-vector denote different circular polarizations, but in the 3-vector representation one should account for the polarization state separately; it actually is a spin quantum number. In both cases the space of possible wave vectors is three-dimensional.

The photon is the gauge boson for electromagnetism,[13]:29-30 and therefore all other quantum numbers of the photon (such as lepton number, baryon number, and flavour quantum numbers) are zero.[14] Also, the photon does not obey the Pauli exclusion principle.[15]:1221

Photons are emitted in many natural processes. For example, when a charge is accelerated it emits synchrotron radiation. During a molecular, atomic or nuclear transition to a lower energy level, photons of various energy will be emitted, from radio waves to gamma rays. A photon can also be emitted when a particle and its corresponding antiparticle are annihilated (for example, electron–positron annihilation).[15]:572, 1114, 1172

In empty space, the photon moves at c (the speed of light) and its energy and momentum are related by $E = pc$, where p is the magnitude of the momentum vector $\mathbf{p}$. This derives from the following relativistic relation, with $m = 0$:[16]

$$E^2 = p^2 c^2 + m^2 c^4.$$

The energy and momentum of a photon depend only on its frequency (ν) or inversely, its wavelength (λ):

$$E = \hbar\omega = h\nu = \frac{hc}{\lambda}$$

$$\boldsymbol{p} = \hbar\boldsymbol{k},$$

where $\boldsymbol{k}$ is the wave vector (where the wave number $k = |\boldsymbol{k}|$ = $2\pi/\lambda$), $\omega = 2\pi\nu$ is the angular frequency, and $\hbar = h/2\pi$ is the reduced Planck constant.[17]

Since $\boldsymbol{p}$ points in the direction of the photon's propagation, the magnitude of the momentum is

$$p = \hbar k = \frac{h\nu}{c} = \frac{h}{\lambda}.$$

The photon also carries spin angular momentum that does not depend on its frequency.[18] The magnitude of its spin

is $\sqrt{2}\hbar$ and the component measured along its direction of motion, its helicity, must be $\pm\hbar$. These two possible helicities, called right-handed and left-handed, correspond to the two possible circular polarization states of the photon.[19]

To illustrate the significance of these formulae, the annihilation of a particle with its antiparticle in free space must result in the creation of at least *two* photons for the following reason. In the center of momentum frame, the colliding antiparticles have no net momentum, whereas a single photon always has momentum (since it is determined, as we have seen, only by the photon's frequency or wavelength—which cannot be zero). Hence, conservation of momentum (or equivalently, translational invariance) requires that at least two photons are created, with zero net momentum. (However, it is possible if the system interacts with another particle or field for annihilation to produce one photon, as when a positron annihilates with a bound atomic electron, it is possible for only one photon to be emitted, as the nuclear Coulomb field breaks translational symmetry.)[20]:64-65 The energy of the two photons, or, equivalently, their frequency, may be determined from conservation of four-momentum. Seen another way, the photon can be considered as its own antiparticle. The reverse process, pair production, is the dominant mechanism by which high-energy photons such as gamma rays lose energy while passing through matter.[21] That process is the reverse of "annihilation to one photon" allowed in the electric field of an atomic nucleus.

The classical formulae for the energy and momentum of electromagnetic radiation can be re-expressed in terms of photon events. For example, the pressure of electromagnetic radiation on an object derives from the transfer of photon momentum per unit time and unit area to that object, since pressure is force per unit area and force is the change in momentum per unit time.[22]

Experimental checks on photon mass

Current commonly accepted physical theories imply or assume the photon to be strictly massless. If the photon is not a strictly massless particle, it would not move at the exact speed of light in vacuum, c. Its speed would be lower and depend on its frequency. Relativity would be unaffected by this; the so-called speed of light, c, would then not be the actual speed at which light moves, but a constant of nature which is the maximum speed that any object could theoretically attain in space-time.[23] Thus, it would still be the speed of space-time ripples (gravitational waves and gravitons), but it would not be the speed of photons.

If a photon did have non-zero mass, there would be other effects as well. Coulomb's law would be modified and the electromagnetic field would have an extra physical degree of freedom. These effects yield more sensitive experimental probes of the photon mass than the frequency dependence of the speed of light. If Coulomb's law is not exactly valid, then that would cause the presence of an electric field inside a hollow conductor when it is subjected to an external electric field. This thus allows one to test Coulomb's law to very high precision.[24] A null result of such an experiment has set a limit of $m \lesssim 10^{-14}$ eV/c^2.[25]

Sharper upper limits have been obtained in experiments designed to detect effects caused by the galactic vector potential. Although the galactic vector potential is very large because the galactic magnetic field exists on very long length scales, only the magnetic field is observable if the photon is massless. In case of a massive photon, the mass term $\frac{1}{2}m^2 A_\mu A^\mu$ would affect the galactic plasma. The fact that no such effects are seen implies an upper bound on the photon mass of $m < 3 \times 10^{-27}$ eV/c^2.[26] The galactic vector potential can also be probed directly by measuring the torque exerted on a magnetized ring.[27] Such methods were used to obtain the sharper upper limit of 10^{-18}eV/c^2 (the equivalent of 1.07×10^{-27} atomic mass units) given by the Particle Data Group.[28]

These sharp limits from the non-observation of the effects caused by the galactic vector potential have been shown to be model dependent.[29] If the photon mass is generated via the Higgs mechanism then the upper limit of $m \lesssim 10^{-14}$ eV/c^2 from the test of Coulomb's law is valid.

Photons inside superconductors do develop a nonzero effective rest mass; as a result, electromagnetic forces become short-range inside superconductors.[30]

See also: Supernova/Acceleration Probe

4.4.3 Historical development

Main article: Light

In most theories up to the eighteenth century, light was

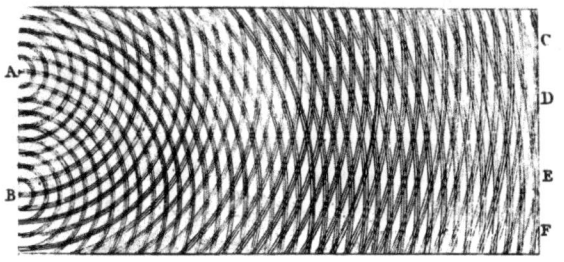

Thomas Young's double-slit experiment in 1801 showed that light can act as a wave, helping to invalidate early particle theories of light.[15]:964

pictured as being made up of particles. Since particle mod-

els cannot easily account for the refraction, diffraction and birefringence of light, wave theories of light were proposed by René Descartes (1637),[31] Robert Hooke (1665),[32] and Christiaan Huygens (1678);[33] however, particle models remained dominant, chiefly due to the influence of Isaac Newton.[34] In the early nineteenth century, Thomas Young and August Fresnel clearly demonstrated the interference and diffraction of light and by 1850 wave models were generally accepted.[35] In 1865, James Clerk Maxwell's prediction[36] that light was an electromagnetic wave—which was confirmed experimentally in 1888 by Heinrich Hertz's detection of radio waves[37]—seemed to be the final blow to particle models of light.

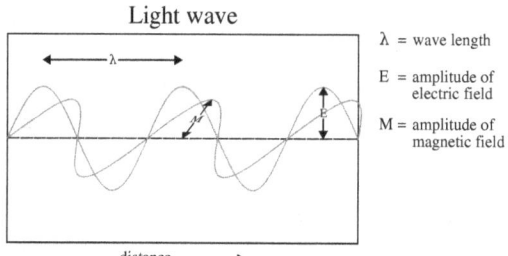

In 1900, Maxwell's theoretical model of light as oscillating electric and magnetic fields seemed complete. However, several observations could not be explained by any wave model of electromagnetic radiation, leading to the idea that light-energy was packaged into quanta *described by E=hv. Later experiments showed that these light-quanta also carry momentum and, thus, can be considered particles: the* photon *concept was born, leading to a deeper understanding of the electric and magnetic fields themselves.*

The Maxwell wave theory, however, does not account for *all* properties of light. The Maxwell theory predicts that the energy of a light wave depends only on its intensity, not on its frequency; nevertheless, several independent types of experiments show that the energy imparted by light to atoms depends only on the light's frequency, not on its intensity. For example, some chemical reactions are provoked only by light of frequency higher than a certain threshold; light of frequency lower than the threshold, no matter how intense, does not initiate the reaction. Similarly, electrons can be ejected from a metal plate by shining light of sufficiently high frequency on it (the photoelectric effect); the energy of the ejected electron is related only to the light's frequency, not to its intensity.[38][Note 4]

At the same time, investigations of blackbody radiation carried out over four decades (1860–1900) by various researchers[39] culminated in Max Planck's hypothesis[5][40] that the energy of *any* system that absorbs or emits electromagnetic radiation of frequency ν is an integer multiple of an energy quantum $E = h\nu$. As shown by Albert Einstein,[6][41] some form of energy quantization *must* be assumed to account for the thermal equilibrium observed between matter and electromagnetic radiation; for this explanation of the photoelectric effect, Einstein received the 1921 Nobel Prize in physics.[42]

Since the Maxwell theory of light allows for all possible energies of electromagnetic radiation, most physicists assumed initially that the energy quantization resulted from some unknown constraint on the matter that absorbs or emits the radiation. In 1905, Einstein was the first to propose that energy quantization was a property of electromagnetic radiation itself.[6] Although he accepted the validity of Maxwell's theory, Einstein pointed out that many anomalous experiments could be explained if the *energy* of a Maxwellian light wave were localized into point-like quanta that move independently of one another, even if the wave itself is spread continuously over space.[6] In 1909[41] and 1916,[43] Einstein showed that, if Planck's law of black-body radiation is accepted, the energy quanta must also carry momentum $p = h/\lambda$, making them full-fledged particles. This photon momentum was observed experimentally[44] by Arthur Compton, for which he received the Nobel Prize in 1927. The pivotal question was then: how to unify Maxwell's wave theory of light with its experimentally observed particle nature? The answer to this question occupied Albert Einstein for the rest of his life,[45] and was solved in quantum electrodynamics and its successor, the Standard Model (see Second quantization and The photon as a gauge boson, below).

4.4.4 Einstein's light quantum

Unlike Planck, Einstein entertained the possibility that there might be actual physical quanta of light—what we now call photons. He noticed that a light quantum with energy proportional to its frequency would explain a number of troubling puzzles and paradoxes, including an unpublished law by Stokes, the ultraviolet catastrophe, and of course the photoelectric effect. Stokes's law said simply that the frequency of fluorescent light cannot be greater than the frequency of the light (usually ultraviolet) inducing it. Einstein eliminated the ultraviolet catastrophe by imagining a gas of photons behaving like a gas of electrons that he had previously considered. He was advised by a colleague to be careful how he wrote up this paper, in order to not challenge Planck too directly, as he was a powerful figure, and indeed the warning was justified, as Planck never forgave him for writing it.[46]

4.4.5 Early objections

Einstein's 1905 predictions were verified experimentally in several ways in the first two decades of the 20th century,

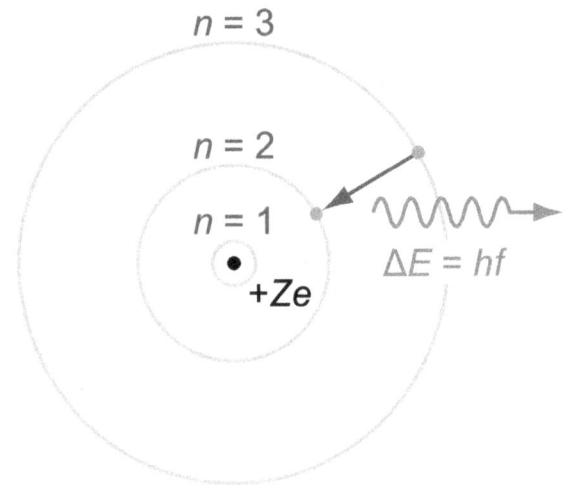

Up to 1923, most physicists were reluctant to accept that light itself was quantized. Instead, they tried to explain photon behavior by quantizing only matter, as in the Bohr model of the hydrogen atom (shown here). Even though these semiclassical models were only a first approximation, they were accurate for simple systems and they led to quantum mechanics.

as recounted in Robert Millikan's Nobel lecture.[47] However, before Compton's experiment[44] showing that photons carried momentum proportional to their wave number (or frequency) (1922), most physicists were reluctant to believe that electromagnetic radiation itself might be particulate. (See, for example, the Nobel lectures of Wien,[39] Planck[40] and Millikan.[47]) Instead, there was a widespread belief that energy quantization resulted from some unknown constraint on the matter that absorbs or emits radiation. Attitudes changed over time. In part, the change can be traced to experiments such as Compton scattering, where it was much more difficult not to ascribe quantization to light itself to explain the observed results.[48]

Even after Compton's experiment, Niels Bohr, Hendrik Kramers and John Slater made one last attempt to preserve the Maxwellian continuous electromagnetic field model of light, the so-called BKS model.[49] To account for the data then available, two drastic hypotheses had to be made:

1. **Energy and momentum are conserved only on the average in interactions between matter and radiation, not in elementary processes such as absorption and emission.** This allows one to reconcile the discontinuously changing energy of the atom (jump between energy states) with the continuous release of energy into radiation.

2. **Causality is abandoned.** For example, spontaneous emissions are merely emissions induced by a "virtual" electromagnetic field.

However, refined Compton experiments showed that energy–momentum is conserved extraordinarily well in elementary processes; and also that the jolting of the electron and the generation of a new photon in Compton scattering obey causality to within 10 ps. Accordingly, Bohr and his co-workers gave their model "as honorable a funeral as possible".[45] Nevertheless, the failures of the BKS model inspired Werner Heisenberg in his development of matrix mechanics.[50]

A few physicists persisted[51] in developing semiclassical models in which electromagnetic radiation is not quantized, but matter appears to obey the laws of quantum mechanics. Although the evidence for photons from chemical and physical experiments was overwhelming by the 1970s, this evidence could not be considered as *absolutely* definitive; since it relied on the interaction of light with matter, a sufficiently complicated theory of matter could in principle account for the evidence. Nevertheless, *all* semiclassical theories were refuted definitively in the 1970s and 1980s by photon-correlation experiments.[Note 5] Hence, Einstein's hypothesis that quantization is a property of light itself is considered to be proven.

4.4.6 Wave–particle duality and uncertainty principles

See also: Wave–particle duality, Squeezed coherent state, Uncertainty principle and De Broglie–Bohm theory

Photons, like all quantum objects, exhibit wave-like and

Photons in a Mach–Zehnder interferometer exhibit wave-like interference and particle-like detection at single-photon detectors.

particle-like properties. Their dual wave–particle nature can be difficult to visualize. The photon displays clearly wave-like phenomena such as diffraction and interference on the length scale of its wavelength. For example, a single photon passing through a double-slit experiment lands on the screen exhibiting interference phenomena but only

if no measure was made on the actual slit being run across. To account for the particle interpretation that phenomenon is called probability distribution but behaves according to Maxwell's equations.[52] However, experiments confirm that the photon is *not* a short pulse of electromagnetic radiation; it does not spread out as it propagates, nor does it divide when it encounters a beam splitter.[53] Rather, the photon seems to be a point-like particle since it is absorbed or emitted *as a whole* by arbitrarily small systems, systems much smaller than its wavelength, such as an atomic nucleus ($\approx 10^{-15}$ m across) or even the point-like electron. Nevertheless, the photon is *not* a point-like particle whose trajectory is shaped probabilistically by the electromagnetic field, as conceived by Einstein and others; that hypothesis was also refuted by the photon-correlation experiments cited above. According to our present understanding, the electromagnetic field itself is produced by photons, which in turn result from a local gauge symmetry and the laws of quantum field theory (see the Second quantization and Gauge boson sections below).

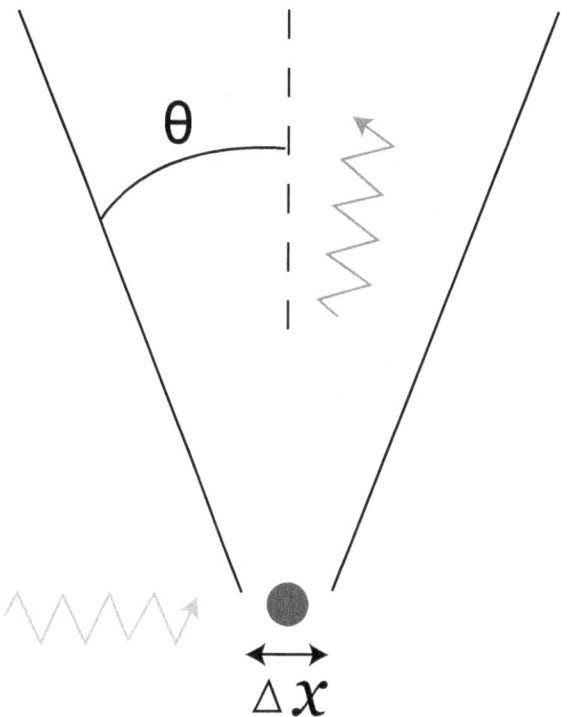

Heisenberg's thought experiment for locating an electron (shown in blue) with a high-resolution gamma-ray microscope. The incoming gamma ray (shown in green) is scattered by the electron up into the microscope's aperture angle θ. The scattered gamma ray is shown in red. Classical optics shows that the electron position can be resolved only up to an uncertainty Δx that depends on θ and the wavelength λ of the incoming light.

A key element of quantum mechanics is Heisenberg's

uncertainty principle, which forbids the simultaneous measurement of the position and momentum of a particle along the same direction. Remarkably, the uncertainty principle for charged, material particles *requires* the quantization of light into photons, and even the frequency dependence of the photon's energy and momentum. An elegant illustration is Heisenberg's thought experiment for locating an electron with an ideal microscope.[54] The position of the electron can be determined to within the resolving power of the microscope, which is given by a formula from classical optics

$$\Delta x \sim \frac{\lambda}{\sin \theta}$$

where θ is the aperture angle of the microscope. Thus, the position uncertainty Δx can be made arbitrarily small by reducing the wavelength λ. The momentum of the electron is uncertain, since it received a "kick" Δp from the light scattering from it into the microscope. If light were *not* quantized into photons, the uncertainty Δp could be made arbitrarily small by reducing the light's intensity. In that case, since the wavelength and intensity of light can be varied independently, one could simultaneously determine the position and momentum to arbitrarily high accuracy, violating the uncertainty principle. By contrast, Einstein's formula for photon momentum preserves the uncertainty principle; since the photon is scattered anywhere within the aperture, the uncertainty of momentum transferred equals

$$\Delta p \sim p_{\text{photon}} \sin \theta = \frac{h}{\lambda} \sin \theta$$

giving the product $\Delta x \Delta p \sim h$, which is Heisenberg's uncertainty principle. Thus, the entire world is quantized; both matter and fields must obey a consistent set of quantum laws, if either one is to be quantized.[55]

The analogous uncertainty principle for photons forbids the simultaneous measurement of the number n of photons (see Fock state and the Second quantization section below) in an electromagnetic wave and the phase ϕ of that wave

$$\Delta n \Delta \phi > 1$$

See coherent state and squeezed coherent state for more details.

Both (photons and material) particles such as electrons create analogous interference patterns when passing through a double-slit experiment. For photons, this corresponds to the interference of a Maxwell light wave whereas, for material particles, this corresponds to the interference of the Schrödinger wave equation. Although this similarity might suggest that Maxwell's equations are simply Schrödinger's equation for photons, most physicists do not agree.[56][57] For one thing, they are mathematically different; most obviously, Schrödinger's one equation solves for a complex field, whereas Maxwell's four equations solve for real fields. More generally, the normal concept of a Schrödinger probability wave function cannot be applied to photons.[58] Being massless, they cannot be localized without being destroyed; technically, photons cannot have a position eigenstate $|\mathbf{r}\rangle$, and, thus, the normal Heisenberg uncertainty principle $\Delta x \Delta p > h/2$ does not pertain to photons. A few substitute wave functions have been suggested for the photon,[59][60][61][62] but they have not come into general use. Instead, physicists generally accept the second-quantized theory of photons described below, quantum electrodynamics, in which photons are quantized excitations of electromagnetic modes.

Another interpretation, that avoids duality, is the De Broglie–Bohm theory: known also as the *pilot-wave model*, the photon in this theory is both, wave and particle.[63] *"This idea seems to me so natural and simple, to resolve the wave-particle dilemma in such a clear and ordinary way, that it is a great mystery to me that it was so generally ignored"*,[64] J.S.Bell.

4.4.7 Bose–Einstein model of a photon gas

Main articles: Bose gas, Bose–Einstein statistics, Spin-statistics theorem and Gas in a box

In 1924, Satyendra Nath Bose derived Planck's law of black-body radiation without using any electromagnetism, but rather a modification of coarse-grained counting of phase space.[65] Einstein showed that this modification is equivalent to assuming that photons are rigorously identical and that it implied a "mysterious non-local interaction",[66][67] now understood as the requirement for a symmetric quantum mechanical state. This work led to the concept of coherent states and the development of the laser. In the same papers, Einstein extended Bose's formalism to material particles (bosons) and predicted that they would condense into their lowest quantum state at low enough temperatures; this Bose–Einstein condensation was observed experimentally in 1995.[68] It was later used by Lene Hau to slow, and then completely stop, light in 1999[69] and 2001.[70]

The modern view on this is that photons are, by virtue of their integer spin, bosons (as opposed to fermions with half-integer spin). By the spin-statistics theorem, all bosons obey Bose–Einstein statistics (whereas all fermions obey Fermi–Dirac statistics).[71]

4.4.8 Stimulated and spontaneous emission

Main articles: Stimulated emission and Laser
In 1916, Einstein showed that Planck's radiation law could

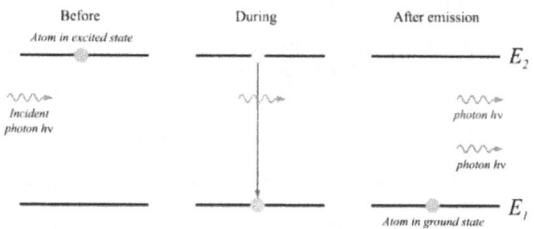

Before | During | After emission

Atom in excited state

E_2

Incident photon hv

photon hv

photon hv

Atom in ground state

E_1

Stimulated emission (in which photons "clone" themselves) was predicted by Einstein in his kinetic analysis, and led to the development of the laser. Einstein's derivation inspired further developments in the quantum treatment of light, which led to the statistical interpretation of quantum mechanics.

be derived from a semi-classical, statistical treatment of photons and atoms, which implies a relation between the rates at which atoms emit and absorb photons. The condition follows from the assumption that light is emitted and absorbed by atoms independently, and that the thermal equilibrium is preserved by interaction with atoms. Consider a cavity in thermal equilibrium and filled with electromagnetic radiation and atoms that can emit and absorb that radiation. Thermal equilibrium requires that the energy density $\rho(\nu)$ of photons with frequency ν (which is proportional to their number density) is, on average, constant in time; hence, the rate at which photons of any particular frequency are *emitted* must equal the rate of *absorbing* them.[72]

Einstein began by postulating simple proportionality relations for the different reaction rates involved. In his model, the rate R_{ji} for a system to *absorb* a photon of frequency ν and transition from a lower energy E_j to a higher energy E_i is proportional to the number N_j of atoms with energy E_j and to the energy density $\rho(\nu)$ of ambient photons with that frequency,

$$R_{ji} = N_j B_{ji} \rho(\nu)$$

where B_{ji} is the rate constant for absorption. For the reverse process, there are two possibilities: spontaneous emission of a photon, and a return to the lower-energy state that is initiated by the interaction with a passing photon. Following Einstein's approach, the corresponding rate R_{ij} for the emission of photons of frequency ν and transition from a higher energy E_i to a lower energy E_j is

$$R_{ij} = N_i A_{ij} + N_i B_{ij} \rho(\nu)$$

where A_{ij} is the rate constant for emitting a photon spontaneously, and B_{ij} is the rate constant for emitting it in response to ambient photons (induced or stimulated emission). In thermodynamic equilibrium, the number of atoms in state i and that of atoms in state j must, on average, be constant; hence, the rates R_{ji} and R_{ij} must be equal. Also, by arguments analogous to the derivation of Boltzmann statistics, the ratio of N_i and N_j is $g_i/g_j \exp\left(E_j - E_i\right)/kT$, where $g_{i,j}$ are the degeneracy of the state i and that of j, respectively, $E_{i,j}$ their energies, k the Boltzmann constant and T the system's temperature. From this, it is readily derived that $g_i B_{ij} = g_j B_{ji}$ and

$$A_{ij} = \frac{8\pi h \nu^3}{c^3} B_{ij}.$$

The A and Bs are collectively known as the *Einstein coefficients*.[73]

Einstein could not fully justify his rate equations, but claimed that it should be possible to calculate the coefficients A_{ij}, B_{ji} and B_{ij} once physicists had obtained "mechanics and electrodynamics modified to accommodate the quantum hypothesis".[74] In fact, in 1926, Paul Dirac derived the B_{ij} rate constants in using a semiclassical approach,[75] and, in 1927, succeeded in deriving *all* the rate constants from first principles within the framework of quantum theory.[76][77] Dirac's work was the foundation of quantum electrodynamics, i.e., the quantization of the electromagnetic field itself. Dirac's approach is also called *second quantization* or quantum field theory;[78][79][80] earlier quantum mechanical treatments only treat material particles as quantum mechanical, not the electromagnetic field.

Einstein was troubled by the fact that his theory seemed incomplete, since it did not determine the *direction* of a spontaneously emitted photon. A probabilistic nature of light-particle motion was first considered by Newton in his treatment of birefringence and, more generally, of the splitting of light beams at interfaces into a transmitted beam and a reflected beam. Newton hypothesized that hidden variables in the light particle determined which path it would follow.[34] Similarly, Einstein hoped for a more complete theory that would leave nothing to chance, beginning his separation[45] from quantum mechanics. Ironically, Max Born's probabilistic interpretation of the wave function[81][82] was inspired by Einstein's later work searching for a more complete theory.[83]

4.4.9 Second quantization and high energy photon interactions

Main article: Quantum field theory
In 1910, Peter Debye derived Planck's law of black-body

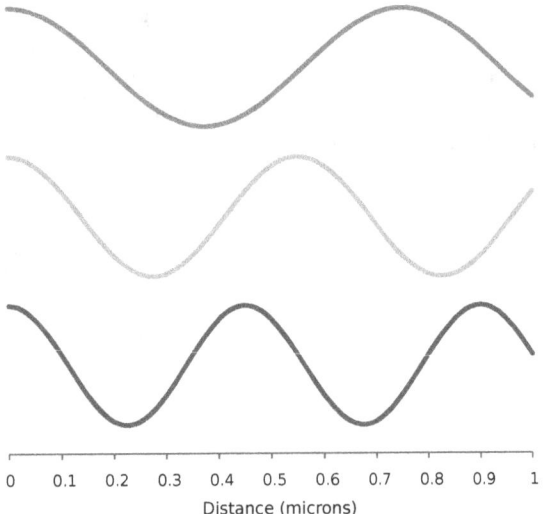

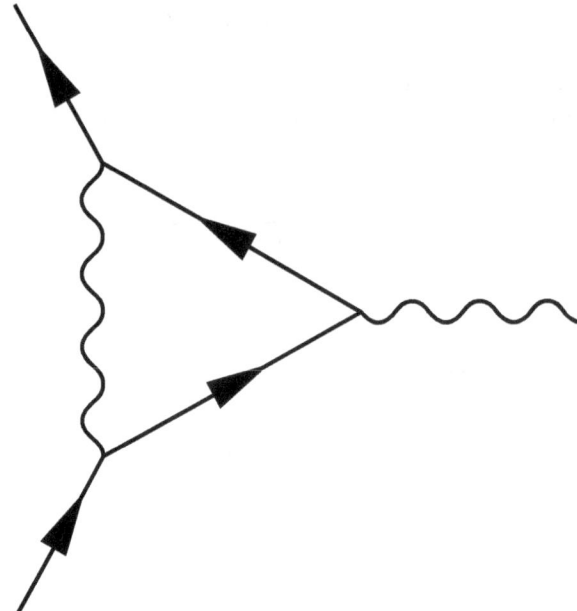

Different electromagnetic modes *(such as those depicted here) can be treated as independent simple harmonic oscillators. A photon corresponds to a unit of energy E=hν in its electromagnetic mode.*

In quantum field theory, the probability of an event is computed by summing the probability amplitude (a complex number) for all possible ways in which the event can occur, as in the Feynman diagram shown here; the probability equals the square of the modulus of the total amplitude.

radiation from a relatively simple assumption.[84] He correctly decomposed the electromagnetic field in a cavity into its Fourier modes, and assumed that the energy in any mode was an integer multiple of $h\nu$, where ν is the frequency of the electromagnetic mode. Planck's law of black-body radiation follows immediately as a geometric sum. However, Debye's approach failed to give the correct formula for the energy fluctuations of blackbody radiation, which were derived by Einstein in 1909.[41]

In 1925, Born, Heisenberg and Jordan reinterpreted Debye's concept in a key way.[85] As may be shown classically, the Fourier modes of the electromagnetic field—a complete set of electromagnetic plane waves indexed by their wave vector **k** and polarization state—are equivalent to a set of uncoupled simple harmonic oscillators. Treated quantum mechanically, the energy levels of such oscillators are known to be $E = nh\nu$, where ν is the oscillator frequency. The key new step was to identify an electromagnetic mode with energy $E = nh\nu$ as a state with n photons, each of energy $h\nu$. This approach gives the correct energy fluctuation formula.

Dirac took this one step further.[76][77] He treated the interaction between a charge and an electromagnetic field as a small perturbation that induces transitions in the photon states, changing the numbers of photons in the modes, while conserving energy and momentum overall. Dirac was able to derive Einstein's A_{ij} and B_{ij} coefficients from first principles, and showed that the Bose–Einstein statistics of photons is a natural consequence of quantizing the electromagnetic field correctly (Bose's reasoning went in the opposite direction; he derived Planck's law of black-body radiation

by *assuming* B–E statistics). In Dirac's time, it was not yet known that all bosons, including photons, must obey Bose–Einstein statistics.

Dirac's second-order perturbation theory can involve virtual photons, transient intermediate states of the electromagnetic field; the static electric and magnetic interactions are mediated by such virtual photons. In such quantum field theories, the probability amplitude of observable events is calculated by summing over *all* possible intermediate steps, even ones that are unphysical; hence, virtual photons are not constrained to satisfy $E = pc$, and may have extra polarization states; depending on the gauge used, virtual photons may have three or four polarization states, instead of the two states of real photons. Although these transient virtual photons can never be observed, they contribute measurably to the probabilities of observable events. Indeed, such second-order and higher-order perturbation calculations can give apparently infinite contributions to the sum. Such unphysical results are corrected for using the technique of renormalization.

Other virtual particles may contribute to the summation as well; for example, two photons may interact indirectly through virtual electron–positron pairs.[86] In fact, such photon-photon scattering (see two-photon physics), as well as electron-photon scattering, is meant to be one of the modes of operations of the planned particle accelerator, the International Linear Collider.[87]

In modern physics notation, the quantum state of the electromagnetic field is written as a Fock state, a tensor product of the states for each electromagnetic mode

$$|n_{k_0}\rangle \otimes |n_{k_1}\rangle \otimes \cdots \otimes |n_{k_n}\rangle \ldots$$

where $|n_{k_i}\rangle$ represents the state in which n_{k_i} photons are in the mode k_i. In this notation, the creation of a new photon in mode k_i (e.g., emitted from an atomic transition) is written as $|n_{k_i}\rangle \rightarrow |n_{k_i} + 1\rangle$. This notation merely expresses the concept of Born, Heisenberg and Jordan described above, and does not add any physics.

4.4.10 The hadronic properties of the photon

Measurements of the interaction between energetic photons and hadrons show that the interaction is much more intense than expected by the interaction of merely photons with the hadron's electric charge. Furthermore, the interaction of energetic photons with protons is similar to the interaction of photons with neutrons[88] in spite of the fact that the electric charge structures of protons and neutrons are substantially different.

A theory called Vector Meson Dominance (VMD) was developed to explain this effect. According to VMD, the photon is a superposition of the pure electromagnetic photon (which interacts only with electric charges) and vector meson.[89]

However, if experimentally probed at very short distances, the intrinsic structure of the photon is recognized as a flux of quark and gluon components, quasi-free according to asymptotic freedom in QCD and described by the photon structure function.[90][91] A comprehensive comparison of data with theoretical predictions is presented in a recent review.[92]

4.4.11 The photon as a gauge boson

Main article: Gauge theory

The electromagnetic field can be understood as a gauge field, i.e., as a field that results from requiring that a gauge symmetry holds independently at every position in spacetime.[93] For the electromagnetic field, this gauge symmetry is the Abelian U(1) symmetry of complex numbers of absolute value 1, which reflects the ability to vary the phase of a complex number without affecting observables or real valued functions made from it, such as the energy or the Lagrangian.

The quanta of an Abelian gauge field must be massless, uncharged bosons, as long as the symmetry is not broken; hence, the photon is predicted to be massless, and to have zero electric charge and integer spin. The particular form of the electromagnetic interaction specifies that the photon must have spin ±1; thus, its helicity must be $\pm\hbar$. These two spin components correspond to the classical concepts of right-handed and left-handed circularly polarized light. However, the transient virtual photons of quantum electrodynamics may also adopt unphysical polarization states.[93]

In the prevailing Standard Model of physics, the photon is one of four gauge bosons in the electroweak interaction; the other three are denoted W^+, W^- and Z^0 and are responsible for the weak interaction. Unlike the photon, these gauge bosons have mass, owing to a mechanism that breaks their SU(2) gauge symmetry. The unification of the photon with W and Z gauge bosons in the electroweak interaction was accomplished by Sheldon Glashow, Abdus Salam and Steven Weinberg, for which they were awarded the 1979 Nobel Prize in physics.[94][95][96] Physicists continue to hypothesize grand unified theories that connect these four gauge bosons with the eight gluon gauge bosons of quantum chromodynamics; however, key predictions of these theories, such as proton decay, have not been observed experimentally.[97]

4.4.12 Contributions to the mass of a system

See also: Mass in special relativity and General relativity

The energy of a system that emits a photon is *decreased* by the energy E of the photon as measured in the rest frame of the emitting system, which may result in a reduction in mass in the amount E/c^2. Similarly, the mass of a system that absorbs a photon is *increased* by a corresponding amount. As an application, the energy balance of nuclear reactions involving photons is commonly written in terms of the masses of the nuclei involved, and terms of the form E/c^2 for the gamma photons (and for other relevant energies, such as the recoil energy of nuclei).[98]

This concept is applied in key predictions of quantum electrodynamics (QED, see above). In that theory, the mass of electrons (or, more generally, leptons) is modified by including the mass contributions of virtual photons, in a technique known as renormalization. Such "radiative corrections" contribute to a number of predictions of QED, such as the magnetic dipole moment of leptons, the Lamb shift, and the hyperfine structure of bound lepton pairs, such as muonium and positronium.[99]

Since photons contribute to the stress–energy tensor, they exert a gravitational attraction on other objects, according to the theory of general relativity. Conversely, photons are themselves affected by gravity; their normally straight trajectories may be bent by warped spacetime, as in gravitational lensing, and their frequencies may be lowered by moving to a higher gravitational potential, as in the Pound–Rebka experiment. However, these effects are not specific to photons; exactly the same effects would be predicted for classical electromagnetic waves.[100]

4.4.13 Photons in matter

See also: Group velocity and Photochemistry

Any 'explanation' of how photons travel through matter has to explain why different arrangements of matter are transparent or opaque at different wavelengths (light through carbon as diamond or not, as graphite) and why individual photons behave in the same way as large groups. Explanations that invoke 'absorption' and 're-emission' have to provide an explanation for the directionality of the photons (diffraction, reflection) and further explain how entangled photon pairs can travel through matter without their quantum state collapsing.

The simplest explanation is that light that travels through transparent matter does so at a lower speed than c, the speed of light in a vacuum. In addition, light can also undergo scattering and absorption. There are circumstances in which heat transfer through a material is mostly radiative, involving emission and absorption of photons within it. An example would be in the core of the Sun. Energy can take about a million years to reach the surface.[101] However, this phenomenon is distinct from scattered radiation passing diffusely through matter, as it involves local equilibrium between the radiation and the temperature. Thus, the time is how long it takes the *energy* to be transferred, not the *photons* themselves. Once in open space, a photon from the Sun takes only 8.3 minutes to reach Earth. The factor by which the speed of light is decreased in a material is called the refractive index of the material. In a classical wave picture, the slowing can be explained by the light inducing electric polarization in the matter, the polarized matter radiating new light, and the new light interfering with the original light wave to form a delayed wave. In a particle picture, the slowing can instead be described as a blending of the photon with quantum excitation of the matter (quasiparticles such as phonons and excitons) to form a polariton; this polariton has a nonzero effective mass, which means that it cannot travel at c.

Alternatively, photons may be viewed as *always* traveling at c, even in matter, but they have their phase shifted (delayed or advanced) upon interaction with atomic scatters: this modifies their wavelength and momentum, but not speed.[102] A light wave made up of these photons does travel slower than the speed of light. In this view the photons are "bare", and are scattered and phase shifted, while in the view of the preceding paragraph the photons are "dressed" by their interaction with matter, and move without scattering or phase shifting, but at a lower speed.

Light of different frequencies may travel through matter at different speeds; this is called dispersion. In some cases, it can result in extremely slow speeds of light in matter. The effects of photon interactions with other quasi-particles may be observed directly in Raman scattering and Brillouin scattering.[103]

Photons can also be absorbed by nuclei, atoms or molecules, provoking transitions between their energy levels. A classic example is the molecular transition of retinal $C_{20}H_{28}O$, which is responsible for vision, as discovered in 1958 by Nobel laureate biochemist George Wald and co-workers. The absorption provokes a cis-trans isomerization that, in combination with other such transitions, is transduced into nerve impulses. The absorption of photons can even break chemical bonds, as in the photodissociation of chlorine; this is the subject of photochemistry.[104][105] Analogously, gamma rays can in some circumstances dissociate atomic nuclei in a process called photodisintegration.

4.4.14 Technological applications

Photons have many applications in technology. These examples are chosen to illustrate applications of photons *per se*, rather than general optical devices such as lenses, etc. that could operate under a classical theory of light. The laser is an extremely important application and is discussed above under stimulated emission.

Individual photons can be detected by several methods. The classic photomultiplier tube exploits the photoelectric effect: a photon landing on a metal plate ejects an electron, initiating an ever-amplifying avalanche of electrons. Charge-coupled device chips use a similar effect in semiconductors: an incident photon generates a charge on a microscopic capacitor that can be detected. Other detectors such as Geiger counters use the ability of photons to ionize gas molecules, causing a detectable change in conductivity.[106]

Planck's energy formula $E = h\nu$ is often used by engineers and chemists in design, both to compute the change in energy resulting from a photon absorption and to predict the frequency of the light emitted for a given energy transition. For example, the emission spectrum of a fluorescent light bulb can be designed using gas molecules with different electronic energy levels and adjusting the typical energy with which an electron hits the gas molecules within the bulb.[Note 6]

Under some conditions, an energy transition can be excited by "two" photons that individually would be insufficient. This allows for higher resolution microscopy, because the sample absorbs energy only in the region where two beams of different colors overlap significantly, which can be made much smaller than the excitation volume of a single beam (see two-photon excitation microscopy). Moreover, these photons cause less damage to the sample, since they are of lower energy.[107]

In some cases, two energy transitions can be coupled so that, as one system absorbs a photon, another nearby system "steals" its energy and re-emits a photon of a different frequency. This is the basis of fluorescence resonance energy transfer, a technique that is used in molecular biology to study the interaction of suitable proteins.[108]

Several different kinds of hardware random number generator involve the detection of single photons. In one example, for each bit in the random sequence that is to be produced, a photon is sent to a beam-splitter. In such a situation, there are two possible outcomes of equal probability. The actual outcome is used to determine whether the next bit in the sequence is "0" or "1".[109][110]

4.4.15 Recent research

See also: Quantum optics

Much research has been devoted to applications of photons in the field of quantum optics. Photons seem well-suited to be elements of an extremely fast quantum computer, and the quantum entanglement of photons is a focus of research. Nonlinear optical processes are another active research area, with topics such as two-photon absorption, self-phase modulation, modulational instability and optical parametric oscillators. However, such processes generally do not require the assumption of photons *per se*; they may often be modeled by treating atoms as nonlinear oscillators. The nonlinear process of spontaneous parametric down conversion is often used to produce single-photon states. Finally, photons are essential in some aspects of optical communication, especially for quantum cryptography.[Note 7]

4.4.16 See also

- Advanced Photon Source at Argonne National Laboratory

- Ballistic photon

- Doppler shift

- Electromagnetic radiation

- HEXITEC

- Laser

- Light

- Luminiferous aether

- Medipix

- Phonons

- Photon counting

- Photon energy

- Photon polarization

- Photonic molecule

- Photography

- Photonics

- Quantum optics

- Single photon sources

- Static forces and virtual-particle exchange

- Two-photon physics

- EPR paradox

- Dirac equation

4.4.17 Notes

[1] Although the 1967 Elsevier translation of Planck's Nobel Lecture interprets Planck's *Lichtquant* as "photon", the more literal 1922 translation by Hans Thacher Clarke and Ludwik Silberstein *The origin and development of the quantum theory*, The Clarendon Press, 1922 (here) uses "light-quantum". No evidence is known that Planck himself used the term "photon" by 1926 (see also this note).

[2] Isaac Asimov credits Arthur Compton with defining quanta of energy as photons in 1923. Asimov, I. (1966). *The Neutrino, Ghost Particle of the Atom*. Garden City (NY): Doubleday. ISBN 0-380-00483-6. LCCN 66017073. and Asimov, I. (1966). *The Universe From Flat Earth To Quasar*. New York (NY): Walker. ISBN 0-8027-0316-X. LCCN 66022515.

[3] The mass of the photon is believed to be exactly zero, based on experiment and theoretical considerations described in the article. Some sources also refer to the *relativistic mass* concept, which is just the energy scaled to units of mass. For a photon with wavelength λ or energy E, this is $h/\lambda c$ or E/c^2. This usage for the term "mass" is no longer common in scientific literature. Further info: What is the mass of a photon? http://math.ucr.edu/home/baez/physics/ ParticleAndNuclear/photon_mass.html

[4] The phrase "no matter how intense" refers to intensities below approximately 10^{13} W/cm^2 at which point perturbation theory begins to break down. In contrast, in the intense regime, which for visible light is above approximately 10^{14} W/cm^2, the classical wave description correctly predicts the energy acquired by electrons, called ponderomotive energy. (See also: Boreham *et al.* (1996). "Photon density and the correspondence principle of electromagnetic interaction".) By comparison, sunlight is only about 0.1 W/cm^2.

[5] These experiments produce results that cannot be explained by any classical theory of light, since they involve anticorrelations that result from the quantum measurement process. In 1974, the first such experiment was carried out by Clauser, who reported a violation of a classical Cauchy–Schwarz inequality. In 1977, Kimble *et al.* demonstrated an analogous anti-bunching effect of photons interacting with a beam splitter; this approach was simplified and sources of error eliminated in the photon-anticorrelation experiment of Grangier *et al.* (1986). This work is reviewed and simplified further in Thorn *et al.* (2004). (These references are listed below under #Additional references.)

[6] An example is US Patent Nr. 5212709.

[7] Introductory-level material on the various sub-fields of quantum optics can be found in Fox, M. (2006). *Quantum Optics: An Introduction*. Oxford University Press. ISBN 0-19-856673-5.

4.4.18   References

[1] Amsler, C. (Particle Data Group); Amsler; Doser; Antonelli; Asner; Babu; Baer; Band; Barnett; Bergren; Beringer; Bernardi; Bertl; Bichsel; Biebel; Bloch; Blucher; Blusk; Cahn; Carena; Caso; Ceccucci; Chakraborty; Chen; Chivukula; Cowan; Dahl; d'Ambrosio; Damour et al. (2008). "Review of Particle Physics: Gauge and Higgs bosons" (PDF). *Physics Letters B* **667**: 1. Bibcode:2008PhLB..667....1P. doi:10.1016/j.physletb.2008.07.018.

[2] Joos, George (1951). *Theoretical Physics*. London and Glasgow: Blackie and Son Limited. p. 679.

[3] Kimble, H.J.; Dagenais, M.; Mandel, L.; Dagenais; Mandel (1977). "Photon Anti-bunching in Resonance Fluorescence". *Physical Review Letters* **39** (11): 691–695. Bibcode:1977PhRvL..39..691K. doi:10.1103/PhysRevLett.39.691.

[4] Grangier, P.; Roger, G.; Aspect, A.; Roger; Aspect (1986). "Experimental Evidence for a Photon Anticorrelation Effect on a Beam Splitter: A New Light on Single-Photon Interferences". *Europhysics Letters* **1** (4): 173–179. Bibcode:1986EL......1..173G. doi:10.1209/0295-5075/1/4/004.

[5] Planck, M. (1901). "On the Law of Distribution of Energy in the Normal Spectrum". *Annalen der Physik* **4** (3): 553–563. Bibcode:1901AnP...309..553P. doi:10.1002/andp.19013090310. Archived from the original on 2008-04-18.

[6] Einstein, A. (1905). "Über einen die Erzeugung und Verwandlung des Lichtes betreffenden heuristischen Gesichtspunkt" (PDF). *Annalen der Physik* (in German) **17** (6): 132–148. Bibcode:1905AnP...322..132E. doi:10.1002/andp.19053220607.. An English translation is available from Wikisource.

[7] "Discordances entre l'expérience et la théorie électromagnétique du rayonnement." In Électrons et Photons. Rapports et Discussions de Cinquième Conseil de Physique, edited by Institut International de Physique Solvay. Paris: Gauthier-Villars, pp. 55-85.

[8] Helge Kragh: *Photon: New light on an old name*. Arxiv, 2014-2-28

[9] Villard, P. (1900). "Sur la réflexion et la réfraction des rayons cathodiques et des rayons déviables du radium". *Comptes Rendus des Séances de l'Académie des Sciences* (in French) **130**: 1010–1012.

[10] Villard, P. (1900). "Sur le rayonnement du radium". *Comptes Rendus des Séances de l'Académie des Sciences* (in French) **130**: 1178–1179.

[11] Rutherford, E.; Andrade, E.N.C. (1914). "The Wavelength of the Soft Gamma Rays from Radium B". *Philosophical Magazine* **27** (161): 854–868. doi:10.1080/14786440508635156.

[12] Kobychev, V.V.; Popov, S.B. (2005). "Constraints on the photon charge from observations of extragalactic sources". *Astronomy Letters* **31** (3): 147–151. arXiv:hep-ph/0411398. Bibcode:2005AstL...31..147K. doi:10.1134/1.1883345.

[13] Role as gauge boson and polarization section 5.1 inAitchison, I.J.R.; Hey, A.J.G. (1993). *Gauge Theories in Particle Physics*. IOP Publishing. ISBN 0-85274-328-9.

[14] See p.31 inAmsler, C. et al. (2008). "Review of Particle Physics". *Physics Letters B* **667**: 1–1340. Bibcode:2008PhLB..667....1P. doi:10.1016/j.physletb.2008.07.018.

[15] Halliday, David; Resnick, Robert; Walker, Jerl (2005), *Fundamental of Physics* (7th ed.), USA: John Wiley and Sons, Inc., ISBN 0-471-23231-9

[16] See section 1.6 in Alonso, M.; Finn, E.J. (1968). *Fundamental University Physics Volume III: Quantum and Statistical Physics*. Addison-Wesley. ISBN 0-201-00262-0.

[17] Davison E. Soper, Electromagnetic radiation is made of photons, Institute of Theoretical Science, University of Oregon

[18] This property was experimentally verified by Raman and Bhagavantam in 1931: Raman, C.V.; Bhagavantam, S. (1931). "Experimental proof of the spin of the photon" (PDF). *Indian Journal of Physics* **6**: 353.

[19] Burgess, C.; Moore, G. (2007). "1.3.3.2". *The Standard Model. A Primer*. Cambridge University Press. ISBN 0-521-86036-9.

[20] Griffiths, David J. (2008), *Introduction to Elementary Particles* (2nd revised ed.), WILEY-VCH, ISBN 978-3-527-40601-2

[21] E.g., section 9.3 in Alonso, M.; Finn, E.J. (1968). *Fundamental University Physics Volume III: Quantum and Statistical Physics*. Addison-Wesley.

[22] E.g., Appendix XXXII in Born, M. (1962). *Atomic Physics*. Blackie & Son. ISBN 0-486-65984-4.

[23] Mermin, David (February 1984). "Relativity without light". *American Journal of Physics* **52** (2): 119–124. Bibcode:1984AmJPh..52..119M. doi:10.1119/1.13917.

[24] Plimpton, S.; Lawton, W. (1936). "A Very Accurate Test of Coulomb's Law of Force Between Charges". *Physical Review* **50** (11): 1066. Bibcode:1936PhRv...50.1066P. doi:10.1103/PhysRev.50.1066.

[25] Williams, E.; Faller, J.; Hill, H. (1971). "New Experimental Test of Coulomb's Law: A Laboratory Upper Limit on the Photon Rest Mass". *Physical Review Letters* **26** (12): 721. Bibcode:1971PhRvL..26..721W. doi:10.1103/PhysRevLett.26.721.

[26] Chibisov, G V (1976). "Astrophysical upper limits on the photon rest mass". *Soviet Physics Uspekhi* **19** (7): 624. Bibcode:1976SvPhU..19..624C. doi:10.1070/PU1976v019n07ABEH005277.

[27] Lakes, Roderic (1998). "Experimental Limits on the Photon Mass and Cosmic Magnetic Vector Potential". *Physical Review Letters* **80** (9): 1826. Bibcode:1998PhRvL..80.1826L. doi:10.1103/PhysRevLett.80.1826.

[28] Amsler, C; Doser, M; Antonelli, M; Asner, D; Babu, K; Baer, H; Band, H; Barnett, R et al. (2008). "Review of Particle Physics*". *Physics Letters B* **667**: 1. Bibcode:2008PhLB..667....1P. doi:10.1016/j.physletb.2008.07.018. Summary Table

[29] Adelberger, Eric; Dvali, Gia; Gruzinov, Andrei (2007). "Photon-Mass Bound Destroyed by Vortices". *Physical Review Letters* **98** (1): 010402. arXiv:hep-ph/0306245. Bibcode:2007PhRvL..98a0402A. doi:10.1103/PhysRevLett.98.010402. PMID 17358459. preprint

[30] Wilczek, Frank (2010). *The Lightness of Being: Mass, Ether, and the Unification of Forces*. Basic Books. p. 212. ISBN 978-0-465-01895-6.

[31] Descartes, R. (1637). *Discours de la méthode (Discourse on Method)* (in French). Imprimerie de Ian Maire. ISBN 0-268-00870-1.

[32] Hooke, R. (1667). *Micrographia: or some physiological descriptions of minute bodies made by magnifying glasses with observations and inquiries thereupon ...* London (UK): Royal Society of London. ISBN 0-486-49564-7.

[33] Huygens, C. (1678). *Traité de la lumière* (in French).. An English translation is available from Project Gutenberg

[34] Newton, I. (1952) [1730]. *Opticks* (4th ed.). Dover (NY): Dover Publications. Book II, Part III, Propositions XII–XX; Queries 25–29. ISBN 0-486-60205-2.

[35] Buchwald, J.Z. (1989). *The Rise of the Wave Theory of Light: Optical Theory and Experiment in the Early Nineteenth Century*. University of Chicago Press. ISBN 0-226-07886-8. OCLC 18069573.

[36] Maxwell, J.C. (1865). "A Dynamical Theory of the Electromagnetic Field". *Philosophical Transactions of the Royal Society* **155**: 459–512. Bibcode:1865RSPT..155..459C. doi:10.1098/rstl.1865.0008. This article followed a presentation by Maxwell on 8 December 1864 to the Royal Society.

[37] Hertz, H. (1888). "Über Strahlen elektrischer Kraft". *Sitzungsberichte der Preussischen Akademie der Wissenschaften (Berlin)* (in German) **1888**: 1297–1307.

[38] Frequency-dependence of luminiscence p. 276f., photoelectric effect section 1.4 in Alonso, M.; Finn, E.J. (1968). *Fundamental University Physics Volume III: Quantum and Statistical Physics*. Addison-Wesley. ISBN 0-201-00262-0.

[39] Wien, W. (1911). "Wilhelm Wien Nobel Lecture".

[40] Planck, M. (1920). "Max Planck's Nobel Lecture".

[41] Einstein, A. (1909). "Über die Entwicklung unserer Anschauungen über das Wesen und die Konstitution der Strahlung" (PDF). *Physikalische Zeitschrift* (in German) **10**: 817–825.. An English translation is available from Wikisource.

[42] Presentation speech by Svante Arrhenius for the 1921 Nobel Prize in Physics, December 10, 1922. Online text from [nobelprize.org], The Nobel Foundation 2008. Access date 2008-12-05.

[43] Einstein, A. (1916). "Zur Quantentheorie der Strahlung". *Mitteilungen der Physikalischen Gesellschaft zu Zürich* **16**: 47. Also *Physikalische Zeitschrift*, **18**, 121–128 (1917). (German)

[44] Compton, A. (1923). "A Quantum Theory of the Scattering of X-rays by Light Elements". *Physical Review* **21** (5): 483–502. Bibcode:1923PhRv...21..483C. doi:10.1103/PhysRev.21.483.

[45] Pais, A. (1982). *Subtle is the Lord: The Science and the Life of Albert Einstein*. Oxford University Press. ISBN 0-19-853907-X.

[46] *Einstein and the Quantum: The Quest of the Valiant Swabian*, A. Douglas Stone, Princeton University Press, 2013.

[47] Millikan, R.A (1924). "Robert A. Millikan's Nobel Lecture".

[48] Hendry, J. (1980). "The development of attitudes to the wave-particle duality of light and quantum theory, 1900–1920". *Annals of Science* **37** (1): 59–79. doi:10.1080/00033798000200121.

[49] Bohr, N.; Kramers, H.A.; Slater, J.C. (1924). "The Quantum Theory of Radiation". *Philosophical Magazine* **47**: 785–802. doi:10.1080/14786442408565262. Also *Zeitschrift für Physik*, **24**, 69 (1924).

[50] Heisenberg, W. (1933). "Heisenberg Nobel lecture".

[51] Mandel, L. (1976). E. Wolf, ed. "The case for and against semiclassical radiation theory". *Progress in Optics*. Progress in Optics (North-Holland) **13**: 27–69. doi:10.1016/S0079-6638(08)70018-0. ISBN 978-0-444-10806-7.

[52] Taylor, G.I. (1909). *Interference fringes with feeble light*. *Proceedings of the Cambridge Philosophical Society* **15**: 114–115.

[53] Saleh, B. E. A. and Teich, M. C. (2007). *Fundamentals of Photonics*. Wiley. ISBN 0-471-35832-0.

[54] Heisenberg, W. (1927). "Über den anschaulichen Inhalt der quantentheoretischen Kinematik und Mechanik". *Zeitschrift für Physik* (in German) **43** (3–4): 172–198. Bibcode:1927ZPhy...43..172H. doi:10.1007/BF01397280.

[55] E.g., p. 10f. in Schiff, L.I. (1968). *Quantum Mechanics* (3rd ed.). McGraw-Hill. ASIN B001B3MINM. ISBN 0-07-055287-8.

[56] Kramers, H.A. (1958). *Quantum Mechanics*. Amsterdam: North-Holland. ASIN B0006AUW5C. ISBN 0-486-49533-7.

[57] Bohm, D. (1989) [1954]. *Quantum Theory*. Dover Publications. ISBN 0-486-65969-0.

[58] Newton, T.D.; Wigner, E.P. (1949). "Localized states for elementary particles". *Reviews of Modern Physics* **21** (3): 400–406. Bibcode:1949RvMP...21..400N. doi:10.1103/RevModPhys.21.400.

[59] Bialynicki-Birula, I. (1994). "On the wave function of the photon" (PDF). *Acta Physica Polonica A* **86**: 97–116.

[60] Sipe, J.E. (1995). "Photon wave functions". *Physical Review A* **52** (3): 1875–1883. Bibcode:1995PhRvA..52.1875S. doi:10.1103/PhysRevA.52.1875.

[61] Bialynicki-Birula, I. (1996). "Photon wave function". *Progress in Optics*. Progress in Optics **36**: 245–294. doi:10.1016/S0079-6638(08)70316-0. ISBN 978-0-444-82530-8.

[62] Scully, M.O.; Zubairy, M.S. (1997). *Quantum Optics*. Cambridge (UK): Cambridge University Press. ISBN 0-521-43595-1.

[63] The best illustration is the Couder experiment, demonstrating the behaviour of a mechanical analog, see https://www.youtube.com/watch?v=W9yWv5dqSKk

[64] Bell, J. S., "Speakable and Unspeakable in Quantum Mechanics", Cambridge: Cambridge University Press, 1987.

[65] Bose, S.N. (1924). "Plancks Gesetz und Lichtquantenhypothese". *Zeitschrift für Physik* (in German) **26**: 178–181. Bibcode:1924ZPhy...26..178B. doi:10.1007/BF01327326.

[66] Einstein, A. (1924). "Quantentheorie des einatomigen idealen Gases". *Sitzungsberichte der Preussischen Akademie der Wissenschaften (Berlin), Physikalisch-mathematische Klasse* (in German) **1924**: 261–267.

[67] Einstein, A. (1925). "Quantentheorie des einatomigen idealen Gases, Zweite Abhandlung". *Sitzungsberichte der Preussischen Akademie der Wissenschaften (Berlin), Physikalisch-mathematische Klasse* (in German) **1925**: 3–14. doi:10.1002/3527608958.ch28. ISBN 978-3-527-60895-9.

[68] Anderson, M.H.; Ensher, J.R.; Matthews, M.R.; Wieman, C.E.; Cornell, E.A. (1995). "Observation of Bose–Einstein Condensation in a Dilute Atomic Vapor". *Science* **269** (5221): 198–201. Bibcode:1995Sci...269..198A. doi:10.1126/science.269.5221.198. JSTOR 2888436. PMID 17789847.

[69] "Physicists Slow Speed of Light". News.harvard.edu (1999-02-18). Retrieved on 2015-05-11.

[70] "Light Changed to Matter, Then Stopped and Moved". photonics.com (February 2007). Retrieved on 2015-05-11.

[71] Streater, R.F.; Wightman, A.S. (1989). *PCT, Spin and Statistics, and All That*. Addison-Wesley. ISBN 0-201-09410-X.

[72] Einstein, A. (1916). "Strahlungs-emission und -absorption nach der Quantentheorie". *Verhandlungen der Deutschen Physikalischen Gesellschaft* (in German) **18**: 318–323. Bibcode:1916DPhyG..18..318E.

[73] Section 1.4 in Wilson, J.; Hawkes, F.J.B. (1987). *Lasers: Principles and Applications*. New York: Prentice Hall. ISBN 0-13-523705-X.

[74] P. 322 in Einstein, A. (1916). "Strahlungs-emission und -absorption nach der Quantentheorie". *Verhandlungen der Deutschen Physikalischen Gesellschaft* (in German) **18**: 318–323. Bibcode:1916DPhyG..18..318E.:

> Die Konstanten A_m^n and B_m^n würden sich direkt berechnen lassen, wenn wir im Besitz einer im Sinne der Quantenhypothese modifizierten Elektrodynamik und Mechanik wären."

[75] Dirac, P.A.M. (1926). "On the Theory of Quantum Mechanics". *Proceedings of the Royal Society A* **112** (762): 661–677. Bibcode:1926RSPSA.112..661D. doi:10.1098/rspa.1926.0133.

[76] Dirac, P.A.M. (1927). "The Quantum Theory of the Emission and Absorption of Radiation" (PDF). *Proceedings of the Royal Society A* **114** (767): 243–265. Bibcode:1927RSPSA.114..243D. doi:10.1098/rspa.1927.0039.

[77] Dirac, P.A.M. (1927b). *The Quantum Theory of Dispersion. Proceedings of the Royal Society A* **114**: 710–728. doi:10.1098/rspa.1927.0071.

[78] Heisenberg, W.; Pauli, W. (1929). "Zur Quantentheorie der Wellenfelder". *Zeitschrift für Physik* (in German) **56**: 1. Bibcode:1929ZPhy...56....1H. doi:10.1007/BF01340129.

[79] Heisenberg, W.; Pauli, W. (1930). "Zur Quantentheorie der Wellenfelder". *Zeitschrift für Physik* (in German) **59** (3–4): 139. Bibcode:1930ZPhy...59..168H. doi:10.1007/BF01341423.

[80] Fermi, E. (1932). "Quantum Theory of Radiation" (PDF). *Reviews of Modern Physics* **4**: 87. Bibcode:1932RvMP....4...87F. doi:10.1103/RevModPhys.4.87.

[81] Born, M. (1926). "Zur Quantenmechanik der Stossvorgänge" (PDF). *Zeitschrift für Physik* (in German) **37** (12): 863–867. Bibcode:1926ZPhy...37..863B. doi:10.1007/BF01397477.

[82] Born, M. (1926). "Quantenmechanik der Stossvorgänge". *Zeitschrift für Physik* (in German) **38** (11–12): 803. Bibcode:1926ZPhy...38..803B. doi:10.1007/BF01397184.

[83] Pais, A. (1986). *Inward Bound: Of Matter and Forces in the Physical World.* Oxford University Press. p. 260. ISBN 0-19-851997-4. Specifically, Born claimed to have been inspired by Einstein's never-published attempts to develop a "ghost-field" theory, in which point-like photons are guided probabilistically by ghost fields that follow Maxwell's equations.

[84] Debye, P. (1910). "Der Wahrscheinlichkeitsbegriff in der Theorie der Strahlung". *Annalen der Physik* (in German) **33** (16): 1427–1434. Bibcode:1910AnP...338.1427D. doi:10.1002/andp.19103381617.

[85] Born, M.; Heisenberg, W.; Jordan, P. (1925). "Quantenmechanik II". *Zeitschrift für Physik* (in German) **35** (8–9): 557–615. Bibcode:1926ZPhy...35..557B. doi:10.1007/BF01379806.

[86] Photon-photon-scattering section 7-3-1, renormalization chapter 8-2 in Itzykson, C.; Zuber, J.-B. (1980). *Quantum Field Theory.* McGraw-Hill. ISBN 0-07-032071-3.

[87] Weiglein, G. (2008). "Electroweak Physics at the ILC". *Journal of Physics: Conference Series* **110** (4): 042033. arXiv:0711.3003. Bibcode:2008JPhCS.110d2033W. doi:10.1088/1742-6596/110/4/042033.

[88] Bauer, T. H.; Spital, R. D.; Yennie, D. R.; Pipkin, F. M. (1978). "The hadronic properties of the photon in high-energy interactions". *Reviews of Modern Physics* **50** (2): 261. Bibcode:1978RvMP...50..261B. doi:10.1103/RevModPhys.50.261.

[89] Sakurai, J. J. (1960). "Theory of strong interactions". *Annals of Physics* **11**: 1. Bibcode:1960AnPhy..11....1S. doi:10.1016/0003-4916(60)90126-3.

[90] Walsh, T. F.; Zerwas, P. (1973). "Two-photon processes in the parton model". *Physics Letters B* **44** (2): 195. Bibcode:1973PhLB...44..195W. doi:10.1016/0370-2693(73)90520-0.

[91] Witten, E. (1977). "Anomalous cross section for photon-photon scattering in gauge theories". *Nuclear Physics B* **120** (2): 189. Bibcode:1977NuPhB.120..189W. doi:10.1016/0550-3213(77)90038-4.

[92] Nisius, R. (2000). "The photon structure from deep inelastic electron–photon scattering". *Physics Reports* **332** (4–6): 165. Bibcode:2000PhR...332..165N. doi:10.1016/S0370-1573(99)00115-5.

[93] Ryder, L.H. (1996). *Quantum field theory* (2nd ed.). Cambridge University Press. ISBN 0-521-47814-6.

[94] Sheldon Glashow Nobel lecture, delivered 8 December 1979.

[95] Abdus Salam Nobel lecture, delivered 8 December 1979.

[96] Steven Weinberg Nobel lecture, delivered 8 December 1979.

[97] E.g., chapter 14 in Hughes, I. S. (1985). *Elementary particles* (2nd ed.). Cambridge University Press. ISBN 0-521-26092-2.

[98] E.g., section 10.1 in Dunlap, R.A. (2004). *An Introduction to the Physics of Nuclei and Particles.* Brooks/Cole. ISBN 0-534-39294-6.

[99] Radiative correction to electron mass section 7-1-2, anomalous magnetic moments section 7-2-1, Lamb shift section 7-3-2 and hyperfine splitting in positronium section 10-3 in Itzykson, C.; Zuber, J.-B. (1980). *Quantum Field Theory.* McGraw-Hill. ISBN 0-07-032071-3.

[100] E. g. sections 9.1 (gravitational contribution of photons) and 10.5 (influence of gravity on light) in Stephani, H.; Stewart, J. (1990). *General Relativity: An Introduction to the Theory of Gravitational Field.* Cambridge University Press. pp. 86 ff, 108 ff. ISBN 0-521-37941-5.

[101] Naeye, R. (1998). *Through the Eyes of Hubble: Birth, Life and Violent Death of Stars.* CRC Press. ISBN 0-7503-0484-7. OCLC 40180195.

[102] Ch 4 in Hecht, Eugene (2001). *Optics.* Addison Wesley. ISBN 978-0-8053-8566-3.

[103] Polaritons section 10.10.1, Raman and Brillouin scattering section 10.11.3 in Patterson, J.D.; Bailey, B.C. (2007). *Solid-State Physics: Introduction to the Theory.* Springer. pp. 569 ff, 580 ff. ISBN 3-540-24115-9.

[104] E.g., section 11-5 C in Pine, S.H.; Hendrickson, J.B.; Cram, D.J.; Hammond, G.S. (1980). *Organic Chemistry* (4th ed.). McGraw-Hill. ISBN 0-07-050115-7.

[105] Nobel lecture given by G. Wald on December 12, 1967, online at nobelprize.org: The Molecular Basis of Visual Excitation.

[106] Photomultiplier section 1.1.10, CCDs section 1.1.8, Geiger counters section 1.3.2.1 in Kitchin, C.R. (2008). *Astrophysical Techniques*. Boca Raton (FL): CRC Press. ISBN 1-4200-8243-4.

[107] Denk, W.; Svoboda, K. (1997). "Photon upmanship: Why multiphoton imaging is more than a gimmick". *Neuron* **18** (3): 351–357. doi:10.1016/S0896-6273(00)81237-4. PMID 9115730.

[108] Lakowicz, J.R. (2006). *Principles of Fluorescence Spectroscopy*. Springer. pp. 529 ff. ISBN 0-387-31278-1.

[109] Jennewein, T.; Achleitner, U.; Weihs, G.; Weinfurter, H.; Zeilinger, A. (2000). "A fast and compact quantum random number generator". *Review of Scientific Instruments* **71** (4): 1675–1680. arXiv:quant-ph/9912118. Bibcode:2000RScI...71.1675J. doi:10.1063/1.1150518.

[110] Stefanov, A.; Gisin, N.; Guinnard, O.; Guinnard, L.; Zbiden, H. (2000). "Optical quantum random number generator". *Journal of Modern Optics* **47** (4): 595–598. doi:10.1080/095003400147908.

4.4.19 Additional references

By date of publication:

- Clauser, J.F. (1974). "Experimental distinction between the quantum and classical field-theoretic predictions for the photoelectric effect". *Physical Review D* **9** (4): 853–860. Bibcode:1974PhRvD...9..853C. doi:10.1103/PhysRevD.9.853.

- Kimble, H.J.; Dagenais, M.; Mandel, L. (1977). "Photon Anti-bunching in Resonance Fluorescence". *Physical Review Letters* **39** (11): 691–695. Bibcode:1977PhRvL..39..691K. doi:10.1103/PhysRevLett.39.691.

- Pais, A. (1982). *Subtle is the Lord: The Science and the Life of Albert Einstein*. Oxford University Press.

- Feynman, Richard (1985). *QED: The Strange Theory of Light and Matter*. Princeton University Press. ISBN 978-0-691-12575-6.

- Grangier, P.; Roger, G.; Aspect, A. (1986). "Experimental Evidence for a Photon Anticorrelation Effect on a Beam Splitter: A New Light on Single-Photon Interferences". *Europhysics Letters* **1** (4): 173–179. Bibcode:1986EL......1..173G. doi:10.1209/0295-5075/1/4/004.

- Lamb, W.E. (1995). "Anti-photon". *Applied Physics B* **60** (2–3): 77–84. Bibcode:1995ApPhB..60...77L. doi:10.1007/BF01135846.

- Special supplemental issue of *Optics and Photonics News* (vol. 14, October 2003) article web link

  - Roychoudhuri, C.; Rajarshi, R. (2003). "The nature of light: what is a photon?". *Optics and Photonics News* **14**: S1 (Supplement).

  - Zajonc, A. "Light reconsidered". *Optics and Photonics News* **14**: S2–S5 (Supplement).

  - Loudon, R. "What is a photon?". *Optics and Photonics News* **14**: S6–S11 (Supplement).

  - Finkelstein, D. "What is a photon?". *Optics and Photonics News* **14**: S12–S17 (Supplement).

  - Muthukrishnan, A.; Scully, M.O.; Zubairy, M.S. "The concept of the photon—revisited". *Optics and Photonics News* **14**: S18–S27 (Supplement).

  - Mack, H.; Schleich, W.P.. "A photon viewed from Wigner phase space". *Optics and Photonics News* **14**: S28–S35 (Supplement).

- Glauber, R. (2005). "One Hundred Years of Light Quanta" (PDF). *2005 Physics Nobel Prize Lecture*.

- Hentschel, K. (2007). "Light quanta: The maturing of a concept by the stepwise accretion of meaning". *Physics and Philosophy* **1** (2): 1–20.

Education with single photons:

- Thorn, J.J.; Neel, M.S.; Donato, V.W.; Bergreen, G.S.; Davies, R.E.; Beck, M. (2004). "Observing the quantum behavior of light in an undergraduate laboratory" (PDF). *American Journal of Physics* **72** (9): 1210–1219. Bibcode:2004AmJPh..72.1210T. doi:10.1119/1.1737397.

- Bronner, P.; Strunz, Andreas; Silberhorn, Christine; Meyn, Jan-Peter (2009). "Interactive screen experiments with single photons". *European Journal of Physics* **30** (2): 345–353. Bibcode:2009EJPh...30..345B. doi:10.1088/0143-0807/30/2/014.

4.4.20 External links

- The dictionary definition of photon at Wiktionary

- Media related to Photon at Wikimedia Commons

Chapter 5

Higgs boson

5.1 Higgs boson

The **Higgs boson** or **Higgs particle** is an elementary particle in the Standard Model of particle physics. It is the quantum excitation of the **Higgs field**[6][7]—a fundamental field of crucial importance to particle physics theory,[7] first suspected to exist in the 1960s, that unlike other known fields such as the electromagnetic field, takes a non-zero constant value almost everywhere. The question of the Higgs field's existence has been the last unverified part of the Standard Model of particle physics and, according to some, "the central problem in particle physics".[8][9] The presence of this field, now believed to be confirmed, explains why some fundamental particles have mass when, based on the symmetries controlling their interactions, they should be massless. The existence of the Higgs field would also resolve several other long-standing puzzles, such as the reason for the weak force's extremely short range.

Although it is hypothesized that the Higgs field permeates the entire Universe, evidence for its existence has been very difficult to obtain. In principle, the Higgs field can be detected through its excitations, manifest as Higgs particles, but these are extremely difficult to produce and detect. The importance of this fundamental question led to a 40 year search, and the construction of one of the world's most expensive and complex experimental facilities to date, CERN's Large Hadron Collider,[10] able to create Higgs bosons and other particles for observation and study. On 4 July 2012, the discovery of a new particle with a mass between 125 and 127 GeV/c^2 was announced; physicists suspected that it was the Higgs boson.[11][12][13] Since then, however, the particle had been shown to behave, interact, and decay in many of the ways predicted by the Standard Model. It was also tentatively confirmed to have even parity and zero spin,[1] two fundamental attributes of a Higgs boson. This appears to be the first elementary scalar particle discovered in nature.[14] More data are needed to verify that the discovered particle has properties matching those predicted for the Higgs boson by the Standard Model, or whether, as predicted by some theories, multiple Higgs bosons exist.[3]

The Higgs boson is named after Peter Higgs, one of six physicists who, in 1964, proposed the mechanism that suggested the existence of such a particle. On December 10, 2013, two of them, Peter Higgs and François Englert, were awarded the Nobel Prize in Physics for their work and prediction (Englert's co-researcher Robert Brout had died in 2011 and the Nobel Prize is not ordinarily given posthumously).[15] Although Higgs's name has come to be associated with this theory, several researchers between about 1960 and 1972 each independently developed different parts of it. In mainstream media the Higgs boson has often been called the "God particle", from a 1993 book on the topic; the nickname is strongly disliked by many physicists, including Higgs, who regard it as sensationalistic.[16][17][18]

In the Standard Model, the Higgs particle is a boson with no spin, electric charge, or colour charge. It is also very unstable, decaying into other particles almost immediately. It is a quantum excitation of one of the four components of the Higgs field. The latter constitutes a scalar field, with two neutral and two electrically charged components that form a complex doublet of the weak isospin SU(2) symmetry. The Higgs field is tachyonic (this does not refer to faster-than-light speeds, it means that symmetry-breaking through condensation of a particle must occur under certain conditions), and has a "Mexican hat" shaped potential with nonzero strength everywhere (including otherwise empty space), which in its vacuum state breaks the weak isospin symmetry of the electroweak interaction. When this happens, three components of the Higgs field are "absorbed" by the SU(2) and U(1) gauge bosons (the "Higgs mechanism") to become the longitudinal components of the now-massive W and Z bosons of the weak force. The remaining electrically neutral component separately couples to other particles known as fermions (via Yukawa couplings), causing these to acquire mass as well. Some versions of the theory predict more than one kind of Higgs fields and bosons. Alternative "Higgsless" models would have been considered if the Higgs boson was not discovered.

5.1.1   A non-technical summary

"Higgs" terminology

Overview

Physicists explain the properties and forces between elementary particles in terms of the Standard Model—a widely accepted and "remarkably" accurate[21] framework based on gauge invariance and symmetries, believed to explain almost everything in the known universe, other than gravity.[22] But by around 1960 all attempts to create a gauge invariant theory for two of the four fundamental forces had consistently failed at one crucial point: although gauge invariance seemed extremely important, it seemed to make any theory of electromagnetism and the weak force go haywire, by demanding that either many particles with mass were massless or that non-existent forces and massless particles had to exist. Scientists had no idea how to get past this point.

In 1962 physicist Philip Anderson wrote a paper that built upon work by Yoichiro Nambu concerning "broken symmetries" in superconductivity and particle physics. He suggested that "broken symmetries" might also be the missing piece needed to solve the problems of gauge invariance. In 1964 a theory was created almost simultaneously by 3 different groups of researchers, that showed Anderson's suggestion was possible - the gauge theory and "mass problems" could indeed be resolved if an unusual kind of field, now generally called the "Higgs field", existed throughout the universe; if the Higgs field did exist, it would apparently cause existing particles to acquire mass instead of new massless particles being formed. Although these ideas did not gain much initial support or attention, by 1972 they had been developed into a comprehensive theory and proved capable of giving "sensible" results that accurately described particles known at the time, and which accurately predicted of several other particles discovered during the following years.[Note 7] During the 1970s these theories rapidly became the "standard model". There was not yet any direct evidence that the Higgs field actually existed, but even without proof of the field, the accuracy of its predictions led scientists to believe the theory might be true. By the 1980s the question whether or not the Higgs field existed had come to be regarded as one of the most important unanswered questions in particle physics.

If Higgs field could be shown to exist, it would be a monumental discovery for science and human knowledge, and would open doorways to new knowledge in many disciplines. If not, then other more complicated theories would need to be considered. The simplest means to test the existence of the Higgs field would be a search for a new elementary particle that the field would have to give off, a particle known as "Higgs bosons" or the "Higgs particle". This particle would be extremely difficult to find. After significant technological advancements, by the 1990s two large experimental installations were being designed and constructed that allowed to search for the Higgs boson.

While several symmetries in nature are spontaneously broken through a form of the Higgs mechanism, in the context of the Standard Model the term "Higgs mechanism" almost always means symmetry breaking of the electroweak field. It is considered confirmed, but revealing the exact cause has been difficult. Various analogies have also been invented to describe the Higgs field and boson, including analogies with well-known symmetry breaking effects such as the rainbow and prism, electric fields, ripples, and resistance of macro objects moving through media, like people moving through crowds or some objects moving through syrup or molasses. However, analogies based on simple resistance to motion are inaccurate as the Higgs field does not work by resisting motion.

5.1.2   Significance

Scientific impact

Evidence of the Higgs field and its properties has been extremely significant scientifically, for many reasons. The Higgs boson's importance is largely that it is able to be examined using existing knowledge and experimental technology, as a way to confirm and study the entire Higgs field theory.[6][7] Conversely, proof that the Higgs field and boson do not exist would also have been significant. In discussion form, the relevance includes:

Practical and technological impact of discovery

As yet, there are no known immediate technological benefits of finding the Higgs particle. However, a common pattern for fundamental discoveries is for practical applications to follow later, once the discovery has been explored further, at which point they become the basis for new technologies of importance to society.[44][45][46]

The challenges in particle physics have furthered major technological of widespread importance. For example, the World Wide Web began as a project to improve CERN's communication system. CERN's requirement to process massive amounts of data produced by the Large Hadron Collider also led to contributions to the fields of distributed and cloud computing.

5.1.3 History

See also: 1964 PRL symmetry breaking papers, Higgs mechanism and History of quantum field theory

Particle physicists study matter made from fundamental

Nobel Prize Laureate Peter Higgs in Stockholm, December 2013

particles whose interactions are mediated by exchange particles - gauge bosons - acting as force carriers. At the beginning of the 1960s a number of these particles had been discovered or proposed, along with theories suggesting how they relate to each other, some of which had already been reformulated as field theories in which the objects of study are not particles and forces, but quantum fields and their symmetries.[47]:150 However, attempts to unify known fundamental forces such as the electromagnetic force and the weak nuclear force were known to be incomplete. One known omission was that gauge invariant approaches, including non-abelian models such as Yang–Mills theory (1954), which held great promise for unified theories, also seemed to predict known massive particles as massless.[48] Goldstone's theorem, relating to continuous symmetries within some theories, also appeared to rule out many obvious solutions,[49] since it appeared to show that zero-mass particles would have to also exist that were "simply not seen".[50] According to Guralnik, physicists had "no understanding" how these problems could be overcome.[50]

Particle physicist and mathematician Peter Woit summarised the state of research at the time:

> "Yang and Mills work on non-abelian gauge theory had one huge problem: in perturbation theory it has massless particles which don't correspond to anything we see. One way of getting rid of

this problem is now fairly well-understood, the phenomenon of confinement realized in QCD, where the strong interactions get rid of the massless "gluon" states at long distances. By the very early sixties, people had begun to understand another source of massless particles: spontaneous symmetry breaking of a continuous symmetry. What Philip Anderson realized and worked out in the summer of 1962 was that, when you have *both* gauge symmetry *and* spontaneous symmetry breaking, the Nambu–Goldstone massless mode can combine with the massless gauge field modes to produce a physical massive vector field. This is what happens in superconductivity, a subject about which Anderson was (and is) one of the leading experts." *[text condensed]* [48]

The Higgs mechanism is a process by which vector bosons can get rest mass *without* explicitly breaking gauge invariance, as a byproduct of spontaneous symmetry breaking.[51][52] The mathematical theory behind spontaneous symmetry breaking was initially conceived and published within particle physics by Yoichiro Nambu in 1960,[53] the concept that such a mechanism could offer a possible solution for the "mass problem" was originally suggested in 1962 by Philip Anderson (who had previously written papers on broken symmetry and its outcomes in superconductivity[54] and concluded in his 1963 paper on Yang-Mills theory that *"considering the superconducting analog... [t]hese two types of bosons seem capable of canceling each other out... leaving finite mass bosons"*),[55]:4–5[56] and Abraham Klein and Benjamin Lee showed in March 1964 that Goldstone's theorem could be avoided this way in at least some non-relativistic cases and speculated it might be possible in truly relativistic cases.[57]

These approaches were quickly developed into a full relativistic model, independently and almost simultaneously, by three groups of physicists: by François Englert and Robert Brout in August 1964;[58] by Peter Higgs in October 1964;[59] and by Gerald Guralnik, Carl Hagen, and Tom Kibble (GHK) in November 1964.[60] Higgs also wrote a short but important[51] response published in September 1964 to an objection by Gilbert,[61] which showed that if calculating within the radiation gauge, Goldstone's theorem and Gilbert's objection would become inapplicable.[Note 11] (Higgs later described Gilbert's objection as prompting his own paper.[62]) Properties of the model were further considered by Guralnik in 1965,[63] by Higgs in 1966,[64] by Kibble in 1967,[65] and further by GHK in 1967.[66] The original three 1964 papers showed that when a gauge theory is combined with an additional field that spontaneously breaks the symmetry, the gauge bosons can consistently acquire a finite mass.[51][52][67] In 1967, Steven Weinberg[68] and Abdus Salam[69] independently showed how a Higgs mechanism could be used to break the electroweak symmetry of Sheldon Glashow's unified model for the weak and electromagnetic interactions[70] (itself an extension of work by Schwinger), forming what became the Standard Model of particle physics. Weinberg was the first to observe that this would also provide mass terms for the fermions.[71] [Note 12]

However, the seminal papers on spontaneous breaking of gauge symmetries were at first largely ignored, because it was widely believed that the (non-Abelian gauge) theories in question were a dead-end, and in particular that they could not be renormalised. In 1971–72, Martinus Veltman and Gerard 't Hooft proved renormalisation of Yang–Mills was possible in two papers covering massless, and then massive, fields.[71] Their contribution, and others' work on the renormalization group - including "substantial" theoretical work by Russian physicists Ludvig Faddeev, Andrei Slavnov, Efim Fradkin and Igor Tyutin[72] - was eventually "enormously profound and influential",[73] but even with all key elements of the eventual theory published there was still almost no wider interest. For example, Coleman found in a study that "essentially no-one paid any attention" to Weinberg's paper prior to 1971[74] and discussed by David Politzer in his 2004 Nobel speech.[73] – now the most cited in particle physics[75] – and even in 1970 according to Politzer, Glashow's teaching of the weak interaction contained no mention of Weinberg's, Salam's, or Glashow's own work.[73] In practice, Politzer states, almost everyone learned of the theory due to physicist Benjamin Lee, who combined the work of Veltman and 't Hooft with insights by others, and popularised the completed theory.[73] In this way, from 1971, interest and acceptance "exploded"[73] and the ideas were quickly absorbed in the mainstream.[71][73]

The resulting electroweak theory and Standard Model have accurately predicted (among other things) weak neutral currents, three bosons, the top and charm quarks, and with great precision, the mass and other properties of some of these.[Note 7] Many of those involved eventually won Nobel Prizes or other renowned awards. A 1974 paper and comprehensive review in Reviews of Modern Physics commented that "while no one doubted the [mathematical] correctness of these arguments, no one quite believed that nature was diabolically clever enough to take advantage of them",[76]:9 adding that the theory had so far produced accurate answers that accorded with experiment, but it was unknown whether the theory was fundamentally correct.[76]:9,36(footnote),43–44,47 By 1986 and again in the 1990s it became possible to write that understanding and proving the Higgs sector of the Standard Model was "the central problem today in particle physics".[8][9]

Summary and impact of the PRL papers

The three papers written in 1964 were each recognised as milestone papers during Physical Review Letters 's 50th anniversary celebration.[67] Their six authors were also awarded the 2010 J. J. Sakurai Prize for Theoretical Particle Physics for this work.[77] (A controversy also arose the same year, because in the event of a Nobel Prize only up to three scientists could be recognised, with six being credited for the papers.[78]) Two of the three PRL papers (by Higgs and by GHK) contained equations for the hypothetical field that eventually would become known as the Higgs field and its hypothetical quantum, the Higgs boson.[59][60] Higgs' subsequent 1966 paper showed the decay mechanism of the boson; only a massive boson can decay and the decays can prove the mechanism.

In the paper by Higgs the boson is massive, and in a closing sentence Higgs writes that "an essential feature" of the theory "is the prediction of incomplete multiplets of scalar and vector bosons".[59] (Frank Close comments that 1960s gauge theorists were focused on the problem of massless vector bosons, and the implied existence of a massive scalar boson was not seen as important; only Higgs directly addressed it.[79]:154, 166, 175) In the paper by GHK the boson is massless and decoupled from the massive states.[60] In reviews dated 2009 and 2011, Guralnik states that in the GHK model the boson is massless only in a lowest-order approximation, but it is not subject to any constraint and acquires mass at higher orders, and adds that the GHK paper was the only one to show that there are no massless Goldstone bosons in the model and to give a complete analysis of the general Higgs mechanism.[50][80] All three reached similar conclusions, despite their very different approaches: Higgs' paper essentially used classical techniques, Englert and Brout's involved calculating vacuum polarization in perturbation theory around an assumed symmetry-breaking vacuum state, and GHK used operator formalism and conservation laws to explore in depth the ways in which Goldstone's theorem may be worked around.[51]

5.1.4 Theoretical properties

Main article: Higgs mechanism

Theoretical need for the Higgs

Gauge invariance is an important property of modern particle theories such as the Standard Model, partly due to its success in other areas of fundamental physics such as electromagnetism and the strong interaction (quantum chromodynamics). However, there were great difficulties

"Symmetry breaking illustrated": – At high energy levels (left) the ball settles in the center, and the result is symmetrical. At lower energy levels (right), the overall "rules" remain symmetrical, but the "Mexican hat" potential comes into effect: "local" symmetry inevitably becomes broken since eventually the ball must at random roll one way or another.

in developing gauge theories for the weak nuclear force or a possible unified electroweak interaction. Fermions with a mass term would violate gauge symmetry and therefore cannot be gauge invariant. (This can be seen by examining the Dirac Lagrangian for a fermion in terms of left and right handed components; we find none of the spin-half particles could ever flip helicity as required for mass, so they must be massless.[Note 13]) W and Z bosons are observed to have mass, but a boson mass term contains terms, which clearly depend on the choice of gauge and therefore these masses too cannot be gauge invariant. Therefore, it seems that *none* of the standard model fermions *or* bosons could "begin" with mass as an inbuilt property except by abandoning gauge invariance. If gauge invariance were to be retained, then these particles had to be acquiring their mass by some other mechanism or interaction. Additionally, whatever was giving these particles their mass, had to not "break" gauge invariance as the basis for other parts of the theories where it worked well, *and* had to not require or predict unexpected massless particles and long-range forces (seemingly an inevitable consequence of Goldstone's theorem) which did not actually seem to exist in nature.

A solution to all of these overlapping problems came from the discovery of a previously unnoticed borderline case hidden in the mathematics of Goldstone's theorem,[Note 11] that under certain conditions it *might* theoretically be possible for a symmetry to be broken *without* disrupting gauge invariance and *without* any new massless particles or forces, and having "sensible" (renormalisable) results mathematically: this became known as the Higgs mechanism.

The Standard Model hypothesizes a field which is responsible for this effect, called the Higgs field (symbol: ϕ), which has the unusual property of a non-zero amplitude in its ground state; i.e., a non-zero vacuum expectation value. It can have this effect because of its unusual "Mexican hat" shaped potential whose lowest "point" is not at its "centre". Below a certain extremely high energy level the existence of this non-zero vacuum expectation spontaneously breaks electroweak gauge symmetry which in turn gives rise to the

Higgs mechanism and triggers the acquisition of mass by those particles interacting with the field. This effect occurs because scalar field components of the Higgs field are "absorbed" by the massive bosons as degrees of freedom, and couple to the fermions via Yukawa coupling, thereby producing the expected mass terms. In effect when symmetry breaks under these conditions, the Goldstone bosons that arise *interact* with the Higgs field (and with other particles capable of interacting with the Higgs field) instead of becoming new massless particles, the intractable problems of both underlying theories "neutralise" each other, and the residual outcome is that elementary particles acquire a consistent mass based on how strongly they interact with the Higgs field. It is the simplest known process capable of giving mass to the gauge bosons while remaining compatible with gauge theories.[81] Its quantum would be a scalar boson, known as the Higgs boson.[82]

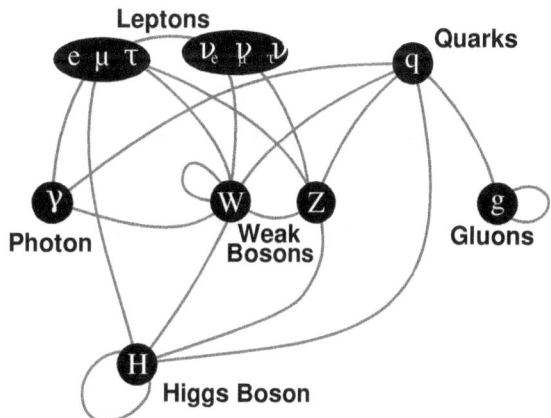

Summary of interactions between certain particles described by the Standard Model.

Properties of the Higgs field

In the Standard Model, the Higgs field is a scalar tachyonic field – 'scalar' meaning it does not transform under Lorentz transformations, and 'tachyonic' meaning the field (but not the particle) has imaginary mass and in certain configurations must undergo symmetry breaking. It consists of four components, two neutral ones and two charged component fields. Both of the charged components and one of the neutral fields are Goldstone bosons, which act as the longitudinal third-polarization components of the massive W+, W−, and Z bosons. The quantum of the remaining neutral component corresponds to (and is theoretically realised as) the massive Higgs boson,[83] this component can interact with fermions via Yukawa coupling to give them mass, as well.

Mathematically, the Higgs field has imaginary mass and is therefore a tachyonic field.[84] While tachyons (particles that move faster than light) are a purely hypothetical concept, fields with imaginary mass have come to play an important role in modern physics.[85][86] Under no circumstances do any excitations ever propagate faster than light in such theories — the presence or absence of a tachyonic mass has no effect whatsoever on the maximum velocity of signals (there is no violation of causality).[87] Instead of faster-than-light particles, the imaginary mass creates an instability:- any configuration in which one or more field excitations are tachyonic must spontaneously decay, and the resulting configuration contains no physical tachyons. This process is known as tachyon condensation, and is now believed to be the explanation for how the Higgs mechanism itself arises in nature, and therefore the reason behind electroweak symmetry breaking.

Although the notion of imaginary mass might seem troubling, it is only the field, and not the mass itself, that is quantized. Therefore, the field operators at spacelike separated points still commute (or anticommute), and information and particles still do not propagate faster than light.[88] Tachyon condensation drives a physical system that has reached a local limit and might naively be expected to produce physical tachyons, to an alternate stable state where no physical tachyons exist. Once a tachyonic field such as the Higgs field reaches the minimum of the potential, its quanta are not tachyons any more but rather are ordinary particles such as the Higgs boson.[89]

Properties of the Higgs boson

Since the Higgs field is scalar, the Higgs boson has no spin. The Higgs boson is also its own antiparticle and is CP-even, and has zero electric and colour charge.[90]

The Minimal Standard Model does not predict the mass of the Higgs boson.[91] If that mass is between 115 and 180 GeV/c^2, then the Standard Model can be valid at energy scales all the way up to the Planck scale (10^{19} GeV).[92] Many theorists expect new physics beyond the Standard Model to emerge at the TeV-scale, based on unsatisfactory properties of the Standard Model.[93] The highest possible mass scale allowed for the Higgs boson (or some other electroweak symmetry breaking mechanism) is 1.4 TeV; beyond this point, the Standard Model becomes inconsistent without such a mechanism, because unitarity is violated in certain scattering processes.[94]

It is also possible, although experimentally difficult, to estimate the mass of the Higgs boson indirectly. In the Standard Model, the Higgs boson has a number of indirect effects; most notably, Higgs loops result in tiny corrections to masses of W and Z bosons. Precision measurements of electroweak parameters, such as the Fermi constant and masses of W/Z bosons, can be used to calculate constraints on the mass of the Higgs. As of July 2011, the preci-

sion electroweak measurements tell us that the mass of the Higgs boson is likely to be less than about 161 GeV/c^2 at 95% confidence level (this upper limit would increase to 185 GeV/c^2 if the lower bound of 114.4 GeV/c^2 from the LEP-2 direct search is allowed for[95]). These indirect constraints rely on the assumption that the Standard Model is correct. It may still be possible to discover a Higgs boson above these masses if it is accompanied by other particles beyond those predicted by the Standard Model.[96]

Production

If Higgs particle theories are valid, then a Higgs particle can be produced much like other particles that are studied, in a particle collider. This involves accelerating a large number of particles to extremely high energies and extremely close to the speed of light, then allowing them to smash together. Protons and lead ions (the bare nuclei of lead atoms) are used at the LHC. In the extreme energies of these collisions, the desired esoteric particles will occasionally be produced and this can be detected and studied; any absence or difference from theoretical expectations can also be used to improve the theory. The relevant particle theory (in this case the Standard Model) will determine the necessary kinds of collisions and detectors. The Standard Model predicts that Higgs bosons could be formed in a number of ways,[97][98][99] although the probability of producing a Higgs boson in any collision is always expected to be very small—for example, only 1 Higgs boson per 10 billion collisions in the Large Hadron Collider.[Note 14] The most common expected processes for Higgs boson production are:

- *Gluon fusion*. If the collided particles are hadrons such as the proton or antiproton—as is the case in the LHC and Tevatron—then it is most likely that two of the gluons binding the hadron together collide. The easiest way to produce a Higgs particle is if the two gluons combine to form a loop of virtual quarks. Since the coupling of particles to the Higgs boson is proportional to their mass, this process is more likely for heavy particles. In practice it is enough to consider the contributions of virtual top and bottom quarks (the heaviest quarks). This process is the dominant contribution at the LHC and Tevatron being about ten times more likely than any of the other processes.[97][98]

- *Higgs Strahlung*. If an elementary fermion collides with an anti-fermion—e.g., a quark with an anti-quark or an electron with a positron—the two can merge to form a virtual W or Z boson which, if it carries sufficient energy, can then emit a Higgs boson. This process was the dominant production mode at the LEP, where an electron and a positron collided to form a virtual Z boson, and it was the second largest contribution for Higgs production at the Tevatron. At the LHC this process is only the third largest, because the LHC collides protons with protons, making a quark–antiquark collision less likely than at the Tevatron. Higgs Strahlung is also known as *associated production*.[97][98][99]

- *Weak boson fusion.* Another possibility when two (anti-)fermions collide is that the two exchange a virtual W or Z boson, which emits a Higgs boson. The colliding fermions do not need to be the same type. So, for example, an up quark may exchange a Z boson with an anti-down quark. This process is the second most important for the production of Higgs particle at the LHC and LEP.[97][99]

- *Top fusion.* The final process that is commonly considered is by far the least likely (by two orders of magnitude). This process involves two colliding gluons, which each decay into a heavy quark–antiquark pair. A quark and antiquark from each pair can then combine to form a Higgs particle.[97][98]

Decay

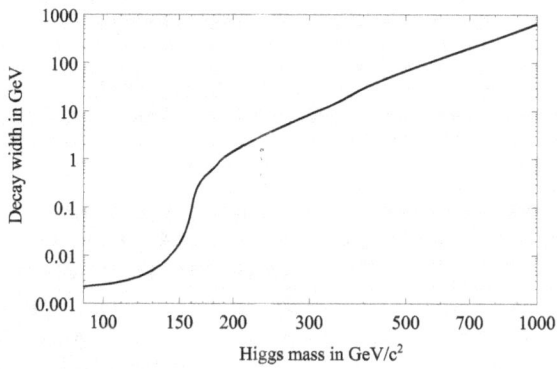

The Standard Model prediction for the decay width of the Higgs particle depends on the value of its mass.

Quantum mechanics predicts that if it is possible for a particle to decay into a set of lighter particles, then it will eventually do so.[101] This is also true for the Higgs boson. The likelihood with which this happens depends on a variety of factors including: the difference in mass, the strength of the interactions, etc. Most of these factors are fixed by the Standard Model, except for the mass of the Higgs boson itself. For a Higgs boson with a mass of 126 GeV/c^2 the SM predicts a mean life time of about 1.6×10^{-22} s.[Note 2]

Since it interacts with all the massive elementary particles of the SM, the Higgs boson has many different processes

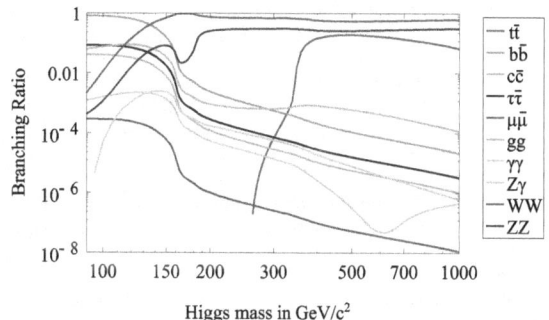

The Standard Model prediction for the branching ratios of the different decay modes of the Higgs particle depends on the value of its mass.

through which it can decay. Each of these possible processes has its own probability, expressed as the *branching ratio*; the fraction of the total number decays that follows that process. The SM predicts these branching ratios as a function of the Higgs mass (see plot).

One way that the Higgs can decay is by splitting into a fermion–antifermion pair. As general rule, the Higgs is more likely to decay into heavy fermions than light fermions, because the mass of a fermion is proportional to the strength of its interaction with the Higgs.[102] By this logic the most common decay should be into a top–antitop quark pair. However, such a decay is only possible if the Higgs is heavier than ~346 GeV/c^2, twice the mass of the top quark. For a Higgs mass of 126 GeV/c^2 the SM predicts that the most common decay is into a bottom–antibottom quark pair, which happens 56.1% of the time.[5] The second most common fermion decay at that mass is a tau–antitau pair, which happens only about 6% of the time.[5]

Another possibility is for the Higgs to split into a pair of massive gauge bosons. The most likely possibility is for the Higgs to decay into a pair of W bosons (the light blue line in the plot), which happens about 23.1% of the time for a Higgs boson with a mass of 126 GeV/c^2.[5] The W bosons can subsequently decay either into a quark and an antiquark or into a charged lepton and a neutrino. However, the decays of W bosons into quarks are difficult to distinguish from the background, and the decays into leptons cannot be fully reconstructed (because neutrinos are impossible to detect in particle collision experiments). A cleaner signal is given by decay into a pair of Z-bosons (which happens about 2.9% of the time for a Higgs with a mass of 126 GeV/c^2),[5] if each of the bosons subsequently decays into a pair of easy-to-detect charged leptons (electrons or muons).

Decay into massless gauge bosons (i.e., gluons or photons) is also possible, but requires intermediate loop of virtual heavy quarks (top or bottom) or massive gauge bosons.[102] The most common such process is the decay into a pair of gluons through a loop of virtual heavy quarks. This process, which is the reverse of the gluon fusion process mentioned above, happens approximately 8.5% of the time for a Higgs boson with a mass of 126 GeV/c^2.[5] Much rarer is the decay into a pair of photons mediated by a loop of W bosons or heavy quarks, which happens only twice for every thousand decays.[5] However, this process is very relevant for experimental searches for the Higgs boson, because the energy and momentum of the photons can be measured very precisely, giving an accurate reconstruction of the mass of the decaying particle.[102]

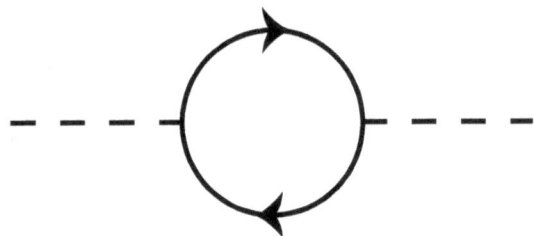

A one-loop Feynman diagram of the first-order correction to the Higgs mass. In the Standard Model the effects of these corrections are potentially enormous, giving rise to the so-called hierarchy problem.

Alternative models

Main article: Alternatives to the Standard Model Higgs

The Minimal Standard Model as described above is the simplest known model for the Higgs mechanism with just one Higgs field. However, an extended Higgs sector with additional Higgs particle doublets or triplets is also possible, and many extensions of the Standard Model have this feature. The non-minimal Higgs sector favoured by theory are the two-Higgs-doublet models (2HDM), which predict the existence of a quintet of scalar particles: two CP-even neutral Higgs bosons h^0 and H^0, a CP-odd neutral Higgs boson A^0, and two charged Higgs particles $H^\pm$. Supersymmetry ("SUSY") also predicts relations between the Higgs-boson masses and the masses of the gauge bosons, and could accommodate a 125 GeV/c^2 neutral Higgs boson.

The key method to distinguish between these different models involves study of the particles' interactions ("coupling") and exact decay processes ("branching ratios"), which can be measured and tested experimentally in particle collisions. In the Type-I 2HDM model one Higgs doublet couples to up and down quarks, while the second doublet does not couple to quarks. This model has two interesting limits, in which the lightest Higgs couples to just fermions ("gauge-phobic") or just gauge bosons ("fermiophobic"), but not both. In the Type-II 2HDM model, one Higgs doublet only couples to up-type quarks, the other only couples to down-type quarks.[103] The heavily researched Minimal Supersymmetric Standard Model (MSSM) includes a Type-II 2HDM Higgs sector, so it could be disproven by evidence of a Type-I 2HDM Higgs.

In other models the Higgs scalar is a composite particle. For example, in technicolor the role of the Higgs field is played by strongly bound pairs of fermions called techniquarks. Other models, feature pairs of top quarks (see top quark condensate). In yet other models, there is no Higgs field at all and the electroweak symmetry is broken using extra dimensions.[104][105]

Further theoretical issues and hierarchy problem

Main articles: Hierarchy problem and Hierarchy problem § The Higgs mass

The Standard Model leaves the mass of the Higgs boson as a parameter to be measured, rather than a value to be calculated. This is seen as theoretically unsatisfactory, particularly as quantum corrections (related to interactions with virtual particles) should apparently cause the Higgs particle to have a mass immensely higher than that observed, but at the same time the Standard Model requires a mass of the order of 100 to 1000 GeV to ensure unitarity (in this case, to unitarise longitudinal vector boson scattering).[106] Reconciling these points appears to require explaining why there is an almost-perfect cancellation resulting in the visible mass of ~ 125 GeV, and it is not clear how to do this. Because the weak force is about 10^{32} times stronger than gravity, and (linked to this) the Higgs boson's mass is so much less than the Planck mass or the grand unification energy, it appears that either there is some underlying connection or reason for these observations which is unknown and not described by the Standard Model, or some unexplained and extremely precise fine-tuning of parameters – however at present neither of these explanations is proven. This is known as a hierarchy problem.[107] More broadly, the hierarchy problem amounts to the worry that a future theory of fundamental particles and interactions should not have excessive fine-tunings or unduly delicate cancellations, and should allow masses of particles such as the Higgs boson to be calculable. The problem is in some ways unique to spin-0 particles (such as the Higgs boson), which can give rise to issues related to quantum corrections that do not affect particles with spin.[106] A number of solutions have been proposed, including supersymmetry, conformal solutions and solutions via extra dimensions such as braneworld models.

There are also issues of quantum triviality, which suggests that it may not be possible to create a consistent quantum field theory involving elementary scalar particles.

5.1.5 Experimental search

Main article: Search for the Higgs boson

To produce Higgs bosons, two beams of particles are accelerated to very high energies and allowed to collide within a particle detector. Occasionally, although rarely, a Higgs boson will be created fleetingly as part of the collision byproducts. Because the Higgs boson decays very quickly, particle detectors cannot detect it directly. Instead the detectors register all the decay products (the *decay signature*) and from the data the decay process is reconstructed. If the observed decay products match a possible decay process (known as a *decay channel*) of a Higgs boson, this indicates that a Higgs boson may have been created. In practice, many processes may produce similar decay signatures. Fortunately, the Standard Model precisely predicts the likelihood of each of these, and each known process, occurring. So, if the detector detects more decay signatures consistently matching a Higgs boson than would otherwise be expected if Higgs bosons did not exist, then this would be strong evidence that the Higgs boson exists.

Because Higgs boson production in a particle collision is likely to be very rare (1 in 10 billion at the LHC),[Note 14] and many other possible collision events can have similar decay signatures, the data of hundreds of trillions of collisions needs to be analysed and must "show the same picture" before a conclusion about the existence of the Higgs boson can be reached. To conclude that a new particle has been found, particle physicists require that the statistical analysis of two independent particle detectors each indicate that there is lesser than a one-in-a-million chance that the observed decay signatures are due to just background random Standard Model events—i.e., that the observed number of events is more than 5 standard deviations (sigma) different from that expected if there was no new particle. More collision data allows better confirmation of the physical properties of any new particle observed, and allows physicists to decide whether it is indeed a Higgs boson as described by the Standard Model or some other hypothetical new particle.

To find the Higgs boson, a powerful particle accelerator was needed, because Higgs bosons might not be seen in lower-energy experiments. The collider needed to have a high luminosity in order to ensure enough collisions were seen for conclusions to be drawn. Finally, advanced computing facilities were needed to process the vast amount of data (25 petabytes per year as at 2012) produced by the collisions.[109] For the announcement of 4 July 2012, a new collider known as the Large Hadron Collider was constructed at CERN with a planned eventual collision energy of 14 TeV—over seven times any previous collider—and over 300 trillion (3×10^{14}) LHC proton–proton col-

lisions were analysed by the LHC Computing Grid, the world's largest computing grid (as of 2012), comprising over 170 computing facilities in a worldwide network across 36 countries.[109][110][111]

Search prior to 4 July 2012

The first extensive search for the Higgs boson was conducted at the Large Electron–Positron Collider (LEP) at CERN in the 1990s. At the end of its service in 2000, LEP had found no conclusive evidence for the Higgs.[Note 15] This implied that if the Higgs boson were to exist it would have to be heavier than 114.4 GeV/c^2.[112]

The search continued at Fermilab in the United States, where the Tevatron—the collider that discovered the top quark in 1995—had been upgraded for this purpose. There was no guarantee that the Tevatron would be able to find the Higgs, but it was the only supercollider that was operational since the Large Hadron Collider (LHC) was still under construction and the planned Superconducting Super Collider had been cancelled in 1993 and never completed. The Tevatron was only able to exclude further ranges for the Higgs mass, and was shut down on 30 September 2011 because it no longer could keep up with the LHC. The final analysis of the data excluded the possibility of a Higgs boson with a mass between 147 GeV/c^2 and 180 GeV/c^2. In addition, there was a small (but not significant) excess of events possibly indicating a Higgs boson with a mass between 115 GeV/c^2 and 140 GeV/c^2.[113]

The Large Hadron Collider at CERN in Switzerland, was designed specifically to be able to either confirm or exclude the existence of the Higgs boson. Built in a 27 km tunnel under the ground near Geneva originally inhabited by LEP, it was designed to collide two beams of protons, initially at energies of 3.5 TeV per beam (7 TeV total), or almost 3.6 times that of the Tevatron, and upgradeable to 2×7 TeV (14 TeV total) in future. Theory suggested if the Higgs boson existed, collisions at these energy levels should be able to reveal it. As one of the most complicated scientific instruments ever built, its operational readiness was delayed for 14 months by a magnet quench event nine days after its inaugural tests, caused by a faulty electrical connection that damaged over 50 superconducting magnets and contaminated the vacuum system.[114][115][116]

Data collection at the LHC finally commenced in March 2010.[117] By December 2011 the two main particle detectors at the LHC, ATLAS and CMS, had narrowed down the mass range where the Higgs could exist to around 116-130 GeV (ATLAS) and 115-127 GeV (CMS).[118][119] There had also already been a number of promising event excesses that had "evaporated" and proven to be nothing but random fluctuations. However, from around May 2011,[120] both experiments had seen among their results, the slow emergence of a small yet consistent excess of gamma and 4-lepton decay signatures and several other particle decays, all hinting at a new particle at a mass around 125 GeV.[120] By around November 2011, the anomalous data at 125 GeV was becoming "too large to ignore" (although still far from conclusive), and the team leaders at both ATLAS and CMS each privately suspected they might have found the Higgs.[120] On November 28, 2011, at an internal meeting of the two team leaders and the director general of CERN, the latest analyses were discussed outside their teams for the first time, suggesting both ATLAS and CMS might be converging on a possible shared result at 125 GeV, and initial preparations commenced in case of a successful finding.[120] While this information was not known publicly at the time, the narrowing of the possible Higgs range to around 115–130 GeV and the repeated observation of small but consistent event excesses across multiple channels at both ATLAS and CMS in the 124-126 GeV region (described as "tantalising hints" of around 2-3 sigma) were public knowledge with "a lot of interest".[121] It was therefore widely anticipated around the end of 2011, that the LHC would provide sufficient data to either exclude or confirm the finding of a Higgs boson by the end of 2012, when their 2012 collision data (with slightly higher 8 TeV collision energy) had been examined.[121][122]

Discovery of candidate boson at CERN

On 22 June 2012 CERN announced an upcoming seminar covering tentative findings for 2012,[126][127] and shortly afterwards (from around 1 July 2012 according to an analysis of the spreading rumour in social media[128]) rumours began to spread in the media that this would include a major announcement, but it was unclear whether this would be a stronger signal or a formal discovery.[129][130] Speculation escalated to a "fevered" pitch when reports emerged that Peter Higgs, who proposed the particle, was to be attending the seminar,[131][132] and that "five leading physicists" had been invited – generally believed to signify the five living 1964 authors – with Higgs, Englert, Guralnik, Hagen attending and Kibble confirming his invitation (Brout having died in 2011).[133][134]

On 4 July 2012 both of the CERN experiments announced they had independently made the same discovery:[135] CMS of a previously unknown boson with mass 125.3 ± 0.6 GeV/c^2[136][137] and ATLAS of a boson with mass 126.0 ± 0.6 GeV/c^2.[138][139] Using the combined analysis of two interaction types (known as 'channels'), both experiments independently reached a local significance of 5 sigma — implying that the probability of getting at least as strong a result by chance alone is less than 1 in 3 million. When additional channels were taken into account, the CMS sig-

nificance was reduced to 4.9 sigma.[137]

The two teams had been working 'blinded' from each other from around late 2011 or early 2012,[120] meaning they did not discuss their results with each other, providing additional certainty that any common finding was genuine validation of a particle.[109] This level of evidence, confirmed independently by two separate teams and experiments, meets the formal level of proof required to announce a confirmed discovery.

On 31 July 2012, the ATLAS collaboration presented additional data analysis on the "observation of a new particle", including data from a third channel, which improved the significance to 5.9 sigma (1 in 588 million chance of obtaining at least as strong evidence by random background effects alone) and mass 126.0 ± 0.4 (stat) ± 0.4 (sys) GeV/c^2, [139] and CMS improved the significance to 5-sigma and mass 125.3 ± 0.4 (stat) ± 0.5 (sys) GeV/c^2.[136]

The new particle tested as a possible Higgs boson

Following the 2012 discovery, it was still unconfirmed whether or not the 125 GeV/c^2 particle was a Higgs boson. On one hand, observations remained consistent with the observed particle being the Standard Model Higgs boson, and the particle decayed into at least some of the predicted channels. Moreover, the production rates and branching ratios for the observed channels broadly matched the predictions by the Standard Model within the experimental uncertainties. However, the experimental uncertainties currently still left room for alternative explanations, meaning an announcement of the discovery of a Higgs boson would have been premature.[102] To allow more opportunity for data collection, the LHC's proposed 2012 shutdown and 2013–14 upgrade were postponed by 7 weeks into 2013.[140]

In November 2012, in a conference in Kyoto researchers said evidence gathered since July was falling into line with the basic Standard Model more than its alternatives, with a range of results for several interactions matching that theory's predictions.[141] Physicist Matt Strassler highlighted "considerable" evidence that the new particle is not a pseudoscalar negative parity particle (consistent with this required finding for a Higgs boson), "evaporation" or lack of increased significance for previous hints of non-Standard Model findings, expected Standard Model interactions with W and Z bosons, absence of "significant new implications" for or against supersymmetry, and in general no significant deviations to date from the results expected of a Standard Model Higgs boson.[142] However some kinds of extensions to the Standard Model would also show very similar results;[143] so commentators noted that based on other particles that are still being understood long after their discovery, it may take years to be sure, and decades to fully understand the particle that has been found.[141][142]

These findings meant that as of January 2013, scientists were very sure they had found an unknown particle of mass ~ 125 GeV/c^2, and had not been misled by experimental error or a chance result. They were also sure, from initial observations, that the new particle was some kind of boson. The behaviours and properties of the particle, so far as examined since July 2012, also seemed quite close to the behaviours expected of a Higgs boson. Even so, it could still have been a Higgs boson or some other unknown boson, since future tests could show behaviours that do not match a Higgs boson, so as of December 2012 CERN still only stated that the new particle was "consistent with" the Higgs boson,[11][13] and scientists did not yet positively say it was the Higgs boson.[144] Despite this, in late 2012, widespread media reports announced (incorrectly) that a Higgs boson had been confirmed during the year.[Note 16]

In January 2013, CERN director-general Rolf-Dieter Heuer stated that based on data analysis to date, an answer could be possible 'towards' mid-2013,[150] and the deputy chair of physics at Brookhaven National Laboratory stated in February 2013 that a "definitive" answer might require "another few years" after the collider's 2015 restart.[151] In early March 2013, CERN Research Director Sergio Bertolucci stated that confirming spin-0 was the major remaining requirement to determine whether the particle is at least some kind of Higgs boson.[152]

Preliminary confirmation of existence and current status

On 14 March 2013 CERN confirmed that:

> "CMS and ATLAS have compared a number of options for the spin-parity of this particle, and these all prefer no spin and even parity [two fundamental criteria of a Higgs boson consistent with the Standard Model]. This, coupled with the measured interactions of the new particle with other particles, strongly indicates that it is a Higgs boson." [1]

This also makes the particle the first elementary scalar particle to be discovered in nature.[14]

Examples of tests used to validate whether the 125 GeV particle is a Higgs boson:[142][153]

5.1.6 Public discussion

Naming

Names used by physicists The name most strongly associated with the particle and field is the Higgs boson[79]:168 and Higgs field. For some time the particle was known by a combination of its PRL author names (including at times Anderson), for example the Brout–Englert–Higgs particle, the Anderson-Higgs particle, or the Englert–Brout–Higgs–Guralnik–Hagen–Kibble mechanism,[Note 17] and these are still used at times.[51][160] Fueled in part by the issue of recognition and a potential shared Nobel Prize,[160][161] the most appropriate name is still occasionally a topic of debate as at 2012.[160] (Higgs himself prefers to call the particle either by an acronym of all those involved, or "the scalar boson", or "the so-called Higgs particle".[161])

A considerable amount has been written on how Higgs' name came to be exclusively used. Two main explanations are offered.

Nickname The Higgs boson is often referred to as the "God particle" in popular media outside the scientific community.[170][171][172][173][174] The nickname comes from the title of the 1993 book on the Higgs boson and particle physics - The God Particle: If the Universe Is the Answer, What Is the Question? by Nobel Physics prizewinner and Fermilab director Leon Lederman.[21] Lederman wrote it in the context of failing US government support for the Superconducting Super Collider,[175] a part-constructed titanic[176][177] competitor to the Large Hadron Collider with planned collision energies of 2 × 20 TeV that was championed by Lederman since its 1983 inception[175][178][179] and shut down in 1993. The book sought in part to promote awareness of the significance and need for such a project in the face of its possible loss of funding.[180] Lederman, a leading researcher in the field, wanted to title his book "The Goddamn Particle: If the Universe is the Answer, What is the Question?" But his editor decided that the title was too controversial and convinced Lederman to change the title to "The God Particle: If the Universe is the Answer, What is the Question?"[181]

And since the Higgs Boson deals with how matter was formed at the time of the big bang, and since newspapers loved the term, the term "God particle" was used.

While media use of this term may have contributed to wider awareness and interest,[182] many scientists feel the name is inappropriate[16][17][183] since it is sensational hyperbole and misleads readers;[184] the particle also has nothing to do with God, leaves open numerous questions in fundamental physics, and does not explain the ultimate origin of the universe. Higgs, an atheist, was reported to be displeased and stated in a 2008 interview that he found it "embarrassing" because it was "the kind of misuse... which I think might offend some people".[184][185][186] Science writer Ian Sample stated in his 2010 book on the search that the nickname is "universally hate[d]" by physicists and perhaps the "worst derided" in the history of physics, but that (according to Lederman) the publisher rejected all titles mentioning "Higgs" as unimaginative and too unknown.[187]

Lederman begins with a review of the long human search for knowledge, and explains that his tongue-in-cheek title draws an analogy between the impact of the Higgs field on the fundamental symmetries at the Big Bang, and the apparent chaos of structures, particles, forces and interactions that resulted and shaped our present universe, with the biblical story of Babel in which the primordial single language of early Genesis was fragmented into many disparate languages and cultures.[188]

> Today ... we have the standard model, which reduces all of reality to a dozen or so particles and four forces. ... It's a hard-won simplicity [...and...] remarkably accurate. But it is also incomplete and, in fact, internally inconsistent... This boson is so central to the state of physics today, so crucial to our final understanding of the structure of matter, yet so elusive, that I have given it a nickname: the God Particle. Why God Particle? Two reasons. One, the publisher wouldn't let us call it the Goddamn Particle, though that might be a more appropriate title, given its villainous nature and the expense it is causing. And two, there is a connection, of sorts, to another book, a *much* older one...
> — Leon M. Lederman and Dick Teresi, *The God Particle: If the Universe is the Answer, What is the Question*[21] p. 22

Lederman asks whether the Higgs boson was added just to perplex and confound those seeking knowledge of the universe, and whether physicists will be confounded by it as recounted in that story, or ultimately surmount the challenge and understand "how beautiful is the universe [God has] made".[189]

Other proposals   A renaming competition by British newspaper *The Guardian* in 2009 resulted in their science correspondent choosing the name "the champagne bottle boson" as the best submission: "The bottom of a champagne bottle is in the shape of the Higgs potential and is often used as an illustration in physics lectures. So it's not an embarrassingly grandiose name, it is memorable, and [it] has some physics connection too."[190] The name *Higgson*

was suggested as well, in an opinion piece in the Institute of Physics' online publication *physicsworld.com*.[191]

Media explanations and analogies

There has been considerable public discussion of analogies and explanations for the Higgs particle and how the field creates mass,[192][193] including coverage of explanatory attempts in their own right and a competition in 1993 for the best popular explanation by then-UK Minister for Science Sir William Waldegrave[194] and articles in newspapers worldwide.

Photograph of light passing through a dispersive prism: the rainbow effect arises because photons are not all affected to the same degree by the dispersive material of the prism.

An educational collaboration involving an LHC physicist and a High School Teachers at CERN educator suggests that dispersion of light – responsible for the rainbow and dispersive prism – is a useful analogy for the Higgs field's symmetry breaking and mass-causing effect.[195]

Matt Strassler uses electric fields as an analogy:[196]

> Some particles interact with the Higgs field while others don't. Those particles that feel the Higgs field act as if they have mass. Something

similar happens in an electric field – charged objects are pulled around and neutral objects can sail through unaffected. So you can think of the Higgs search as an attempt to make waves in the Higgs field *[create Higgs bosons]* to prove it's really there.

A similar explanation was offered by *The Guardian*:[197]

> The Higgs boson is essentially a ripple in a field said to have emerged at the birth of the universe and to span the cosmos to this day ... The particle is crucial however: it is the smoking gun, the evidence required to show the theory is right.

The Higgs field's effect on particles was famously described by physicist David Miller as akin to a room full of political party workers spread evenly throughout a room: the crowd gravitates to and slows down famous people but does not slow down others.[Note 18] He also drew attention to well-known effects in solid state physics where an electron's effective mass can be much greater than usual in the presence of a crystal lattice.[198]

Analogies based on drag effects, including analogies of "syrup" or "molasses" are also well known, but can be somewhat misleading since they may be understood (incorrectly) as saying that the Higgs field simply resists some particles' motion but not others' – a simple resistive effect could also conflict with Newton's third law.[200]

Recognition and awards

There has been considerable discussion of how to allocate the credit if the Higgs boson is proven, made more pointed as a Nobel prize had been expected, and the very wide basis of people entitled to consideration. These include a range of theoreticians who made the Higgs mechanism theory possible, the theoreticians of the 1964 PRL papers (including Higgs himself), the theoreticians who derived from these, a working electroweak theory and the Standard Model itself, and also the experimentalists at CERN and other institutions who made possible the proof of the Higgs field and boson in reality. The Nobel prize has a limit of 3 persons to share an award, and some possible winners are already prize holders for other work, or are deceased (the prize is only awarded to persons in their lifetime). Existing prizes for works relating to the Higgs field, boson, or mechanism include:

- Nobel Prize in Physics (1979) – Glashow, Salam, and Weinberg, *for contributions to the theory of the unified weak and electromagnetic interaction between elementary particles* [201]

- Nobel Prize in Physics (1999) – 't Hooft and Veltman, *for elucidating the quantum structure of electroweak interactions in physics* [202]

- Nobel Prize in Physics (2008) – Nambu (shared), *for the discovery of the mechanism of spontaneous broken symmetry in subatomic physics* [53]

- J. J. Sakurai Prize for Theoretical Particle Physics (2010) – Hagen, Englert, Guralnik, Higgs, Brout, and Kibble, *for elucidation of the properties of spontaneous symmetry breaking in four-dimensional relativistic gauge theory and of the mechanism for the consistent generation of vector boson masses* [77] (for the 1964 papers described above)

- Wolf Prize (2004) – Englert, Brout, and Higgs

- Nobel Prize in Physics (2013) - Peter Higgs and François Englert, *for the theoretical discovery of a mechanism that contributes to our understanding of the origin of mass of subatomic particles, and which recently was confirmed through the discovery of the predicted fundamental particle, by the ATLAS and CMS experiments at CERN's Large Hadron Collider* [203]

Additionally Physical Review Letters' 50-year review (2008) recognized the 1964 PRL symmetry breaking papers and Weinberg's 1967 paper *A model of Leptons* (the most cited paper in particle physics, as of 2012) "milestone Letters".[75]

Following reported observation of the Higgs-like particle in July 2012, several Indian media outlets reported on the supposed neglect of credit to Indian physicist Satyendra Nath Bose after whose work in the 1920s the class of particles "bosons" is named[204][205] (although physicists have described Bose's connection to the discovery as tenuous).[206]

5.1.7 Technical aspects and mathematical formulation

See also: Standard Model (mathematical formulation)

In the Standard Model, the Higgs field is a four-component scalar field that forms a complex doublet of the weak isospin SU(2) symmetry:

while the field has charge $+1/2$ under the weak hypercharge U(1) symmetry (in the convention where the electric charge, Q, the weak isospin, I_3, and the weak hypercharge, Y, are related by $Q = I_3 + Y$).[207]

The Higgs part of the Lagrangian is[207]

where $(d, u, e, \nu)^i_{L,R}$ are left-handed and right-handed quarks and leptons of the ith generation, $\lambda^{ij}_{u,d,e}$ are matrices of Yukawa couplings where h.c. denotes the hermitian conjugate terms. In the symmetry breaking ground state, only the terms containing ϕ^0 remain, giving rise to mass terms for the fermions. Rotating the quark and lepton fields to the basis where the matrices of Yukawa couplings are diagonal, one gets

where the masses of the fermions are $m^i_{u,d,e} = \lambda^i_{u,d,e} v / \sqrt{2}$, and $\lambda^i_{u,d,e}$ denote the eigenvalues of the Yukawa matrices.[207]

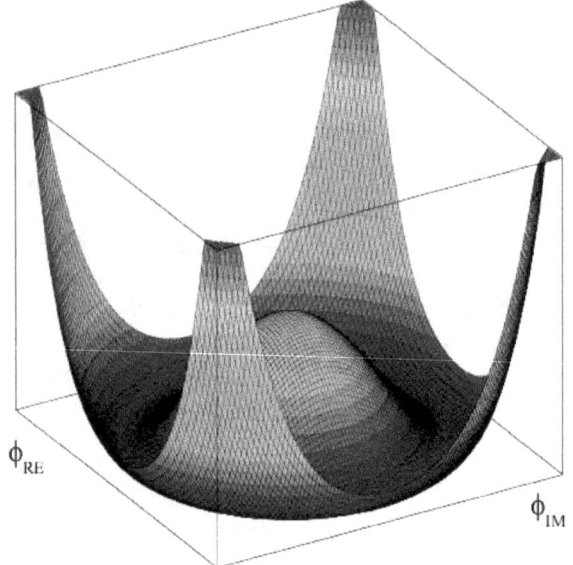

The potential for the Higgs field, plotted as function of ϕ^0 and ϕ^3. It has a Mexican-hat *or* champagne-bottle profile *at the ground.*

where W^a_μ and B_μ are the gauge bosons of the SU(2) and U(1) symmetries, g and g' their respective coupling constants, $\tau^a = \sigma^a/2$ (where σ^a are the Pauli matrices) a complete set generators of the SU(2) symmetry, and $\lambda > 0$ and $\mu^2 > 0$, so that the ground state breaks the SU(2) symmetry (see figure). The ground state of the Higgs field (the bottom of the potential) is degenerate with different ground states related to each other by a SU(2) gauge transformation. It is always possible to pick a gauge such that in the ground state $\phi^1 = \phi^2 = \phi^3 = 0$. The expectation value of ϕ^0 in the ground state (the vacuum expectation value or vev) is then $\langle \phi^0 \rangle = v$, where $v = \frac{|\mu|}{\sqrt{\lambda}}$. The measured value of this parameter is ~246 GeV/c^2.[102] It has units of mass, and is the only free parameter of the Standard Model that is not a dimensionless number. Quadratic terms in W_μ and B_μ arise, which give masses to the W and Z bosons:[207]

with their ratio determining the Weinberg angle, $\cos\theta_W = \frac{M_W}{M_Z} = \frac{|g|}{\sqrt{g^2 + g'^2}}$, and leave a massless U(1) photon, γ.

The quarks and the leptons interact with the Higgs field through Yukawa interaction terms:

5.1.8   See also

Standard Model

- Quantum gauge theory

- History of quantum field theory

- Introduction to quantum mechanics

- Noncommutative standard model and noncommutative geometry generally

- Standard Model (mathematical formulation) (and especially Standard Model fields overview and mass terms and the Higgs mechanism)

Other

- Bose–Einstein statistics

- Dalitz plot

- Higgs boson in fiction

- Quantum triviality

- ZZ diboson

- Scalar boson

- Stueckelberg action

- Tachyonic field

5.1.9 Notes

[1] Note that such events also occur due to other processes. Detection involves a statistically significant excess of such events at specific energies.

[2] In the Standard Model, the total decay width of a Higgs boson with a mass of 126 GeV/c^2 is predicted to be 4.21×10^{-3} GeV.[5] The mean lifetime is given by $\tau = \hbar / \Gamma$.

[3] The range of a force is inversely proportional to the mass of the particles transmitting it.[19] In the Standard Model, forces are carried by virtual particles. These particles' movement and interactions with each other are limited by the energy–time uncertainty principle. As a result, the more massive a single virtual particle is, the greater its energy, and therefore the shorter the distance it can travel. A particle's mass therefore determines the maximum distance at which it can interact with other particles and on any force it mediates. By the same token, the reverse is also true: massless and near-massless particles can carry long distance forces. *(See also: Compton wavelength and Static forces and virtual-particle exchange)* Since experiments have shown that the weak force acts over only a very short range, this implies that there must exist massive gauge bosons. And indeed, their masses have since been confirmed by measurement.

[4] It is quite common for a law of physics to hold true only if certain assumptions held true or only under certain conditions. For example, Newton's laws of motion apply only at speeds where relativistic effects are negligible; and laws related to conductivity, gases, and classical physics (as opposed to quantum mechanics) may apply only within certain ranges of size, temperature, pressure, or other conditions.

[5] Electroweak symmetry is broken by the Higgs field in its lowest energy state, called its "ground state". At high energy levels this does not happen, and the gauge bosons of the weak force would therefore be expected to be massless.

[6] By the 1960s, many had already started to see gauge theories as failing to explain particle physics because theorists had been unable to solve the mass problem or even explain how gauge theory could provide a solution. So the idea that the Standard Model – which relied on a Higgs field, not yet proved to exist – could be fundamentally incorrect. Against this, once the model was developed around 1972, no better theory existed, and its predictions and solutions were so accurate, that it became the preferred theory anyway. It then became crucial to science, to know whether it was *correct*.

[7] The success of the Higgs-based electroweak theory and Standard Model is illustrated by their predictions of the mass of two particles later detected: the W boson (predicted mass: 80.390 ± 0.018 GeV, experimental measurement: 80.387 ± 0.019 GeV), and the Z boson (predicted mass: 91.1874 ± 0.0021, experimental measurement: 91.1876 ± 0.0021 GeV). The existence of the Z boson was itself another prediction. Other accurate predictions included the weak neutral current, the gluon, and the top and charm quarks, all later proven to exist as the theory said.

[8] For example, Huffington Post/Reuters[35] and others[36][37]

[9] The bubble's effects would be expected to propagate across the universe at the speed of light from wherever it occurred. However space is vast – with even the nearest galaxy being over 2 million lightyears from us, and others being many billions of lightyears distant, so the effect of such an event would be unlikely to arise here for billions of years after first occurring.[39][40]

[10] If the Standard Model is valid, then the particles and forces we observe in our universe exist as they do, because of underlying quantum fields. Quantum fields can have states of differing stability, including 'stable', 'unstable' and 'metastable' states (the latter remain stable unless sufficiently perturbed). If a more stable vacuum state were able to arise, then existing particles and forces would no longer arise as they presently do. Different particles or forces would arise from (and be shaped by) whatever new quantum states arose. The world we know depends upon these particles and forces, so if this happened, everything around us, from subatomic particles to galaxies, and all fundamental forces, would be reconstituted into new fundamental particles and forces and structures. The universe would potentially lose all of its present structures and become inhabited by new ones (depending upon the exact states involved) based upon the same quantum fields.

[11] Goldstone's theorem only applies to gauges having manifest Lorentz covariance, a condition that took time to become questioned. But the process of quantisation requires a gauge to be fixed and at this point it becomes possible to choose a gauge such as the 'radiation' gauge which is not invariant over time, so that these problems can be avoided. According to Bernstein (1974, p.8):

> "the "radiation gauge" condition $\nabla \cdot A(x) = 0$ is clearly noncovariant, which means that if we wish to maintain transversality of the photon in all Lorentz frames, the photon field $A\mu(x)$ cannot transform like a four-vector. This is no catastrophe, since the photon *field* is not an observable, and one can readily show that the S-matrix elements, which *are* observable have covariant structures .... in gauge theories one might arrange things so that one had a symmetry breakdown because of the noninvariance of the vacuum; but, because the Goldstone *et al.* proof breaks down, the zero mass Goldstone mesons need not appear." [Emphasis in original]

Bernstein (1974) contains an accessible and comprehensive background and review of this area, see external links

[12] A field with the "Mexican hat" potential $V(\phi) = \mu^2 \phi^2 + \lambda \phi^4$ and $\mu^2 < 0$ has a minimum not at zero but at some non-zero value ϕ_0 . By expressing the action in terms of the field $\tilde{\phi} = \phi - \phi_0$ (where ϕ_0 is a constant independent of position), we find the Yukawa term has a component $g\phi_0 \bar{\psi}\psi$. Since both g and ϕ_0 are constants, this looks exactly like

the mass term for a fermion of mass $g\phi_0$. The field $\tilde{\phi}$ is then the Higgs field.

[13] In the Standard Model, the mass term arising from the Dirac Lagrangian for any fermion ψ is $-m\bar{\psi}\psi$. This is *not* invariant under the electroweak symmetry, as can be seen by writing ψ in terms of left and right handed components:

$$-m\bar{\psi}\psi \;=\; -m(\bar{\psi}_L\psi_R + \bar{\psi}_R\psi_L)$$

i.e., contributions from $\bar{\psi}_L\psi_L$ and $\bar{\psi}_R\psi_R$ terms do not appear. We see that the mass-generating interaction is achieved by constant flipping of particle chirality. Since the spin-half particles have no right/left helicity pair with the same SU(2) and SU(3) representation and the same weak hypercharge, then assuming these gauge charges are conserved in the vacuum, none of the spin-half particles could ever swap helicity. Therefore, in the absence of some other cause, all fermions must be massless.

[14] The example is based on the production rate at the LHC operating at 7 TeV. The total cross-section for producing a Higgs boson at the LHC is about 10 picobarn,[97] while the total cross-section for a proton–proton collision is 110 millibarn.[100]

[15] Just before LEP's shut down, some events that hinted at a Higgs were observed, but it was not judged significant enough to extend its run and delay construction of the LHC.

[16] Announced in articles in *Time*,[145] Forbes,[146] *Slate*,[147] *NPR*,[148] and others.[149]

[17] Other names have included: the "Anderson–Higgs" mechanism,[159] "Higgs–Kibble" mechanism (by Abdus Salam)[79] and "ABEGHHK'tH" mechanism [for Anderson, Brout, Englert, Guralnik, Hagen, Higgs, Kibble and 't Hooft] (by Peter Higgs).[79]

[18] In Miller's analogy, the Higgs field is compared to political party workers spread evenly throughout a room. There will be some people (in Miller's example an anonymous person) who pass through the crowd with ease, paralleling the interaction between the field and particles that do not interact with it, such as massless photons. There will be other people (in Miller's example the British prime minister) who would find their progress being continually slowed by the swarm of admirers crowding around, paralleling the interaction for particles that do interact with the field and by doing so, acquire a finite mass.[198][199]

5.1.10 References

[1] O'Luanaigh, C. (14 March 2013). "New results indicate that new particle is a Higgs boson". CERN. Retrieved 2013-10-09.

[2] Bryner, J. (14 March 2013). "Particle confirmed as Higgs boson". *NBC News*. Retrieved 2013-03-14.

[3] Heilprin, J. (14 March 2013). "Higgs Boson Discovery Confirmed After Physicists Review Large Hadron Collider Data at CERN". *The Huffington Post*. Retrieved 2013-03-14.

[4] ATLAS; CMS (26 March 2015). "Combined Measurement of the Higgs Boson Mass in pp Collisions at √s=7 and 8 TeV with the ATLAS and CMS Experiments". arXiv:1503.07589.

[5] LHC Higgs Cross Section Working Group; Dittmaier; Mariotti; Passarino; Tanaka; Alekhin; Alwall; Bagnaschi; Banfi (2012). "Handbook of LHC Higgs Cross Sections: 2. Differential Distributions". *CERN Report 2 (Tables A.1 – A.20)* **1201**: 3084. arXiv:1201.3084. Bibcode:2012arXiv1201.3084L.

[6] Onyisi, P. (23 October 2012). "Higgs boson FAQ". University of Texas ATLAS group. Retrieved 2013-01-08.

[7] Strassler, M. (12 October 2012). "The Higgs FAQ 2.0". *ProfMattStrassler.com*. Retrieved 2013-01-08. [Q] Why do particle physicists care so much about the Higgs particle? [A] Well, actually, they don't. What they really care about is the Higgs *field*, because it is *so* important. [emphasis in original]

[8] José Luis Lucio and Arnulfo Zepeda (1987). *Proceedings of the II Mexican School of Particles and Fields, Cuernavaca-Morelos, 1986*. World Scientific. p. 29. ISBN 9971504340.

[9] Gunion, Dawson, Kane, and Haber (199). *The Higgs Hunter's Guide (1st ed.)*. pp. 11 (?). ISBN 9780786743186. – quoted as being in the first (1990) edition of the book by Peter Higgs in his talk "My Life as a Boson", 2001, ref#25.

[10] Strassler, M. (8 October 2011). "The Known Particles – If The Higgs Field Were Zero". *ProfMattStrassler.com*. Retrieved 13 November 2012. The Higgs field: so important it merited an entire experimental facility, the Large Hadron Collider, dedicated to understanding it.

[11] Biever, C. (6 July 2012). "It's a boson! But we need to know if it's the Higgs". *New Scientist*. Retrieved 2013-01-09. 'As a layman, I would say, I think we have it,' said Rolf-Dieter Heuer, director general of CERN at Wednesday's seminar announcing the results of the search for the Higgs boson. But when pressed by journalists afterwards on what exactly 'it' was, things got more complicated. 'We have discovered a boson – now we have to find out what boson it is' Q: 'If we don't know the new particle is a Higgs, what do we know about it?' We know it is some kind of boson, says Vivek Sharma of CMS [...] Q: 'are the CERN scientists just being too cautious? What would be enough evidence to call it a Higgs boson?' As there could be many different kinds of Higgs bosons, there's no straight answer. [emphasis in original]

[12] Siegfried, T. (20 July 2012). "Higgs Hysteria". *Science News*. Retrieved 2012-12-09. In terms usually reserved for athletic achievements, news reports described the finding as a monumental milestone in the history of science.

[13] Del Rosso, A. (19 November 2012). "Higgs: The beginning of the exploration". *CERN Bulletin* (47–48). Retrieved 2013-01-09. Even in the most specialized circles, the new particle discovered in July is not yet being called the "Higgs boson". Physicists still hesitate to call it that before they have determined that its properties fit with those the Higgs theory predicts the Higgs boson has.

[14] Naik, G. (14 March 2013). "New Data Boosts Case for Higgs Boson Find". *The Wall Street Journal*. Retrieved 2013-03-15. 'We've never seen an elementary particle with spin zero,' said Tony Weidberg, a particle physicist at the University of Oxford who is also involved in the CERN experiments.

[15] Overbye, D. (8 October 2013). "For Nobel, They Can Thank the 'God Particle'". *The New York Times*. Retrieved 2013-11-03.

[16] Sample, I. (29 May 2009). "Anything but the God particle". *The Guardian*. Retrieved 2009-06-24.

[17] Evans, R. (14 December 2011). "The Higgs boson: Why scientists hate that you call it the 'God particle'". *National Post*. Retrieved 2013-11-03.

[18] The nickname occasionally has been satirised in mainstream media as well. Borowitz, Andy (July 13, 2012). "5 questions for the Higgs boson". *The New Yorker*.

[19] Shu, F. H. (1982). *The Physical Universe: An Introduction to Astronomy*. University Science Books. pp. 107–108. ISBN 978-0-935702-05-7.

[20] Shu, F. H. (1982). *The Physical Universe: An Introduction to Astronomy*. University Science Books. pp. 107–108. ISBN 978-0-935702-05-7.

[21] Leon M. Lederman and Dick Teresi (1993). *The God Particle: If the Universe is the Answer, What is the Question*. Houghton Mifflin Company.

[22] Heath, Nick, *The Cern tech that helped track down the God particle*, TechRepublic, 4 July 2012

[23] Rao, Achintya (2 July 2012). "Why would I care about the Higgs boson?". *CMS Public Website*. CERN. Retrieved 18 July 2012.

[24] Max Jammer, *Concepts of Mass in Contemporary Physics and Philosophy* (Princeton, NJ: Princeton University Press, 2000) pp.162–163, who provides many references in support of this statement.

[25] The Large Hadron Collider: Shedding Light on the Early Universe – lecture by R.-D. Heuer, CERN, Chios, Greece, 28 September 2011

[26] Alekhin, Djouadi and Moch, S.; Djouadi, A.; Moch, S. (2012-08-13). "The top quark and Higgs boson masses and the stability of the electroweak vacuum". *Physics Letters B* **716**: 214. arXiv:1207.0980. Bibcode:2012PhLB..716..214A. doi:10.1016/j.physletb.2012.08.024. Retrieved 20 February 2013.

[27] M.S. Turner, F. Wilczek (1982). "Is our vacuum metastable?". *Nature* **298** (5875): 633–634. Bibcode:1982Natur.298..633T. doi:10.1038/298633a0.

[28] S. Coleman and F. De Luccia (1980). "Gravitational effects on and of vacuum decay". *Physical Review* **D21** (12): 3305. Bibcode:1980PhRvD..21.3305C. doi:10.1103/PhysRevD.21.3305.

[29] M. Stone (1976). "Lifetime and decay of excited vacuum states". *Phys. Rev. D* **14** (12): 3568–3573. Bibcode:1976PhRvD..14.3568S. doi:10.1103/PhysRevD.14.3568.

[30] P.H. Frampton (1976). "Vacuum Instability and Higgs Scalar Mass". *Phys. Rev. Lett.* **37** (21): 1378–1380. Bibcode:1976PhRvL..37.1378F. doi:10.1103/PhysRevLett.37.1378.

[31] P.H. Frampton (1977). "Consequences of Vacuum Instability in Quantum Field Theory". *Phys. Rev. D15* (10): 2922–28. Bibcode:1977PhRvD..15.2922F. doi:10.1103/PhysRevD.15.2922.

[32] Ellis, Espinosa, Giudice, Hoecker, & Riotto, J.; Espinosa, J.R.; Giudice, G.F.; Hoecker, A.; Riotto, A. (2009). "The Probable Fate of the Standard Model". *Phys. Lett. B* **679** (4): 369–375. arXiv:0906.0954. Bibcode:2009PhLB..679..369E. doi:10.1016/j.physletb.2009.07.054.

[33] Masina, Isabella (2013-02-12). "Higgs boson and top quark masses as tests of electroweak vacuum stability". *Phys. Rev. D* **87** (5): 53001. arXiv:1209.0393. Bibcode:2013PhRvD..87e3001M. doi:10.1103/PhysRevD.87.053001.

[34] Buttazzo, Degrassi, Giardino, Giudice, Sala, Salvio, Strumia (2013-07-12). "Investigating the near-criticality of the Higgs boson". *JHEP 1312 (2013) 089*. arXiv:1307.3536. Bibcode:2013JHEP...12..089B. doi:10.1007/JHEP12(2013)089.

[35] Irene Klotz (editing by David Adams and Todd Eastham) (2013-02-18). "Universe Has Finite Lifespan, Higgs Boson Calculations Suggest". *Huffington Post*. Reuters. Retrieved 21 February 2013. Earth will likely be long gone before any Higgs boson particles set off an apocalyptic assault on the universe

[36] Hoffman, Mark (2013-02-19). "Higgs Boson Will Destroy The Universe Eventually". *ScienceWorldReport*. Retrieved 21 February 2013.

[37] "Higgs boson will aid in creation of the universe – and how it will end". *Catholic Online/NEWS CONSORTIUM*. 2013-02-20. Retrieved 21 February 2013. [T]he Earth will likely be long gone before any Higgs boson particles set off an apocalyptic assault on the universe

[38] Salvio, Alberto (2015-04-09). "A Simple Motivated Completion of the Standard Model below the Planck Scale: Axions and Right-Handed Neutrinos". *Physics Letters B* **743**: 428. arXiv:1501.03781. Bibcode:2015PhLB..743..428S. doi:10.1016/j.physletb.2015.03.015.

[39] Boyle, Alan (2013-02-19). "Will our universe end in a 'big slurp'? Higgs-like particle suggests it might". *NBC News' Cosmic log*. Retrieved 21 February 2013. [T]he bad news is that its mass suggests the universe will end in a fast-spreading bubble of doom. The good news? It'll probably be tens of billions of years. The article quotes Fermilab's Joseph Lykken: "[T]he parameters for our universe, including the Higgs [and top quark's masses] suggest that we're just at the edge of stability, in a "metastable" state. Physicists have been contemplating such a possibility for more than 30 years. Back in 1982, physicists Michael Turner and Frank Wilczek wrote in Nature that "without warning, a bubble of true vacuum could nucleate somewhere in the universe and move outwards..."

[40] Peralta, Eyder (2013-02-19). "If Higgs Boson Calculations Are Right, A Catastrophic 'Bubble' Could End Universe". *npr – two way*. Retrieved 21 February 2013. Article cites Fermilab's Joseph Lykken: "The bubble forms through an unlikely quantum fluctuation, at a random time and place," Lykken tells us. "So in principle it could happen tomorrow, but then most likely in a very distant galaxy, so we are still safe for billions of years before it gets to us."

[41] Bezrukov; Shaposhnikov (2007-10-19). "The Standard Model Higgs boson as the inflaton". *Phys.Lett.* *B659 (2008) 703-706.* arXiv:0710.3755. Bibcode:2008PhLB..659..703B. doi:10.1016/j.physletb.2007.11.072.

[42] Salvio, Alberto (2013-08-09). "Higgs Inflation at NNLO after the Boson Discovery". *Phys.Lett.* *B727 (2013) 234-239.* arXiv:1308.2244. Bibcode:2013PhLB..727..234S. doi:10.1016/j.physletb.2013.10.042.

[43] Cole, K. (2000-12-14). "One Thing Is Perfectly Clear: Nothingness Is Perfect". *Los Angeles Times*. p. 'Science File'. Retrieved 17 January 2013. [T]he Higgs' influence (or the influence of something like it) could reach much further. For example, something like the Higgs—if not exactly the Higgs itself—may be behind many other unexplained "broken symmetries" in the universe as well ... In fact, something very much like the Higgs may have been behind the collapse of the symmetry that led to the Big Bang, which created the universe. When the forces first began to separate from their primordial sameness—taking on the distinct characters they have today—they released energy in the same way as water releases energy when it turns to ice. Except in this case, the freezing packed enough energy to blow up the universe. ... However it happened, the moral is clear: Only when the perfection shatters can everything else be born.

[44] Higgs Matters – Kathy Sykes, 30 Nove 2012

[45] Why the public should care about the Higgs Boson – Jodi Lieberman, American Physical Society (APS)

[46] Matt Strassler's blog – Why the Higgs particle matters 2 July 2012

[47] Sean Carroll (13 November 2012). *The Particle at the End of the Universe: How the Hunt for the Higgs Boson Leads Us to the Edge of a New World*. Penguin Group US. ISBN 978-1-101-60970-5.

[48] Woit, Peter (13 November 2010). "The Anderson–Higgs Mechanism". Dr. Peter Woit (Senior Lecturer in Mathematics Columbia University and Ph.D. particle physics). Retrieved 12 November 2012.

[49] Goldstone, J; Salam, Abdus; Weinberg, Steven (1962). "Broken Symmetries". *Physical Review* **127** (3): 965–970. Bibcode:1962PhRv..127..965G. doi:10.1103/PhysRev.127.965.

[50] Guralnik, G. S. (2011). "The Beginnings of Spontaneous Symmetry Breaking in Particle Physics". arXiv:1110.2253 [physics.hist-ph].

[51] Kibble, T. W. B. (2009). "Englert–Brout–Higgs–Guralnik–Hagen–Kibble Mechanism". *Scholarpedia* **4** (1): 6441. Bibcode:2009SchpJ...4.6441K. doi:10.4249/scholarpedia.6441. Retrieved 2012-11-23.

[52] Kibble, T. W. B. "History of Englert–Brout–Higgs–Guralnik–Hagen–Kibble Mechanism (history)". *Scholarpedia* **4** (1): 8741. Bibcode:2009SchpJ...4.8741K. doi:10.4249/scholarpedia.8741. Retrieved 2012-11-23.

[53] The Nobel Prize in Physics 2008 – official Nobel Prize website.

[54] List of Anderson 1958–1959 papers referencing 'symmetry', at APS Journals

[55] Higgs, Peter (2010-11-24). "My Life as a Boson" (PDF). Talk given by Peter Higgs at Kings College, London, Nov 24 2010, expanding on a paper originally presented in 2001. Retrieved 17 January 2013. – the original 2001 paper can be found at: Duff and Liu, ed. (2003) [year of publication]. *2001 A Spacetime Odyssey: Proceedings of the Inaugural Conference of the Michigan Center for Theoretical Physics, Michigan, USA, 21–25 May 2001*. World Scientific. pp. 86–88. ISBN 9812382313. Retrieved 17 January 2013.

[56] Anderson, P. (1963). "Plasmons, gauge invariance and mass". *Physical Review* **130**: 439. Bibcode:1963PhRv..130..439A. doi:10.1103/PhysRev.130.439.

[57] Klein, A.; Lee, B. (1964). "Does Spontaneous Breakdown of Symmetry Imply Zero-Mass Particles?". *Physical Review Letters* **12** (10): 266. Bibcode:1964PhRvL..12..266K. doi:10.1103/PhysRevLett.12.266.

[58] Englert, François; Brout, Robert (1964). "Broken Symmetry and the Mass of Gauge Vector Mesons". *Physical Review Letters* **13** (9): 321–23. Bibcode:1964PhRvL..13..321E. doi:10.1103/PhysRevLett.13.321.

[59] Higgs, Peter (1964). "Broken Symmetries and the Masses of Gauge Bosons". *Physical Review Letters* **13** (16): 508–509. Bibcode:1964PhRvL..13..508H. doi:10.1103/PhysRevLett.13.508.

[60] Guralnik, Gerald; Hagen, C. R.; Kibble, T. W. B. (1964). "Global Conservation Laws and Massless Particles". *Physical Review Letters* **13** (20): 585–587. Bibcode:1964PhRvL..13..585G. doi:10.1103/PhysRevLett.13.585.

[61] Higgs, Peter (1964). "Broken symmetries, massless particles and gauge fields". *Physics Letters* **12** (2): 132–133. Bibcode:1964PhL....12..132H. doi:10.1016/0031-9163(64)91136-9.

[62] Higgs, Peter (2010-11-24). "My Life as a Boson" (PDF). Talk given by Peter Higgs at Kings College, London, Nov 24 2010. Retrieved 17 January 2013. Gilbert ... wrote a response to [Klein and Lee's paper] saying 'No, you cannot do that in a relativistic theory. You cannot have a preferred unit time-like vector like that.' This is where I came in, because the next month was when I responded to Gilbert's paper by saying 'Yes, you can have such a thing' but only in a gauge theory with a gauge field coupled to the current.

[63] G.S. Guralnik (2011). "Gauge invariance and the Goldstone theorem – 1965 Feldafing talk". *Modern Physics Letters A* **26** (19): 1381–1392. arXiv:1107.4592. Bibcode:2011MPLA...26.1381G. doi:10.1142/S0217732311036188.

[64] Higgs, Peter (1966). "Spontaneous Symmetry Breakdown without Massless Bosons". *Physical Review* **145** (4): 1156–1163. Bibcode:1966PhRv..145.1156H. doi:10.1103/PhysRev.145.1156.

[65] Kibble, Tom (1967). "Symmetry Breaking in Non-Abelian Gauge Theories". *Physical Review* **155** (5): 1554–1561. Bibcode:1967PhRv..155.1554K. doi:10.1103/PhysRev.155.1554.

[66] "Guralnik, G S; Hagen, C R and Kibble, T W B (1967). Broken Symmetries and the Goldstone Theorem. Advances in Physics, vol. 2" (PDF).

[67] "Physical Review Letters – 50th Anniversary Milestone Papers". Physical Review Letters.

[68] S. Weinberg (1967). "A Model of Leptons". *Physical Review Letters* **19** (21): 1264–1266. Bibcode:1967PhRvL..19.1264W. doi:10.1103/PhysRevLett.19.1264.

[69] A. Salam (1968). N. Svartholm, ed. *Elementary Particle Physics: Relativistic Groups and Analyticity*. Eighth Nobel Symposium. Stockholm: Almquvist and Wiksell. p. 367.

[70] S.L. Glashow (1961). "Partial-symmetries of weak interactions". *Nuclear Physics* **22** (4): 579–588. Bibcode:1961NucPh..22..579G. doi:10.1016/0029-5582(61)90469-2.

[71] Ellis, John; Gaillard, Mary K.; Nanopoulos, Dimitri V. (2012). "A Historical Profile of the Higgs Boson". arXiv:1201.6045 [hep-ph].

[72] "Martin Veltman Nobel Lecture, December 12, 1999, p.391" (PDF). Retrieved 2013-10-09.

[73] Politzer, David. "The Dilemma of Attribution". *Nobel Prize lecture, 2004*. Nobel Prize. Retrieved 22 January 2013. Sidney Coleman published in Science magazine in 1979 a citation search he did documenting that essentially no one paid any attention to Weinberg's Nobel Prize winning paper until the work of 't Hooft (as explicated by Ben Lee). In 1971 interest in Weinberg's paper exploded. I had a parallel personal experience: I took a one-year course on weak interactions from Shelly Glashow in 1970, and he never even mentioned the Weinberg–Salam model or his own contributions.

[74] Coleman, Sidney (1979-12-14). "The 1979 Nobel Prize in Physics". *Science* **206** (4424): 1290–1292. Bibcode:1979Sci...206.1290C. doi:10.1126/science.206.4424.1290.

[75] Letters from the Past – A PRL Retrospective (50 year celebration, 2008)

[76] Jeremy Bernstein (January 1974). "Spontaneous symmetry breaking, gauge theories, the Higgs mechanism and all that" (PDF). *Reviews of Modern Physics* **46** (1): 7. Bibcode:1974RvMP...46....7B. doi:10.1103/RevModPhys.46.7. Retrieved 2012-12-10.

[77] American Physical Society – "J. J. Sakurai Prize for Theoretical Particle Physics".

[78] Merali, Zeeya (4 August 2010). "Physicists get political over Higgs". *Nature Magazine*. Retrieved 28 December 2011.

[79] Close, Frank (2011). *The Infinity Puzzle: Quantum Field Theory and the Hunt for an Orderly Universe*. Oxford: Oxford University Press. ISBN 978-0-19-959350-7.

[80] G.S. Guralnik (2009). "The History of the Guralnik, Hagen and Kibble development of the Theory of Spontaneous Symmetry Breaking and Gauge Particles". *International Journal of Modern Physics A* **24** (14): 2601–2627. arXiv:0907.3466. Bibcode:2009IJMPA..24.2601G. doi:10.1142/S0217751X09045431.

[81] Peskin, Michael E.; Schroeder, Daniel V. (1995). *Introduction to Quantum Field Theory*. Reading, MA: Addison-Wesley Publishing Company. pp. 717–719 and 787–791. ISBN 0-201-50397-2.

[82] Peskin & Schroeder 1995, pp. 715–716

[83] Gunion, John (2000). *The Higgs Hunter's Guide* (illustrated, reprint ed.). Westview Press. pp. 1–3. ISBN 9780738203058.

[84] Lisa Randall, *Warped Passages: Unraveling the Mysteries of the Universe's Hidden Dimensions*, p.286: "People initially thought of tachyons as particles travelling faster than the

speed of light...But we now know that a tachyon indicates an instability in a theory that contains it. Regrettably for science fiction fans, tachyons are not real physical particles that appear in nature."

[85] Sen, Ashoke (April 2002). "Rolling Tachyon". *J. High Energy Phys.* **2002** (0204): 048. arXiv:hep-th/0203211. Bibcode:2002JHEP...04..048S. doi:10.1088/1126-6708/2002/04/048.

[86] Kutasov, David; Marino, Marcos & Moore, Gregory W. (2000). "Some exact results on tachyon condensation in string field theory". *JHEP* **0010**: 045.

[87] Aharonov, Y.; Komar, A.; Susskind, L. (1969). "Superluminal Behavior, Causality, and Instability". *Phys. Rev.* (American Physical Society) **182** (5): 1400–1403. Bibcode:1969PhRv..182.1400A. doi:10.1103/PhysRev.182.1400.

[88] Feinberg, Gerald (1967). "Possibility of Faster-Than-Light Particles". *Physical Review* **159** (5): 1089–1105. Bibcode:1967PhRv..159.1089F. doi:10.1103/PhysRev.159.1089.

[89] Michael E. Peskin and Daniel V. Schroeder (1995). *An Introduction to Quantum Field Theory*, Perseus books publishing.

[90] Flatow, Ira (6 July 2012). "At Long Last, The Higgs Particle... Maybe". *NPR*. Retrieved 10 July 2012.

[91] "Explanatory Figures for the Higgs Boson Exclusion Plots". *ATLAS News*. CERN. Retrieved 6 July 2012.

[92] Bernardi, G.; Carena, M.; Junk, T. (2012). "Higgs Bosons: Theory and Searches" (PDF). p. 7.

[93] Lykken, Joseph D. (2009). "Beyond the Standard Model". *Proceedings of the 2009 European School of High-Energy Physics, Bautzen, Germany, 14 – 27 June 2009*. arXiv:1005.1676.

[94] Plehn, Tilman (2012). *Lectures on LHC Physics*. Lecture Notes is Physics **844**. Springer. Sec. 1.2.2. arXiv:0910.4122. ISBN 3642240399.

[95] "LEP Electroweak Working Group".

[96] Peskin, Michael E.; Wells, James D. (2001). "How Can a Heavy Higgs Boson be Consistent with the Precision Electroweak Measurements?". *Physical Review D* **64** (9): 093003. arXiv:hep-ph/0101342. Bibcode:2001PhRvD..64i3003P. doi:10.1103/PhysRevD.64.093003.

[97] Baglio, Julien; Djouadi, Abdelhak (2011). "Higgs production at the lHC". *Journal of High Energy Physics* **1103** (3): 055. arXiv:1012.0530. Bibcode:2011JHEP...03..055B. doi:10.1007/JHEP03(2011)055.

[98] Baglio, Julien; Djouadi, Abdelhak (2010). "Predictions for Higgs production at the Tevatron and the associated uncertainties". *Journal of High Energy Physics* **1010** (10): 063. arXiv:1003.4266. Bibcode:2010JHEP...10..064B. doi:10.1007/JHEP10(2010)064.

[99] Teixeira-Dias (LEP Higgs working group), P. (2008). "Higgs boson searches at LEP". *Journal of.Physics: Conference Series* **110** (4): 042030. arXiv:0804.4146. Bibcode:2008JPhCS.110d2030T. doi:10.1088/1742-6596/110/4/042030.

[100] "Collisions". *LHC Machine Outreach*. CERN. Retrieved 26 July 2012.

[101] Asquith, Lily (22 June 2012). "Why does the Higgs decay?". *Life and Physics* (London: The Guardian). Retrieved 14 August 2012.

[102] "Higgs bosons: theory and searches" (PDF). *PDGLive*. Particle Data Group. 12 July 2012. Retrieved 15 August 2012.

[103] Branco, G. C.; Ferreira, P.M.; Lavoura, L.; Rebelo, M.N.; Sher, Marc; Silva, João P. (July 2012). "Theory and phenomenology of two-Higgs-doublet models". *Physics Reports* (Elsevier) **516** (1): 1–102. arXiv:1106.0034. Bibcode:2012PhR...516....1B. doi:10.1016/j.physrep.2012.02.002.

[104] Csaki, C.; Grojean, C.; Pilo, L.; Terning, J. (2004). "Towards a realistic model of Higgsless electroweak symmetry breaking". *Physical Review Letters* **92** (10): 101802. arXiv:hep-ph/0308038. Bibcode:2004PhRvL..92j1802C. doi:10.1103/PhysRevLett.92.101802. PMID 15089195.

[105] Csaki, C.; Grojean, C.; Pilo, L.; Terning, J.; Terning, John (2004). "Gauge theories on an interval: Unitarity without a Higgs". *Physical Review D* **69** (5): 055006. arXiv:hep-ph/0305237. Bibcode:2004PhRvD..69e5006C. doi:10.1103/PhysRevD.69.055006.

[106] "The Hierarchy Problem: why the Higgs has a snowball's chance in hell". Quantum Diaries. 2012-07-01. Retrieved 19 March 2013.

[107] "The Hierarchy Problem | Of Particular Significance". Prof-mattstrassler.com. Retrieved 2013-10-09.

[108] "Collisions". *LHC Machine Outreach*. CERN. Retrieved 26 July 2012.

[109] "Hunt for Higgs boson hits key decision point". MSNBC. 2012-12-06. Retrieved 2013-01-19.

[110] Worldwide LHC Computing Grid main page 14 November 2012: *"[A] global collaboration of more than 170 computing centres in 36 countries ... to store, distribute and analyse the ~25 Petabytes (25 million Gigabytes) of data annually generated by the Large Hadron Collider"*

[111] What is the Worldwide LHC Computing Grid? (Public 'About' page) 14 November 2012: *"Currently WLCG is made up of more than 170 computing centers in 36 countries...The WLCG is now the world's largest computing grid"*

[112] W.-M. Yao et al. (2006). "Review of Particle Physics" (PDF). *Journal of Physics G* **33**: 1. arXiv:astro-ph/0601168. Bibcode:2006JPhG...33....1Y. doi:10.1088/0954-3899/33/1/001.

[113] The CDF Collaboration, the D0 Collaboration, the Tevatron New Physics, Higgs Working Group (2012). "Updated Combination of CDF and D0 Searches for Standard Model Higgs Boson Production with up to 10.0 fb^{-1} of Data". arXiv:1207.0449 [hep-ex].

[114] "Interim Summary Report on the Analysis of the 19 September 2008 Incident at the LHC" (PDF). CERN. 15 October 2008. EDMS 973073. Retrieved 28 September 2009.

[115] "CERN releases analysis of LHC incident" (Press release). CERN Press Office. 16 October 2008. Retrieved 28 September 2009.

[116] "LHC to restart in 2009" (Press release). CERN Press Office. 5 December 2008. Retrieved 8 December 2008.

[117] "LHC progress report". *The Bulletin*. CERN. 3 May 2010. Retrieved 7 December 2011.

[118] "ATLAS experiment presents latest Higgs search status". *ATLAS homepage*. CERN. 13 December 2011. Retrieved 13 December 2011.

[119] Taylor, Lucas (13 December 2011). "CMS search for the Standard Model Higgs Boson in LHC data from 2010 and 2011". *CMS public website*. CERN. Retrieved 13 December 2011.

[120] Overbye, D. (5 March 2013). "Chasing The Higgs Boson". *The New York Times*. Retrieved 2013-03-05.

[121] "ATLAS and CMS experiments present Higgs search status" (Press release). CERN Press Office. 13 December 2011. Retrieved 14 September 2012. the statistical significance is not large enough to say anything conclusive. As of today what we see is consistent either with a background fluctuation or with the presence of the boson. Refined analyses and additional data delivered in 2012 by this magnificent machine will definitely give an answer

[122] "WLCG Public Website". CERN. Retrieved 29 October 2012.

[123] CMS collaboration (2014). "Precise determination of the mass of the Higgs boson and tests of compatibility of its couplings with the standard model predictions using proton collisions at 7 and 8 TeV". arXiv:1412.8662.

[124] ATLAS collaboration (2014). "Measurements of Higgs boson production and couplings in the four-lepton channel in pp collisions at center-of-mass energies of 7 and 8 TeV with the ATLAS detector". arXiv:1408.5191.

[125] ATLAS collaboration (2014). "Measurement of Higgs boson production in the diphoton decay channel in pp collisions at center-of-mass energies of 7 and 8 TeV with the ATLAS detector". arXiv:1408.7084.

[126] "Press Conference: Update on the search for the Higgs boson at CERN on 4 July 2012". Indico.cern.ch. 22 June 2012. Retrieved 4 July 2012.

[127] "CERN to give update on Higgs search". CERN. 22 June 2012. Retrieved 2 July 2011.

[128] "Scientists analyse global Twitter gossip around Higgs boson discovery". *phys.org (from arXiv)*. 2013-01-23. Retrieved 6 February 2013. – stated to be *"the first time scientists have been able to analyse the dynamics of social media on a global scale before, during and after the announcement of a major scientific discovery."* For the paper itself see: De Domenico, M.; Lima, A.; Mougel, P.; Musolesi, M. (2013). "The Anatomy of a Scientific Gossip". arXiv:1301.2952. Bibcode:2013NatSR...3E2980D. doi:10.1038/srep02980.

[129] "Higgs boson particle results could be a quantum leap". Times LIVE. 28 June 2012. Retrieved 4 July 2012.

[130] CERN prepares to deliver Higgs particle findings, Australian Broadcasting Corporation. Retrieved 4 July 2012.

[131] "God Particle Finally Discovered? Higgs Boson News At Cern Will Even Feature Scientist It's Named After". Huffingtonpost.co.uk. Retrieved 2013-01-19.

[132] Our Bureau (2012-07-04). "Higgs on way, theories thicken". Calcutta, India: Telegraphindia.com. Retrieved 2013-01-19.

[133] Thornhill, Ted (2013-07-03). "God Particle Finally Discovered? Higgs Boson News At Cern Will Even Feature Scientist It's Named After". *Huffington Post*. Retrieved 23 July 2013.

[134] Cooper, Rob (2013-07-01) [updated subsequently]. "God particle is 'found': Scientists at Cern expected to announce on Wednesday Higgs boson particle has been discovered". *Daily Mail* (London). Retrieved 23 July 2013. - States that *""Five leading theoretical physicists have been invited to the event on Wednesday - sparking speculation that the particle has been discovered."*, including Higgs and Englert, and that Kibble - who was invited but unable to attend - "told the Sunday Times: 'My guess is that is must be a pretty positive result for them to be asking us out there'."

[135] Adrian Cho (13 July 2012). "Higgs Boson Makes Its Debut After Decades-Long Search". *Science* **337** (6091): 141–143. doi:10.1126/science.337.6091.141. PMID 22798574.

[136] CMS collaboration (2012). "Observation of a new boson at a mass of 125 GeV with the CMS experiment at the LHC". *Physics Letters B* **716** (1): 30–61. arXiv:1207.7235. Bibcode:2012PhLB..716...30C. doi:10.1016/j.physletb.2012.08.021.

[137] Taylor, Lucas (4 July 2012). "Observation of a New Particle with a Mass of 125 GeV". *CMS Public Website*. CERN. Retrieved 4 July 2012.

[138] "Latest Results from ATLAS Higgs Search". *ATLAS News*. CERN. 4 July 2012. Retrieved 4 July 2012.

[139] ATLAS collaboration (2012). "Observation of a New Particle in the Search for the Standard Model Higgs Boson with the ATLAS Detector at the LHC". *Physics Letters B* **716** (1): 1–29. arXiv:1207.7214. Bibcode:2012PhLB..716....1A. doi:10.1016/j.physletb.2012.08.020.

[140] Gillies, James (23 July 2012). "LHC 2012 proton run extended by seven weeks". *CERN bulletin*. Retrieved 29 August 2012.

[141] "Higgs boson behaving as expected". *3 News NZ*. 15 November 2012.

[142] Strassler, Matt (2012-11-14). "Higgs Results at Kyoto". *Of Particular Significance: Conversations About Science with Theoretical Physicist Matt Strassler*. Prof. Matt Strassler's personal particle physics website. Retrieved 10 January 2013. ATLAS and CMS only just co-discovered this particle in July ... We will not know after today whether it is a Higgs at all, whether it is a Standard Model Higgs or not, or whether any particular speculative idea...is now excluded. [...] Knowledge about nature does not come easy. We discovered the top quark in 1995, and we are still learning about its properties today... we will still be learning important things about the Higgs during the coming few decades. We've no choice but to be patient.

[143] Sample, Ian (14 November 2012). "Higgs particle looks like a bog Standard Model boson, say scientists". *The Guardian* (London). Retrieved 15 November 2012.

[144] "CERN experiments observe particle consistent with long-sought Higgs boson". CERN press release. 4 July 2012. Retrieved 4 July 2012.

[145] "Person Of The Year 2012". *Time*. 19 December 2012.

[146] "Higgs Boson Discovery Has Been Confirmed". Forbes. Retrieved 2013-10-09.

[147] Slate Video Staff (2012-09-11). "Higgs Boson Confirmed; CERN Discovery Passes Test". Slate.com. Retrieved 2013-10-09.

[148] "The Year Of The Higgs, And Other Tiny Advances In Science". NPR. 2013-01-01. Retrieved 2013-10-09.

[149] "Confirmed: the Higgs boson does exist". *The Sydney Morning Herald*. 4 July 2012.

[150] "AP CERN chief: Higgs boson quest could wrap up by midyear". *MSNBC*. Associated Press. 2013-01-27. Retrieved 20 February 2013. Rolf Heuer, director of [CERN], said he is confident that "towards the middle of the year, we will be there." – Interview by AP, at the World Economic Forum, 26 Jan 2013.

[151] Boyle, Alan (2013-02-16). "Will our universe end in a 'big slurp'? Higgs-like particle suggests it might". *NBCNews.com – cosmic log*. Retrieved 20 February 2013. 'it's going to take another few years' after the collider is restarted to confirm definitively that the newfound particle is the Higgs boson.

[152] Gillies, James (2013-03-06). "A question of spin for the new boson". CERN. Retrieved 7 March 2013.

[153] Adam Falkowski (writing as 'Jester') (2013-02-27). "When shall we call it Higgs?". Résonaances particle physics blog. Retrieved 7 March 2013.

[154] CMS Collaboration (February 2013). "Study of the Mass and Spin-Parity of the Higgs Boson Candidate via Its Decays to Z Boson Pairs". *Phys. Rev. Lett.* (American Physical Society) **110** (8): 081803. arXiv:1212.6639. Bibcode:2013PhRvL.110h1803C. doi:10.1103/PhysRevLett.110.081803. Retrieved 15 September 2014.

[155] ATLAS Collaboration (7 October 2013). "Evidence for the spin-0 nature of the Higgs boson using ATLAS data". *Phys. Lett. B* (American Physical Society) **726** (1-3): 120–144. Bibcode:2013PhLB..726..120A. doi:10.1016/j.physletb.2013.08.026. Retrieved 15 September 2014.

[156] "Higgs-like Particle in a Mirror". American Physical Society. Retrieved 26 February 2013.

[157] The CMS Collaboration (2014-06-22). "Evidence for the direct decay of the 125 GeV Higgs boson to fermions". Nature Publishing Group doi= 10.1038/nphys3005.

[158] Adam Falkowski (writing as 'Jester') (2012-12-13). "Twin Peaks in ATLAS". Résonaances particle physics blog. Retrieved 24 February 2013.

[159] Liu, G. Z.; Cheng, G. (2002). "Extension of the Anderson-Higgs mechanism". *Physical Review B* **65** (13): 132513. arXiv:cond-mat/0106070. Bibcode:2002PhRvB..65m2513L. doi:10.1103/PhysRevB.65.132513.

[160] Editorial (2012-03-21). "Mass appeal: As physicists close in on the Higgs boson, they should resist calls to change its name". *Nature*. 483, 374 (7390): 374. Bibcode:2012Natur.483..374.. doi:10.1038/483374a. Retrieved 21 January 2013.

[161] Becker, Kate (2012-03-29). "A Higgs by Any Other Name". "NOVA" (PBS) physics. Retrieved 21 January 2013.

[162] "Frequently Asked Questions: The Higgs!". *The Bulletin*. CERN. Retrieved 18 July 2012.

[163] Woit's physics blog *"Not Even Wrong"*: Anderson on Anderson-Higgs 2013-04-13

[164] Sample, Ian (2012-07-04). "Higgs boson's many great minds cause a Nobel prize headache". *The Guardian* (London). Retrieved 23 July 2013.

[165] "Rochester's Hagen Sakurai Prize Announcement" (Press release). University of Rochester. 2010.

[166] *C.R. Hagen Sakurai Prize Talk* (YouTube). 2010.

[167] Cho, A (2012-09-14). "Particle physics. Why the 'Higgs'?" (PDF). *Science* **337** (6100): 1287. doi:10.1126/science.337.6100.1287. PMID 22984044. Lee ... apparently used the term 'Higgs Boson' as early as 1966... but what may have made the term stick is a seminal paper Steven Weinberg...published in 1967...Weinberg acknowledged the mix-up in an essay in the *New York Review of Books* in May 2012. (See also the original article in *New York Review of Books*[168] and Frank Close's 2011 book *The Infinity Puzzle*[79]:372)

[168] Weinberg, Steven (2012-05-10). "The Crisis of Big Science". *The New York Review of Books* (footnote 1). Retrieved 12 February 2013.

[169] Examples of early papers using the term "Higgs boson" include 'A phenomenological profile of the Higgs boson' (Ellis, Gaillard and Nanopoulos, 1976), 'Weak interaction theory and neutral currents' (Bjorken, 1977), and 'Mass of the Higgs boson' (Wienberg, received 1975)

[170] Leon Lederman; Dick Teresi (2006). *The God Particle: If the Universe Is the Answer, What Is the Question?*. Houghton Mifflin Harcourt. ISBN 0-547-52462-5.

[171] Kelly Dickerson (September 8, 2014). "Stephen Hawking Says 'God Particle' Could Wipe Out the Universe". livescience.com.

[172] Jim Baggott (2012). *Higgs: The invention and discovery of the 'God Particle'*. Oxford University Press. ISBN 978-0-19-165003-1.

[173] Scientific American Editors (2012). *The Higgs Boson: Searching for the God Particle*. Macmillan. ISBN 978-1-4668-2413-3.

[174] Ted Jaeckel (2007). *The God Particle: The Discovery and Modeling of the Ultimate Prime Particle*. Universal-Publishers. ISBN 978-1-58112-959-5.

[175] Aschenbach, Joy (1993-12-05). "No Resurrection in Sight for Moribund Super Collider : Science: Global financial partnerships could be the only way to salvage such a project. Some feel that Congress delivered a fatal blow". *Los Angeles Times*. Retrieved 16 January 2013. 'We have to keep the momentum and optimism and start thinking about international collaboration,' said Leon M. Lederman, the Nobel Prize-winning physicist who was the architect of the super collider plan

[176] "A Supercompetition For Illinois". *Chicago Tribune*. 1986-10-31. Retrieved 16 January 2013. The SSC, proposed by the U.S. Department of Energy in 1983, is a mind-bending project ... this gigantic laboratory ... this titanic project

[177] Diaz, Jesus (2012-12-15). "This Is [The] World's Largest Super Collider That Never Was". *Gizmodo*. Retrieved 16 January 2013. ...this titanic complex...

[178] Abbott, Charles (June 1987). "Illinois Issues journal, June 1987". p. 18. Lederman, who considers himself an unofficial propagandist for the super collider, said the SSC could reverse the physics brain drain in which bright young physicists have left America to work in Europe and elsewhere.

[179] Kevles, Dan. "Good-bye to the SSC: On the Life and Death of the Superconducting Super Collider" (PDF). *California Institute of Technology: "Engineering & Science"*. 58 no. 2 (Winter 1995): 16–25. Retrieved 16 January 2013. Lederman, one of the principal spokesmen for the SSC, was an accomplished high-energy experimentalist who had made Nobel Prize-winning contributions to the development of the Standard Model during the 1960s (although the prize itself did not come until 1988). He was a fixture at congressional hearings on the collider, an unbridled advocate of its merits.

[180] Calder, Nigel (2005). *Magic Universe:A Grand Tour of Modern Science*. pp. 369–370. ISBN 9780191622359. The possibility that the next big machine would create the Higgs became a carrot to dangle in front of funding agencies and politicians. A prominent American physicist, Leon lederman *[sic]*, advertised the Higgs as The God Particle in the title of a book published in 1993 ...Lederman was involved in a campaign to persuade the US government to continue funding the Superconducting Super Collider... the ink was not dry on Lederman's book before the US Congress decided to write off the billions of dollars already spent

[181] Lederman, Leon (1993). *The God Particle If the Universe Is the Answer, What Is the Question?* (PDF). Dell Publishing. p. Chapter 2, Page 2. ISBN 0-385-31211-3. Retrieved 30 July 2015.

[182] Alister McGrath, Higgs boson: the particle of faith, *The Daily Telegraph*, Published 15 December 2011. Retrieved 15 December 2011.

[183] Sample, Ian (3 March 2009). "Father of the God particle: Portrait of Peter Higgs unveiled". London: The Guardian. Retrieved 24 June 2009.

[184] Chivers, Tom (2011-12-13). "How the 'God particle' got its name". *The Telegraph* (London). Retrieved 2012-12-03.

[185] Key scientist sure "God particle" will be found soon Reuters news story. 7 April 2008.

[186] "Interview: the man behind the 'God particle'", New Scientist 13 September 2008, pp. 44–5 (original interview in the Guardian: Father of the 'God Particle', June 30, 2008)

[187] Sample, Ian (2010). *Massive: The Hunt for the God Particle*. pp. 148–149 and 278–279. ISBN 9781905264957.

[188] Cole, K. (2000-12-14). "One Thing Is Perfectly Clear: Nothingness Is Perfect". *Los Angeles Times*. p. 'Science File'. Retrieved 17 January 2013. Consider the early universe–a state of pure, perfect nothingness; a formless fog of undifferentiated stuff ... 'perfect symmetry' ... What shattered this primordial perfection? One likely culprit is the so-called Higgs field ... Physicist Leon Lederman compares the

way the Higgs operates to the biblical story of Babel [whose citizens] all spoke the same language ... Like God, says Lederman, the Higgs differentiated the perfect sameness, confusing everyone (physicists included) ... [Nobel Prizewinner Richard] Feynman wondered why the universe we live in was so obviously askew ... Perhaps, he speculated, total perfection would have been unacceptable to God. And so, just as God shattered the perfection of Babel, 'God made the laws only nearly symmetrical'

[189] Lederman, p. 22 *et seq*:

> "Something we cannot yet detect and which, one might say, has been put there to test and confuse us ... The issue is whether physicists will be confounded by this puzzle or whether, in contrast to the unhappy Babylonians, we will continue to build the tower and, as Einstein put it, 'know the mind of God'."

> "And the Lord said, Behold the people are unconfounding my confounding. And the Lord sighed and said, Go to, let us go down, and there give them the God Particle so that they may see how beautiful is the universe I have made".

[190] Sample, Ian (12 June 2009). "Higgs competition: Crack open the bubbly, the God particle is dead". *The Guardian* (London). Retrieved 4 May 2010.

[191] Gordon, Fraser (5 July 2012). "Introducing the higgson". *physicsworld.com*. Retrieved 25 August 2012.

[192] Wolchover, Natalie (2012-07-03). "Higgs Boson Explained: How 'God Particle' Gives Things Mass". *Huffington Post*. Retrieved 21 January 2013.

[193] Oliver, Laura (2012-07-04). "Higgs boson: how would you explain it to a seven-year-old?". *The Guardian* (London). Retrieved 21 January 2013.

[194] Zimmer, Ben (2012-07-15). "Higgs boson metaphors as clear as molasses". *The Boston Globe*. Retrieved 21 January 2013.

[195] "The Higgs particle: an analogy for Physics classroom (section)". www.lhc-closer.es (a collaboration website of LHCb physicist Xabier Vidal and High School Teachers at CERN educator Ramon Manzano). Retrieved 2013-01-09.

[196] Flam, Faye (2012-07-12). "Finally – A Higgs Boson Story Anyone Can Understand". *The Philadelphia Inquirer (philly.com)*. Retrieved 21 January 2013.

[197] Sample, Ian (2011-04-28). "How will we know when the Higgs particle has been detected?". *The Guardian* (London). Retrieved 21 January 2013.

[198] Miller, David. "A quasi-political Explanation of the Higgs Boson; for Mr Waldegrave, UK Science Minister 1993". Retrieved 10 July 2012.

[199] Kathryn Grim. "Ten things you may not know about the Higgs boson". Symmetry Magazine. Retrieved 10 July 2012.

[200] David Goldberg, Associate Professor of Physics, Drexel University (2010-10-17). "What's the Matter with the Higgs Boson?". io9.com "Ask a physicist". Retrieved 21 January 2013.

[201] The Nobel Prize in Physics 1979 – official Nobel Prize website.

[202] The Nobel Prize in Physics 1999 – official Nobel Prize website.

[203] – official Nobel Prize website.

[204] Daigle, Katy (10 July 2012). "India: Enough about Higgs, let's discuss the boson". *AP News*. Retrieved 10 July 2012.

[205] Bal, Hartosh Singh (19 September 2012). "The Bose in the Boson". New York Times. Retrieved 21 September 2012.

[206] Alikhan, Anvar (16 July 2012). "The Spark In A Crowded Field". *Outlook India*. Retrieved 10 July 2012.

[207] Peskin & Schroeder 1995, Chapter 20

5.1.11 Further reading

- Nambu, Yoichiro; Jona-Lasinio, Giovanni (1961). "Dynamical Model of Elementary Particles Based on an Analogy with Superconductivity". *Physical Review* **122**: 345–358. Bibcode:1961PhRv..122..345N. doi:10.1103/PhysRev.122.345.

- Klein, Abraham; Lee, Benjamin W. (1964). "Does Spontaneous Breakdown of Symmetry Imply Zero-Mass Particles?". *Physical Review Letters* **12** (10): 266. Bibcode:1964PhRvL..12..266K. doi:10.1103/PhysRevLett.12.266.

- Anderson, Philip W. (1963). "Plasmons, Gauge Invariance, and Mass". *Physical Review* **130**: 439. Bibcode:1963PhRv..130..439A. doi:10.1103/PhysRev.130.439.

- Gilbert, Walter (1964). "Broken Symmetries and Massless Particles". *Physical Review Letters* **12** (25): 713. Bibcode:1964PhRvL..12..713G. doi:10.1103/PhysRevLett.12.713.

- Higgs, Peter (1964). "Broken Symmetries, Massless Particles and Gauge Fields". *Physics Letters* **12** (2): 132–133. Bibcode:1964PhL....12..132H. doi:10.1016/0031-9163(64)91136-9.

- Guralnik, Gerald S.; Hagen, C.R.; Kibble, Tom W.B. (1968). "Broken Symmetries and the Goldstone Theorem". In R.L. Cool and R.E. Marshak. *Advances in Physics, Vol. 2*. Interscience Publishers. pp. 567–708. ISBN 978-0470170571.

5.1.12 External links

Popular science, mass media, and general coverage

- Hunting the Higgs Boson at C.M.S. Experiment, at CERN

- The Higgs Boson" by the CERN exploratorium.

- "Particle Fever", documentary film about the search for the Higgs Boson.

- "The Atom Smashers", documentary film about the search for the Higgs Boson at Fermilab.

- Collected Articles at the *Guardian*

- Video (04:38) – CERN Announcement on 4 July 2012, of the discovery of a particle which is suspected will be a Higgs Boson.

- Video1 (07:44) + Video2 (07:44) – Higgs Boson Explained by CERN Physicist, Dr. Daniel Whiteson (16 June 2011).

- HowStuffWorks: What exactly is the Higgs Boson?

- Carroll, Sean. "Higgs Boson with Sean Carroll". *Sixty Symbols*. University of Nottingham.

- Overbye, Dennis (2013-03-05). "Chasing the Higgs Boson: How 2 teams of rivals at CERN searched for physics' most elusive particle". *New York Times Science pages*. Retrieved 22 July 2013. - New York Times "behind the scenes" style article on the Higgs' search at ATLAS and CMS

- The story of the Higgs theory by the authors of the PRL papers and others closely associated:

  - Higgs, Peter (2010). "My Life as a Boson" (PDF). Talk given at Kings College, London, Nov 24 2010. Retrieved 17 January 2013. (also:)

  - Kibble, Tom (2009). "Englert–Brout–Higgs–Guralnik–Hagen–Kibble mechanism (history)". Scholarpedia. Retrieved 17 January 2013. (also:)

- Guralnik, Gerald (2009). "The History of the Guralnik, Hagen and Kibble development of the Theory of Spontaneous Symmetry Breaking and Gauge Particles". *International Journal of Modern Physics A* **24** (14): 2601–2627. arXiv:0907.3466. Bibcode:2009IJMPA..24.2601G. doi:10.1142/S0217751X09045431., Guralnik, Gerald (2011). "The Beginnings of Spontaneous Symmetry Breaking in Particle Physics. Proceedings of the DPF-2011 Conference, Providence, RI, 8–13 August 2011". arXiv:1110.2253v1 [physics.hist-ph]., and Guralnik, Gerald (2013). "Heretical Ideas that Provided the Cornerstone for the Standard Model of Particle Physics". SPG MITTEILUNGEN March 2013, No. 39, (p. 14), and Talk at Brown University about the 1964 PRL papers

- Philip Anderson (not one of the PRL authors) on symmetry breaking in superconductivity and its migration into particle physics and the PRL papers

- Cartoon about the search

- Cham, Jorge (2014-02-19). "True Tales from the Road: The Higgs Boson Re-Explained". *Piled Higher and Deeper*. Retrieved 2014-02-25.

Significant papers and other

- Observation of a new particle in the search for the Standard Model Higgs Boson with the ATLAS detector at the LHC

- Observation of a new Boson at a mass of 125 GeV with the CMS experiment at the LHC

- Particle Data Group: Review of searches for Higgs Bosons.

- 2001, a spacetime odyssey: proceedings of the Inaugural Conference of the Michigan Center for Theoretical Physics : Michigan, USA, 21–25 May 2001, (p.86 – 88), ed. Michael J. Duff, James T. Liu, ISBN 978-981-238-231-3, containing Higgs' story of the Higgs Boson.

- A.A. Migdal & A.M. Polyakov, *Spontaneous Breakdown of Strong Interaction Symmetry and the Absence of Massless Particles*, Sov.J.-JETP 24,91 (1966) - example of a 1966 Russian paper on the subject.

Introductions to the field

- Spontaneous symmetry breaking, gauge theories, the Higgs mechanism and all that (Bernstein, *Reviews of Modern Physics* Jan 1974) - an introduction of 47 pages covering the development, history and mathematics of Higgs theories from around 1950 to 1974.

Chapter 6

Composite particles

6.1 Hadron

In particle physics, a **hadron** ◀))[i] /ˈhædrɒn/ (Greek: ἁδρός, *hadrós*, "stout, thick") is a composite particle made of quarks held together by the strong force (in a similar way as molecules are held together by the electromagnetic force).

Hadrons are categorized into two families: baryons, made of three quarks, and mesons, made of one quark and one antiquark. Protons and neutrons are examples of baryons; pions are an example of a meson. Hadrons containing more than three valence quarks (exotic hadrons) have been discovered in recent years. A tetraquark state (an exotic meson), named the Z(4430)⁻, was discovered in 2007 by the Belle Collaboration [1] and confirmed as a resonance in 2014 by the LHCb collaboration.[2] Two pentaquark states (exotic baryons), named P+
c(4380) and P+
c(4450), were discovered in 2015 by the LHCb collaboration.[3] There are several more exotic hadron candidates, and other colour-singlet quark combinations may also exist.

Of the hadrons, protons are stable, and neutrons bound within atomic nuclei are stable. Other hadrons are unstable under ordinary conditions; free neutrons decay with a half-life of about 611 seconds. Experimentally, hadron physics is studied by colliding protons or nuclei of heavy elements such as lead, and detecting the debris in the produced particle showers.

6.1.1 Etymology

The term "hadron" was introduced by Lev B. Okun in a plenary talk at the 1962 International Conference on High Energy Physics.[4] In this talk he said:

> Not withstanding the fact that this report deals with weak interactions, we shall frequently have to speak of strongly interacting particles. These particles pose not only numerous scien-

tific problems, but also a terminological problem. The point is that "strongly interacting particles" is a very clumsy term which does not yield itself to the formation of an adjective. For this reason, to take but one instance, decays into strongly interacting particles are called non-leptonic. This definition is not exact because "non-leptonic" may also signify "photonic". In this report I shall call strongly interacting particles "hadrons", and the corresponding decays "hadronic" (the Greek ἁδρός signifies "large", "massive", in contrast to λεπτός which means "small", "light"). I hope that this terminology will prove to be convenient.
> — Lev B. Okun, 1962

6.1.2 Properties

According to the quark model,[5] the properties of hadrons are primarily determined by their so-called *valence quarks*. For example, a proton is composed of two up quarks (each with electric charge $+2/3$, for a total of $+4/3$ together) and one down quark (with electric charge $-1/3$). Adding these together yields the proton charge of +1. Although quarks also carry color charge, hadrons must have zero total color charge because of a phenomenon called color confinement. That is, hadrons must be "colorless" or "white". These are the simplest of the two ways: three quarks of different colors, or a quark of one color and an antiquark carrying the corresponding anticolor. Hadrons with the first arrangement are called baryons, and those with the second arrangement are mesons.

Hadrons, however, are not composed of just three or two quarks, because of the strength of the strong force. More accurately, strong force gluons have enough energy (E) to have resonances composed of massive (m) quarks ($E > mc^2$). Thus, virtual quarks and antiquarks, in a 1:1 ratio, form the majority of massive particles inside a hadron. The two or three quarks are the excess of quarks vs. antiquarks in hadrons, and vice versa in anti-hadrons. Because the virtual quarks are not stable wave packets (quanta), but irregular and transient phenomena, it is not meaningful to ask which quark is real and which virtual; only the excess is apparent from the outside. Massless virtual gluons compose the numerical majority of particles inside hadrons.

Like all subatomic particles, hadrons are assigned quantum numbers corresponding to the representations of the Poincaré group: $J^{PC}(m)$, where J is the spin quantum number, P the intrinsic parity (or P-parity), and C, the charge conjugation (or C-parity), and the particle's mass, m. Note that the mass of a hadron has very little to do with the mass of its valence quarks; rather, due to mass–energy equivalence, most of the mass comes from the large amount of energy associated with the strong interaction. Hadrons may also carry flavor quantum numbers such as isospin (or G parity), and strangeness. All quarks carry an additive, conserved quantum number called a baryon number (B), which is $+\frac{1}{3}$ for quarks and $-\frac{1}{3}$ for antiquarks. This means that baryons (groups of three quarks) have $B = 1$ whereas mesons have $B = 0$.

Hadrons have excited states known as resonances. Each ground state hadron may have several excited states; several hundreds of resonances have been observed in particle physics experiments. Resonances decay extremely quickly (within about 10^{-24} seconds) via the strong nuclear force.

In other phases of matter the hadrons may disappear. For example, at very high temperature and high pressure, unless there are sufficiently many flavors of quarks, the theory of quantum chromodynamics (QCD) predicts that quarks and gluons will no longer be confined within hadrons, "because the strength of the strong interaction diminishes with energy". This property, which is known as asymptotic freedom, has been experimentally confirmed in the energy range between 1 GeV (gigaelectronvolt) and 1 TeV (teraelectronvolt).[6]

All free hadrons except the proton (and antiproton) are unstable.

6.1.3 Baryons

Main article: Baryon

All known baryons are made of three valence quarks, so they are fermions, *i.e.*, they have odd half-integral spin, because they have an odd number of quarks. As quarks possess baryon number $B = \frac{1}{3}$, baryons have baryon number $B = 1$. The best-known baryons are the proton and the neutron.

One can hypothesise baryons with further quark-antiquark pairs in addition to their three quarks. Hypothetical baryons with one extra quark-antiquark pair (5 quarks in all) are called pentaquarks. As of August 2015, there are two known pentaquarks, P+
c(4380) and P+
c(4450), both discovered in 2015 by the LHCb collaboration.[3]

Each type of baryon has a corresponding antiparticle (antibaryon) in which quarks are replaced by their corresponding antiquarks. For example, just as a proton is made of two up-quarks and one down-quark, its corresponding antiparticle, the antiproton, is made of two up-antiquarks and one down-antiquark.

6.1.4 Mesons

Main article: Meson

Mesons are hadrons composed of a quark-antiquark pair. They are bosons, meaning they have integral spin, *i.e.*, 0, 1, or −1, as they have an even number of quarks. They have baryon number $B = 0$. Examples of mesons commonly produced in particle physics experiments include pions and kaons. Pions also play a role in holding atomic nuclei together via the residual strong force.

In principle, mesons with more than one quark-antiquark pair may exist; a hypothetical meson with two pairs is called a tetraquark. Several tetraquark candidates were found in the 2000s, but their status is under debate.[7] Several other hypothetical "exotic" mesons lie outside the quark model of classification. These include glueballs and hybrid mesons (mesons bound by excited gluons).

6.1.5 See also

- Hadronization, the formation of hadrons out of quarks and gluons

- Large Hadron Collider (LHC)

- List of particles

- Standard model

- Subatomic particles

- Hadron therapy, a.k.a. particle therapy

- Exotic hadrons

6.1.6 References

[1] Choi, S.-K.; Belle Collaboration et al. (2007). "Observation of a resonance-like structure in the π±Ψ′ mass distribution in exclusive B→Kπ±Ψ′ decays". *Physical Review Letters* **100**

(14). arXiv:0708.1790. Bibcode:2008PhRvL.100n2001C. doi:10.1103/PhysRevLett.100.142001.

[2] LHCb collaboration (2014): Observation of the resonant character of the Z(4430)⁻ state

[3] R. Aaij et al. (LHCb collaboration) (2015). "Observation of J/ψp resonances consistent with pentaquark states in Λ0 b→J/ψK–

p decays". *Physical Review Letters* **115** (7). doi:10.1103/PhysRevLett.115.072001.

[4] Lev B. Okun (1962). "The Theory of Weak Interaction". *Proceedings of 1962 International Conference on High-Energy Physics at CERN*. Geneva. p. 845. Bibcode:1962hep..conf..845O.

[5] C. Amsler *et al.* (Particle Data Group) (2008). "Review of Particle Physics – Quark Model" (PDF). *Physics Letters B* **667**: 1. Bibcode:2008PhLB..667....1P. doi:10.1016/j.physletb.2008.07.018.

[6] S. Bethke (2007). "Experimental tests of asymptotic freedom". *Progress in Particle and Nuclear Physics* **58** (2): 351. arXiv:hep-ex/0606035. Bibcode:2007PrPNP..58..351B. doi:10.1016/j.ppnp.2006.06.001.

[7] Mysterious Subatomic Particle May Represent Exotic New Form of Matter

6.2 Meson

In particle physics, **mesons** (/ˈmiːzɒnz/ or /ˈmɛzɒnz/) are hadronic subatomic particles composed of one quark and one antiquark, bound together by the strong interaction. Because mesons are composed of sub-particles, they have a physical size, with a diameter of roughly one fermi, which is about $2/3$ the size of a proton or neutron. All mesons are unstable, with the longest-lived lasting for only a few hundredths of a microsecond. Charged mesons decay (sometimes through intermediate particles) to form electrons and neutrinos. Uncharged mesons may decay to photons.

Mesons are not produced by radioactive decay, but appear in nature only as short-lived products of very high-energy interactions in matter, between particles made of quarks. In cosmic ray interactions, for example, such particles are ordinary protons and neutrons. Mesons are also frequently produced artificially in high-energy particle accelerators that collide protons, anti-protons, or other particles.

In nature, the importance of lighter mesons is that they are the associated quantum-field particles that transmit the nuclear force, in the same way that photons are the particles that transmit the electromagnetic force. The higher energy (more massive) mesons were created momentarily in the Big Bang, but are not thought to play a role in nature today. However, such particles are regularly created in experiments, in order to understand the nature of the heavier types of quark that compose the heavier mesons.

Mesons are part of the hadron particle family, defined simply as particles composed of two quarks. The other members of the hadron family are the baryons: subatomic particles composed of three quarks rather than two. Some experiments show evidence of exotic mesons, which don't have the conventional valence quark content of one quark and one antiquark.

Because quarks have a spin of $1/2$, the difference in quark-number between mesons and baryons results in conventional two-quark mesons being bosons, whereas baryons are fermions.

Each type of meson has a corresponding antiparticle (antimeson) in which quarks are replaced by their corresponding antiquarks and vice versa. For example, a positive pion (π+) is made of one up quark and one down antiquark; and its corresponding antiparticle, the negative pion (π−), is made of one up antiquark and one down quark.

Because mesons are composed of quarks, they participate in both the weak and strong interactions. Mesons with net electric charge also participate in the electromagnetic interaction. They are classified according to their quark content, total angular momentum, parity and various other properties, such as C-parity and G-parity. Although no meson is stable, those of lower mass are nonetheless more stable than the most massive mesons, and are easier to observe and study in particle accelerators or in cosmic ray experiments. They are also typically less massive than baryons, meaning that they are more easily produced in experiments, and thus exhibit certain higher energy phenomena more readily than baryons composed of the same quarks would. For example, the charm quark was first seen in the J/Psi meson (J/ψ) in 1974,[1][2] and the bottom quark in the upsilon meson (Υ) in 1977.[3]

6.2.1 History

From theoretical considerations, in 1934 Hideki Yukawa[4][5] predicted the existence and the approximate mass of the "meson" as the carrier of the nuclear force that holds atomic nuclei together. If there were no nuclear force, all nuclei with two or more protons would fly apart because of the electromagnetic repulsion. Yukawa called his carrier particle the meson, from μέσος *mesos*, the Greek word for "intermediate," because its predicted mass was between that of the electron and that of the proton, which has about 1,836 times the mass of the electron. Yukawa had originally named his particle the "mesotron", but he was corrected by the physicist Werner Heisenberg (whose father was a professor of Greek at the University of Munich). Heisenberg pointed out that there is no "tr" in the Greek word "mesos".[6]

The first candidate for Yukawa's meson, now known in modern terminology as the muon, was discovered in 1936 by Carl David Anderson and others in the decay products of cosmic ray interactions. The mu meson had about the right mass to be Yukawa's carrier of the strong nuclear force, but over the course of the next decade, it became evident that it was not the right particle. It was eventually found that the "mu meson" did not participate in the strong nuclear interaction at all, but rather behaved like a heavy version of the electron, and was eventually classed as a lepton like the electron, rather than a meson. Physicists in making this choice decided that properties other than particle mass should control their classification.

There were years of delays in the subatomic particle research during World War II in 1939–45, with most physicists working in applied projects for wartime necessities. When the war ended in August 1945, many physicists gradually returned to peacetime research. The first true meson to be discovered was what would later be called the "pi meson" (or pion). This discovery was made in 1947, by Cecil Powell, César Lattes, and Giuseppe Occhialini, who were investigating cosmic ray products at the University of Bristol in England, based on photographic films placed in the Andes mountains. Some mesons in these films had about the same mass as the already-known meson, yet seemed to decay into it, leading physicist Robert Marshak to hypothesize in 1947 that it was actually a new and different meson. Over the next few years, more experiments showed that the pion was indeed involved in strong interactions. The pion (as a virtual particle) is the primary force carrier for the nuclear force in atomic nuclei. Other mesons, such as the rho mesons are involved in mediating this force as well, but to lesser extents. Following the discovery of the pion, Yukawa was awarded the 1949 Nobel Prize in Physics for his predictions.

The word *meson* has at times been used to mean *any* force carrier, such as the "Z^0 meson", which is involved in mediating the weak interaction.[7] However, this spurious usage has fallen out of favor. Mesons are now defined as particles composed of pairs of quarks and antiquarks.

6.2.2 Overview

Spin, orbital angular momentum, and total angular momentum

Main articles: Spin (physics), angular momentum operator, Total angular momentum and Quantum numbers

Spin (quantum number S) is a vector quantity that represents the "intrinsic" angular momentum of a particle. It comes in increments of $\frac{1}{2}$ ħ. The ħ is often dropped because it is the "fundamental" unit of spin, and it is implied that "spin 1" means "spin 1 ħ". (In some systems of natural units, ħ is chosen to be 1, and therefore does not appear in equations).

Quarks are fermions—specifically in this case, particles having spin $\frac{1}{2}$ ($S = \frac{1}{2}$). Because spin projections vary in increments of 1 (that is 1 ħ), a single quark has a spin vector of length $\frac{1}{2}$, and has two spin projections ($S_z = +\frac{1}{2}$ and $S_z = -\frac{1}{2}$). Two quarks can have their spins aligned, in which case the two spin vectors add to make a vector of length $S = 1$ and three spin projections ($S_z = +1$, $S_z = 0$, and $S_z = -1$), called the spin-1 triplet. If two quarks have unaligned spins, the spin vectors add up to make a vector of length $S = 0$ and only one spin projection ($S_z = 0$), called the spin-0 singlet. Because mesons are made of one quark and one antiquark, they can be found in triplet and singlet spin states.

There is another quantity of quantized angular momentum, called the orbital angular momentum (quantum number L), that comes in increments of 1 ħ, which represent the angular momentum due to quarks orbiting around each other. The total angular momentum (quantum number J) of a particle is therefore the combination of intrinsic angular momentum (spin) and orbital angular momentum. It can take any value from $J = |L - S|$ to $J = |L + S|$, in increments of 1.

Particle physicists are most interested in mesons with no orbital angular momentum ($L = 0$), therefore the two groups of mesons most studied are the $S = 1$; $L = 0$ and $S = 0$; $L = 0$, which corresponds to $J = 1$ and $J = 0$, although they are not the only ones. It is also possible to obtain $J = 1$ particles from $S = 0$ and $L = 1$. How to distinguish between the $S = 1$, $L = 0$ and $S = 0$, $L = 1$ mesons is an active area of research in meson spectroscopy.

Parity

Main article: Parity (physics)

If the universe were reflected in a mirror, most of the laws of physics would be identical—things would behave the same way regardless of what we call "left" and what we call "right". This concept of mirror reflection is called parity (P). Gravity, the electromagnetic force, and the strong interaction all behave in the same way regardless of whether or not the universe is reflected in a mirror, and thus are said to conserve parity (P-symmetry). However, the weak interaction does distinguish "left" from "right", a phenomenon called parity violation (P-violation).

Based on this, one might think that, if the wavefunction for each particle (more precisely, the quantum field for each particle type) were simultaneously mirror-reversed, then the new set of wavefunctions would perfectly satisfy the laws of physics (apart from the weak interaction). It turns out that this is not quite true: In order for the equations to be satisfied, the wavefunctions of certain types of particles have to be multiplied by −1, in addition to being mirror-reversed. Such particle types are said to have *negative* or *odd* parity ($P = -1$, or alternatively $P = -$), whereas the other particles are said to have *positive* or *even* parity ($P = +1$, or alternatively $P = +$).

For mesons, the parity is related to the orbital angular momentum by the relation:[8]

$$P = (-1)^{L+1}$$

where the L is a result of the parity of the corresponding spherical harmonic of the wavefunction. The '+1' in the exponent comes from the fact that, according to the Dirac equation, a quark and an antiquark have opposite intrinsic parities. Therefore, the intrinsic parity of a meson is the product of the intrinsic parities of the quark (+1) and antiquark (−1). As these are different, their product is −1, and so it contributes a +1 in the exponent.

As a consequence, mesons with no orbital angular momentum ($L = 0$) all have odd parity ($P = -1$).

C-parity

Main article: C-parity

C-parity is only defined for mesons that are their own antiparticle (i.e. neutral mesons). It represents whether or not the wavefunction of the meson remains the same under the interchange of their quark with their antiquark.[9] If

$$|q\bar{q}\rangle = |\bar{q}q\rangle$$

then, the meson is "C even" (C = +1). On the other hand, if

$$|q\bar{q}\rangle = -|\bar{q}q\rangle$$

then the meson is "C odd" (C = −1).

C-parity rarely is studied on its own, but more commonly in combination with P-parity into CP-parity. CP-parity was thought to be conserved, but was later found to be violated in weak interactions.[10][11][12]

G-parity

Main article: G-parity

G parity is a generalization of the C-parity. Instead of simply comparing the wavefunction after exchanging quarks and antiquarks, it compares the wavefunction after exchanging the meson for the corresponding antimeson, regardless of quark content.[13] In the case of neutral meson, G-parity is equivalent to C-parity because neutral mesons are their own antiparticles.

If

$$|q_1\bar{q}_2\rangle = |\bar{q}_1 q_2\rangle$$

then, the meson is "G even" (G = +1). On the other hand, if

$$|q_1\bar{q}_2\rangle = -|\bar{q}_1 q_2\rangle$$

then the meson is "G odd" (G = −1).

Isospin and charge

Main article: Isospin

The concept of isospin was first proposed by Werner Heisenberg in 1932 to explain the similarities between protons and neutrons under the strong interaction.[14] Although they had different electric charges, their masses were so similar that physicists believed that they were actually the same particle. The different electric charges were explained as being the result of some unknown excitation similar to spin. This unknown excitation was later dubbed *isospin* by Eugene Wigner in 1937.[15] When the first mesons were discovered, they too were seen through the eyes of isospin and so the three pions were believed to be the same particle, but in different isospin states.

This belief lasted until Murray Gell-Mann proposed the quark model in 1964 (containing originally only the u, d, and s quarks).[16] The success of the isospin model is now understood to be the result of the similar masses of the u and d quarks. Because the u and d quarks have similar masses, particles made of the same number of them also have similar masses. The exact specific u and d quark composition determines the charge, because u quarks carry charge $+\frac{2}{3}$ whereas d quarks carry charge $-\frac{1}{3}$. For example the three pions all have different charges (π+ (ud), π0 (a quantum superposition of uu and dd states), π− (du)), but have similar masses (~140 MeV/c^2) as they are each made of a same number of total of up and down quarks and antiquarks. Under the isospin model, they were considered to be a single particle in different charged states.

The mathematics of isospin was modeled after that of spin. Isospin projections varied in increments of 1 just like those of spin, and to each projection was associated a "charged state". Because the "pion particle" had three "charged states", it was said to be of isospin $I = 1$. Its "charged states" $\pi+$, $\pi0$, and $\pi-$, corresponded to the isospin projections I_3 = +1, I_3 = 0, and I_3 = −1 respectively. Another example is the "rho particle", also with three charged states. Its "charged states" $\rho+$, $\rho0$, and $\rho-$, corresponded to the isospin projections I_3 = +1, I_3 = 0, and I_3 = −1 respectively. It was later noted that the isospin projections were related to the up and down quark content of particles by the relation

$$I_3 = \frac{1}{2}[(n_u - n_{\bar{u}}) - (n_d - n_{\bar{d}})],$$

where the n's are the number of up and down quarks and antiquarks.

In the "isospin picture", the three pions and three rhos were thought to be the different states of two particles. However, in the quark model, the rhos are excited states of pions. Isospin, although conveying an inaccurate picture of things, is still used to classify hadrons, leading to unnatural and often confusing nomenclature. Because mesons are hadrons, the isospin classification is also used, with I_3 = +$\frac{1}{2}$ for up quarks and down antiquarks, and I_3 = −$\frac{1}{2}$ for up antiquarks and down quarks.

Flavour quantum numbers

Main article: Flavour (particle physics) § Flavour quantum numbers

The strangeness quantum number S (not to be confused with spin) was noticed to go up and down along with particle mass. The higher the mass, the lower the strangeness (the more s quarks). Particles could be described with isospin projections (related to charge) and strangeness (mass) (see the uds nonet figures). As other quarks were discovered, new quantum numbers were made to have similar description of udc and udb nonets. Because only the u and d mass are similar, this description of particle mass and charge in terms of isospin and flavour quantum numbers only works well for the nonets made of one u, one d and one other quark and breaks down for the other nonets (for example ucb nonet). If the quarks all had the same mass, their behaviour would be called *symmetric*, because they would all behave in exactly the same way with respect to the strong interaction. However, as quarks do not have the same mass, they do not interact in the same way (exactly like an electron placed in an electric field will accelerate more than a proton placed in the same field because of its lighter mass), and the symmetry is said to be broken.

It was noted that charge (Q) was related to the isospin projection (I_3), the baryon number (B) and flavour quantum numbers (S, C, B', T) by the Gell-Mann–Nishijima formula:[17]

$$Q = I_3 + \frac{1}{2}(B + S + C + B' + T),$$

where S, C, B', and T represent the strangeness, charm, bottomness and topness flavour quantum numbers respectively. They are related to the number of strange, charm, bottom, and top quarks and antiquark according to the relations:

$$S = -(n_s - n_{\bar{s}})$$
$$C = +(n_c - n_{\bar{c}})$$
$$B' = -(n_b - n_{\bar{b}})$$
$$T = +(n_t - n_{\bar{t}}),$$

meaning that the Gell-Mann–Nishijima formula is equivalent to the expression of charge in terms of quark content:

$$Q = \frac{2}{3}[(n_u - n_{\bar{u}}) + (n_c - n_{\bar{c}}) + (n_t - n_{\bar{t}})] - \frac{1}{3}[(n_d - n_{\bar{d}}) + (n_s - n_{\bar{s}}) + (n_b$$

6.2.3   Classification

Mesons are classified into groups according to their isospin (I), total angular momentum (J), parity (P), G-parity (G) or C-parity (C) when applicable, and quark (q) content. The rules for classification are defined by the Particle Data Group, and are rather convoluted.[18] The rules are presented below, in table form for simplicity.

Types of meson

Mesons are classified into types according to their spin configurations. Some specific configurations are given special names based on the mathematical properties of their spin configuration.

Nomenclature

Flavourless mesons   Flavourless mesons are mesons made of pair of quark and antiquarks of the same flavour (all their flavour quantum numbers are zero: $S = 0$, $C = 0$, $B' = 0$, $T = 0$).[20] The rules for flavourless mesons are:[18]

[†] ∧ The C parity is only relevant to neutral mesons.
[††] ∧ For $J^{PC}=1^{--}$, the ψ is called the J/ψ

In addition:

- When the spectroscopic state of the meson is known, it is added in parentheses.

- When the spectroscopic state is unknown, mass (in MeV/c^2) is added in parentheses.

- When the meson is in its ground state, nothing is added in parentheses.

Flavoured mesons   Flavoured mesons are mesons made of pair of quark and antiquarks of different flavours. The rules are simpler in this case: the main symbol depends on the heavier quark, the superscript depends on the charge, and the subscript (if any) depends on the lighter quark. In table form, they are:[18]

In addition:

- If J^P is in the "normal series" (i.e., $J^P = 0^+, 1^-, 2^+, 3^-$, ...), a superscript $*$ is added.

- If the meson is not pseudoscalar ($J^P = 0^-$) or vector ($J^P = 1^-$), J is added as a subscript.

- When the spectroscopic state of the meson is known, it is added in parentheses.

- When the spectroscopic state is unknown, mass (in MeV/c^2) is added in parentheses.

- When the meson is in its ground state, nothing is added in parentheses.

6.2.4 Exotic mesons

Main article: Exotic meson

There is experimental evidence for particles that are hadrons (i.e., are composed of quarks) and are color-neutral with zero baryon number, and thus by conventional definition are mesons. Yet, these particles do not consist of a single quark-antiquark pair, as all the other conventional mesons discussed above do. A tentative category for these particles is exotic mesons.

There are at least five exotic meson resonances that have been experimentally confirmed to exist by two or more independent experiments. The most statistically significant

of these is the Z(4430), discovered by the Belle experiment in 2007 and confirmed by LHCb in 2014. It is a candidate for being a tetraquark: a particle composed of two quarks and two antiquarks.[21] See the main article above for other particle resonances that are candidates for being exotic mesons.

6.2.5 List

Main article: List of mesons

6.2.6 See also

- Standard Model

6.2.7 Notes

[1] J.J. Aubert *et al.* (1974)

[2] J.E. Augustin *et al.* (1974)

[3] S.W. Herb *et al.* (1977)

[4] The Noble Foundation (1949) Nobel Prize in Physics 1949 – Presentation Speech

[5] H. Yukawa (1935)

[6] G. Gamow (1961)

[7] J. Steinberger (1998)

[8] C. Amsler *et al.* (2008): Quark Model

[9] M.S. Sozzi (2008b)

[10] J.W. Cronin (1980)

[11] V.L. Fitch (1980)

[12] M.S. Sozzi (2008c)

[13] K. Gottfried, V.F. Weisskopf (1986)

[14] W. Heisenberg (1932)

[15] E. Wigner (1937)

[16] M. Gell-Mann (1964)

[17] S.S.M Wong (1998)

[18] C. Amsler *et al.* (2008): Naming scheme for hadrons

[19] W.E. Burcham, M. Jobes (1995)

[20] For the purpose of nomenclature, the isospin projection I_3 isn't considered a flavour quantum number. This means that the charged pion-like mesons ($\pi^\pm$, $a^\pm$, $b^\pm$, and $\rho^\pm$ mesons) follow the rules of flavourless mesons, even if they aren't truly "flavourless".

[21] LHCb collaborators (2014): Observation of the resonant character of the Z(4430)– state

6.2.8   References

- M.S. Sozzi (2008a). "Parity". *Discrete Symmetries and CP Violation: From Experiment to Theory*. Oxford University Press. pp. 15–87. ISBN 0-19-929666-9.

- M.S. Sozzi (2008b). "Charge Conjugation". *Discrete Symmetries and CP Violation: From Experiment to Theory*. Oxford University Press. pp. 88–120. ISBN 0-19-929666-9.

- M.S. Sozzi (2008c). "CP-Symmetry". *Discrete Symmetries and CP Violation: From Experiment to Theory*. Oxford University Press. pp. 231–275. ISBN 0-19-929666-9.

- C. Amsler *et al.* (Particle Data Group) (2008). "Review of Particle Physics". *Physics Letters B* **667** (1): 1–1340. Bibcode:2008PhLB..667....1P. doi:10.1016/j.physletb.2008.07.018.

- S.S.M. Wong (1998). "Nucleon Structure". *Introductory Nuclear Physics* (2nd ed.). New York (NY): John Wiley & Sons. pp. 21–56. ISBN 0-471-23973-9.

- W.E. Burcham, M. Jobes (1995). *Nuclear and Particle Physics* (2nd ed.). Longman Publishing. ISBN 0-582-45088-8.

- R. Shankar (1994). *Principles of Quantum Mechanics* (2nd ed.). New York (NY): Plenum Press. ISBN 0-306-44790-8.

- J. Steinberger (1989). "Experiments with high-energy neutrino beams". *Reviews of Modern Physics* **61** (3): 533–545. Bibcode:1989RvMP...61..533S. doi:10.1103/RevModPhys.61.533.

- K. Gottfried, V.F. Weisskopf (1986). "Hadronic Spectroscopy: G-parity". *Concepts of Particle Physics* **2**. Oxford University Press. pp. 303–311. ISBN 0-19-503393-0.

- J.W. Cronin (1980). "CP Symmetry Violation—The Search for its origin" (PDF). The Nobel Foundation.

- V.L. Fitch (1980). "The Discovery of Charge—Conjugation Parity Asymmetry" (PDF). The Nobel Foundation.

- S.W. Herb; Hom, D.; Lederman, L.; Sens, J.; Snyder, H.; Yoh, J.; Appel, J.; Brown, B. et al. (1977). "Observation of a Dimuon Resonance at 9.5 Gev in 400-GeV Proton-Nucleus Collisions". *Physical Review Letters* **39** (5): 252–255. Bibcode:1977PhRvL..39..252H. doi:10.1103/PhysRevLett.39.252.

- J.J. Aubert; Becker, U.; Biggs, P.; Burger, J.; Chen, M.; Everhart, G.; Goldhagen, P.; Leong, J. et al. (1974). "Experimental Observation of a Heavy Particle J". *Physical Review Letters* **33** (23): 1404–1406. Bibcode:1974PhRvL..33.1404A. doi:10.1103/PhysRevLett.33.1404.

- J.E. Augustin; Boyarski, A.; Breidenbach, M.; Bulos, F.; Dakin, J.; Feldman, G.; Fischer, G.; Fryberger, D. et al. (1974). "Discovery of a Narrow Resonance in e^+e^- Annihilation". *Physical Review Letters* **33** (23): 1406–1408. Bibcode:1974PhRvL..33.1406A. doi:10.1103/PhysRevLett.33.1406.

- M. Gell-Mann (1964). "A Schematic of Baryons and Mesons". *Physics Letters* **8** (3): 214–215. Bibcode:1964PhL.....8..214G. doi:10.1016/S0031-9163(64)92001-3.

- Ishfaq Ahmad (1965). "the Interactions of 200 MeV $\pi\pm$ -Mesons with Complex Nuclei Proposal to Study the Interactions of 200 MeV $\pi\pm$ -Mesons with Complex Nuclei" (PDF). *CERN documents* **3** (5).

- G. Gamow (1988) [1961]. *The Great Physicists from Galileo to Einstein* (Reprint ed.). Dover Publications. p. 315. ISBN 978-0-486-25767-9.

- E. Wigner (1937). "On the Consequences of the Symmetry of the Nuclear Hamiltonian on the Spectroscopy of Nuclei". *Physical Review* **51** (2): 106–119. Bibcode:1937PhRv...51..106W. doi:10.1103/PhysRev.51.106.

- H. Yukawa (1935). "On the Interaction of Elementary Particles" (PDF). *Proc. Phys. Math. Soc. Jap.* **17** (48).

- W. Heisenberg (1932). "Über den Bau der Atomkerne I". *Zeitschrift für Physik* (in German) **77**: 1–11. Bibcode:1932ZPhy...77....1H. doi:10.1007/BF01342433.

- W. Heisenberg (1932). "Über den Bau der Atomkerne II". *Zeitschrift für Physik* (in German) **78** (3–4): 156–164. Bibcode:1932ZPhy...78..156H. doi:10.1007/BF01337585.

- W. Heisenberg (1932). "Über den Bau der Atomkerne III". *Zeitschrift für Physik* (in German) **80** (9–10): 587–596. Bibcode:1933ZPhy...80..587H. doi:10.1007/BF01335696.

6.2.9 External links

- A table of some mesons and their properties

- *Particle Data Group*—Compiles authoritative information on particle properties

- hep-ph/0211411: The light scalar mesons within quark models

- Naming scheme for hadrons (a PDF file)

- Mesons made thinkable, an interactive visualisation allowing physical properties to be compared

Recent findings

- What Happened to the Antimatter? Fermilab's DZero Experiment Finds Clues in Quick-Change Meson

- CDF experiment's definitive observation of matter-antimatter oscillations in the Bs meson

6.3 List of mesons

This list is of all known and predicted scalar, pseudoscalar and vector mesons. See list of particles for a more detailed list of particles found in particle physics.

Mesons are unstable subatomic particles composed of one quark and one antiquark. They are part of the hadron particle family – particles made of quarks. The other members of the hadron family are the baryons – subatomic particles composed of three quarks. The main difference between mesons and baryons is that mesons have integer spin (thus are bosons) while baryons are fermions (half-integer spin). Because mesons are bosons, the Pauli exclusion principle does not apply to them. Because of this, they can act as force mediating particles on short distances, and thus play a part in processes such as the nuclear interaction.

Since mesons are composed of quarks, they participate in both the weak and strong interactions. Mesons with net electric charge also participate in the electromagnetic interaction. They are classified according to their quark content, total angular momentum, parity, and various other properties such as C-parity and G-parity. While no meson is stable, those of lower mass are nonetheless more stable than the most massive mesons, and are easier to observe and study in particle accelerators or in cosmic ray experiments. They are also typically less massive than baryons, meaning that they are more easily produced in experiments, and will exhibit higher-energy phenomena sooner than baryons

would. For example, the charm quark was first seen in the J/Psi meson (J/ψ) in 1974,[1][2] and the bottom quark in the upsilon meson (Υ) in 1977.[3]

Each meson has a corresponding antiparticle (antimeson) where quarks are replaced by their corresponding antiquarks and vice versa. For example, a positive pion (π+) is made of one up quark and one down antiquark; and its corresponding antiparticle, the negative pion (π−), is made of one up antiquark and one down quark. Some experiments show the evidence of *tetraquarks* – "exotic" mesons made of two quarks and two antiquarks, but the particle physics community as a whole does not view their existence as likely, although still possible.[4]

The symbols encountered in these lists are: I (*isospin*), J (*total angular momentum*), P (*parity*), C (*C-parity*), G (*G-parity*), u (*up quark*), d (*down quark*), s (*strange quark*), c (*charm quark*), b (*bottom quark*), Q (*charge*), B (*baryon number*), S (*strangeness*), C (*charm*), and B′ (*bottomness*), as well as a wide array of subatomic particles (hover for name).

6.3.1 Summary table

Because this table was initially derived from published results and many of those results were preliminary, as many as 64 of the mesons in the following table may not exist or have the wrong mass or quantum numbers.

6.3.2 Meson properties

The following lists detail all known and predicted pseudoscalar ($J^P = 0^-$) and vector ($J^P = 1^-$) mesons.

The properties and quark content of the particles are tabulated below; for the corresponding antiparticles, simply change quarks into antiquarks (and vice versa) and flip the sign of Q, B, S, C, and B′. Particles with † next to their names have been predicted by the standard model but not yet observed. Values in red have not been firmly established by experiments, but are predicted by the quark model and are consistent with the measurements.

Pseudoscalar mesons

[a] ^ Makeup inexact due to non-zero quark masses.
[b] ^ PDG reports the resonance width (Γ). Here the conversion $τ = \hbar/Γ$ is given instead.
[c] ^ Strong eigenstate. No definite lifetime (see kaon notes below)
[d] ^ The mass of the K0
L and K0
S are given as that of the K0. However, it is known that a

difference between the masses of the K0
L and K0
S on the order of 2.2×10^{-11} MeV/c^2 exists.[15]

[e] ∧ Weak eigenstate. Makeup is missing small CP–violating term (see notes on neutral kaons below).

Vector mesons

[f] ∧ PDG reports the resonance width (Γ). Here the conversion $\tau = {}^{\hbar}\!/\Gamma$ is given instead.

[g] ∧ The exact value depends on the method used. See the given reference for detail.

Notes on neutral kaons

There are two complications with neutral kaons:[34]

- Due to neutral kaon mixing, the K0
  S and K0
  L are not eigenstates of strangeness. However, they *are* eigenstates of the weak force, which determines how they decay, so these are the particles with definite lifetime.

- The linear combinations given in the table for the K0
  S and K0
  L are not exactly correct, since there is a small correction due to CP violation. See CP violation in kaons.

Note that these issues also exist in principle for other neutral flavored mesons; however, the weak eigenstates are considered separate particles only for kaons because of their dramatically different lifetimes.[34]

6.3.3   See also

- List of baryons
- List of particles
- Timeline of particle discoveries

6.3.4   References

[1] J.J. Aubert *et al.* (1974)

[2] J.E. Augustin *et al.* (1974)

[3] S.W. Herb *et al.* (1977)

[4] C. Amsler *et al.* (2008): Charmonium States

[5] K.A. Olive *et al.* (2014): Meson Summary Table

[6] K.A. Olive *et al.* (2014): Particle listings – π±

[7] K.A. Olive *et al.* (2014): Particle listings – π0

[8] K.A. Olive *et al.* (2014): Particle listings – η

[9] K.A. Olive *et al.* (2014): Particle listings – η′

[10] K.A. Olive *et al.* (2014): Particle listings – η
c

[11] K.A. Olive *et al.* (2014): Particle listings – η
b

[12] K.A. Olive *et al.* (2014): Particle listings – K±

[13] K.A. Olive *et al.* (2014): Particle listings – K0

[14] K.A. Olive *et al.* (2014): Particle listings – K0
S

[15] K.A. Olive *et al.* (2014): Particle listings – K0
L

[16] K.A. Olive *et al.* (2014): Particle listings – D±

[17] K.A. Olive *et al.* (2014): Particle listings – D0

[18] K.A. Olive *et al.* (2014): Particle listings – D±
s

[19] K.A. Olive *et al.* (2014): Particle listings – B±

[20] K.A. Olive *et al.* (2014): Particle listings – B0

[21] K.A. Olive *et al.* (2014): Particle listings – B0
s

[22] K.A. Olive *et al.* (2014): Particle listings – B±
c

[23] K.A. Olive *et al.* (2014): Particle listings – ρ

[24] K.A. Olive *et al.* (2014): Particle listings – ω(782)

[25] K.A. Olive *et al.* (2014): Particle listings – φ

[26] K.A. Olive *et al.* (2014): Particle listings – J/Ψ

[27] K.A. Olive *et al.* (2014): Particle listings – Υ(1S)

[28] K.A. Olive *et al.* (2014): Particle listings – K∗(892)

[29] K.A. Olive *et al.* (2014): Particle listings – D∗±(2010)

[30] K.A. Olive *et al.* (2014): Particle listings – D∗0(2007)

[31] K.A. Olive *et al.* (2014): Particle listings – D∗±
s

[32] K.A. Olive *et al.* (2014): Particle listings – B∗

[33] K.A. Olive *et al.* (2014): Particle listings – B∗
s

[34] J.W. Cronin (1980)

Bibliography

- K.A. Olive *et al.* (Particle Data Group) (2014). "Review of Particle Physics". *Chinese Physics C* **38** (9): 090001.

- M.S. Sozzi (2008a). "Parity". *Discrete Symmetries and CP Violation: From Experiment to Theory*. Oxford University Press. pp. 15–87. ISBN 0-19-929666-9.

- M.S. Sozzi (2008a). "Charge Conjugation". *Discrete Symmetries and CP Violation: From Experiment to Theory*. Oxford University Press. pp. 88–120. ISBN 0-19-929666-9.

- M.S. Sozzi (2008c). "CP-Symmetry". *Discrete Symmetries and CP Violation: From Experiment to Theory*. Oxford University Press. pp. 231–275. ISBN 0-19-929666-9.

- C. Amsler *et al.* (Particle Data Group); Amsler; Doser; Antonelli; Asner; Babu; Baer; Band; Barnett; Bergren; Beringer; Bernardi; Bertl; Bichsel; Biebel; Bloch; Blucher; Blusk; Cahn; Carena; Caso; Ceccucci; Chakraborty; Chen; Chivukula; Cowan; Dahl; d'Ambrosio; Damour et al. (2008). "Review of Particle Physics". *Physics Letters B* **667** (1): 1–1340. Bibcode:2008PhLB..667....1P. doi:10.1016/j.physletb.2008.07.018.

- S.S.M. Wong (1998). "Nucleon Structure". *Introductory Nuclear Physics* (2nd ed.). John Wiley & Sons. pp. 21–56. ISBN 0-471-23973-9.

- R. Shankar (1994). *Principles of Quantum Mechanics* (2nd ed.). Plenum Press. ISBN 0-306-44790-8.

- K. Gottfried, V.F. Weisskopf (1986). "Hadronic Spectroscopy: G-parity". *Concepts of Particle Physics* **2**. Oxford University Press. pp. 303–311. ISBN 0-19-503393-0.

- J.W. Cronin (1980). "CP Symmetry Violation – The Search for its origin" (PDF). *Nobel Lecture*. The Nobel Foundation.

- V.L. Fitch (1980). "The Discovery of Charge – Conjugation Parity Asymmetry" (PDF). *Nobel Lecture*. The Nobel Foundation.

- S.W. Herb; Hom, D.; Lederman, L.; Sens, J.; Snyder, H.; Yoh, J.; Appel, J.; Brown, B. et al. (1977). "Observation of a Dimuon Resonance at 9.5 Gev in 400-GeV Proton-Nucleus Collisions". *Physical Review Letters* **39** (5): 252–255. Bibcode:1977PhRvL..39..252H. doi:10.1103/PhysRevLett.39.252.

- J.J. Aubert; Becker, U.; Biggs, P.; Burger, J.; Chen, M.; Everhart, G.; Goldhagen, P.; Leong, J. et al. (1974). "Experimental Observation of a Heavy Particle J". *Physical Review Letters* **33** (23): 1404–1406. Bibcode:1974PhRvL..33.1404A. doi:10.1103/PhysRevLett.33.1404.

- J.E. Augustin; Boyarski, A.; Breidenbach, M.; Bulos, F.; Dakin, J.; Feldman, G.; Fischer, G.; Fryberger, D. et al. (1974). "Discovery of a Narrow Resonance in e^+e^- Annihilation". *Physical Review Letters* **33** (23): 1406–1408. Bibcode:1974PhRvL..33.1406A. doi:10.1103/PhysRevLett.33.1406.

- M. Gell-Mann (1964). "A Schematic of Baryons and Mesons". *Physics Letters* **8** (3): 214–215. Bibcode:1964PhL......8..214G. doi:10.1016/S0031-9163(64)92001-3.

- E. Wigner (1937). "On the Consequences of the Symmetry of the Nuclear Hamiltonian on the Spectroscopy of Nuclei". *Physical Review* **51** (2): 106–119. Bibcode:1937PhRv...51..106W. doi:10.1103/PhysRev.51.106.

- W. Heisenberg (1932). "Über den Bau der Atomkerne I". *Zeitschrift für Physik* (in German) **77** (1–2): 1–11. Bibcode:1932ZPhy...77....1H. doi:10.1007/BF01342433.

- W. Heisenberg (1932). "Über den Bau der Atomkerne II". *Zeitschrift für Physik* (in German) **78** (3–4): 156–164. Bibcode:1932ZPhy...78..156H. doi:10.1007/BF01337585.

- W. Heisenberg (1932). "Über den Bau der Atomkerne III". *Zeitschrift für Physik* (in German) **80** (9–10): 587–596. Bibcode:1933ZPhy...80..587H. doi:10.1007/BF01335696.

6.3.5 External links

- Particle Data Group – The Review of Particle Physics (2008)

- Mesons made thinkable, an interactive visualisation allowing physical properties to be compared

6.4 Baryon

Not to be confused with Baryonyx.

A **baryon** is a composite subatomic particle made up of three quarks (as distinct from mesons, which are composed of one quark and one antiquark). Baryons and mesons belong to the hadron family of particles, which are the quark-based particles. The name "baryon" comes from the Greek word for "heavy" (βαρύς, *barys*), because, at the time of their naming, most known elementary particles had lower masses than the baryons.

As quark-based particles, baryons participate in the strong interaction, whereas leptons, which are not quark-based, do not. The most familiar baryons are the protons and neutrons that make up most of the mass of the visible matter in the universe. Electrons (the other major component of the atom) are leptons.

Each baryon has a corresponding antiparticle (antibaryon) where quarks are replaced by their corresponding antiquarks. For example, a proton is made of two up quarks and one down quark; and its corresponding antiparticle, the antiproton, is made of two up antiquarks and one down antiquark.

6.4.1   Background

Baryons are strongly interacting fermions that is, they experience the strong nuclear force and are described by Fermi–Dirac statistics, which apply to all particles obeying the Pauli exclusion principle. This is in contrast to the bosons, which do not obey the exclusion principle.

Baryons, along with mesons, are hadrons, meaning they are particles composed of quarks. Quarks have baryon numbers of $B = \frac{1}{3}$ and antiquarks have baryon number of $B = -\frac{1}{3}$. The term "baryon" usually refers to *triquarks*—baryons made of three quarks ($B = \frac{1}{3} + \frac{1}{3} + \frac{1}{3} = 1$).

Other exotic baryons have been proposed, such as pentaquarks—baryons made of four quarks and one antiquark ($B = \frac{1}{3} + \frac{1}{3} + \frac{1}{3} + \frac{1}{3} - \frac{1}{3} = 1$), but their existence is not generally accepted. In theory, heptaquarks (5 quarks, 2 antiquarks), nonaquarks (6 quarks, 3 antiquarks), etc. could also exist. Until recently, it was believed that some experiments showed the existence of pentaquarks—baryons made of four quarks and one antiquark.[1][2] The particle physics community as a whole did not view their existence as likely in 2006,[3] and in 2008, considered evidence to be overwhelmingly against the existence of the reported pentaquarks.[4] However, in July 2015, the LHCb experiment observed two resonances consistent with pentaquark states in the Λ0

b → J/ψK−

p decay, with a combined statistical significance of 15σ.[5][6]

6.4.2   Baryonic matter

Nearly all matter that may be encountered or experienced in everyday life is baryonic matter, which includes atoms of any sort, and provides those with the quality of mass. Non-baryonic matter, as implied by the name, is any sort of matter that is not composed primarily of baryons. Those might include neutrinos or free electrons, dark matter, such as supersymmetric particles, axions, or black holes.

The very existence of baryons is also a significant issue in cosmology because it is assumed that the Big Bang produced a state with equal amounts of baryons and antibaryons. The process by which baryons came to outnumber their antiparticles is called baryogenesis.

6.4.3   Baryogenesis

Main article: Baryogenesis

Experiments are consistent with the number of quarks in the universe being a constant and, to be more specific, the number of baryons being a constant ; in technical language, the total baryon number appears to be *conserved*. Within the prevailing Standard Model of particle physics, the number of baryons may change in multiples of three due to the action of sphalerons, although this is rare and has not been observed under experiment. Some grand unified theories of particle physics also predict that a single proton can decay, changing the baryon number by one; however, this has not yet been observed under experiment. The excess of baryons over antibaryons in the present universe is thought to be due to non-conservation of baryon number in the very early universe, though this is not well understood.

6.4.4   Properties

Isospin and charge

Main article: Isospin

The concept of isospin was first proposed by Werner Heisenberg in 1932 to explain the similarities between protons and neutrons under the strong interaction.[7] Although they had different electric charges, their masses were so similar that physicists believed they were actually the same particle. The different electric charges were explained as being the result of some unknown excitation similar to spin. This unknown excitation was later dubbed *isospin* by Eugene Wigner in 1937.[8]

This belief lasted until Murray Gell-Mann proposed the quark model in 1964 (containing originally only the u, d, and s quarks).[9] The success of the isospin model is now understood to be the result of the similar masses of the u and d quarks. Since the u and d quarks have similar masses, particles made of the same number then also have similar masses. The exact specific u and d quark composition determines the charge, as u quarks carry charge $+^2/_3$ while d quarks carry charge $-^1/_3$. For example the four Deltas all have different charges (Δ++ (uuu), Δ+ (uud), Δ0 (udd), Δ– (ddd)), but have similar masses (~1,232 MeV/c^2) as they are each made of a combination of three u and d quarks. Under the isospin model, they were considered to be a single particle in different charged states.

The mathematics of isospin was modeled after that of spin. Isospin projections varied in increments of 1 just like those of spin, and to each projection was associated a "charged state". Since the "Delta particle" had four "charged states", it was said to be of isospin $I = {}^3/_2$. Its "charged states" Δ++, Δ+, Δ0, and Δ–, corresponded to the isospin projections $I_3 = +^1/_2$, $I_3 = +^1/_2$, $I_3 = -^1/_2$, and $I_3 = -^3/_2$, respectively. Another example is the "nucleon particle". As there were two nucleon "charged states", it was said to be of isospin $^1/_2$. The positive nucleon N+ (proton) was identified with $I_3 = +^1/_2$ and the neutral nucleon N0 (neutron) with $I_3 = -^1/_2$.[10] It was later noted that the isospin projections were related to the up and down quark content of particles by the relation:

$$I_3 = \frac{1}{2}[(n_u - n_{\bar{u}}) - (n_d - n_{\bar{d}})],$$

where the n's are the number of up and down quarks and antiquarks.

In the "isospin picture", the four Deltas and the two nucleons were thought to be the different states of two particles. However in the quark model, Deltas are different states of nucleons (the N^{++} or N$^-$ are forbidden by Pauli's exclusion principle). Isospin, although conveying an inaccurate picture of things, is still used to classify baryons, leading to unnatural and often confusing nomenclature.

Flavour quantum numbers

Main article: Flavour (particle physics) § Flavour quantum numbers

The strangeness flavour quantum number S (not to be confused with spin) was noticed to go up and down along with particle mass. The higher the mass, the lower the strangeness (the more s quarks). Particles could be described with isospin projections (related to charge) and strangeness (mass) (see the uds octet and decuplet figures on the right). As other quarks were discovered, new quantum numbers were made to have similar description of udc and udb octets and decuplets. Since only the u and d mass are similar, this description of particle mass and charge in terms of isospin and flavour quantum numbers works well only for octet and decuplet made of one u, one d, and one other quark, and breaks down for the other octets and decuplets (for example, ucb octet and decuplet). If the quarks all had the same mass, their behaviour would be called *symmetric*, as they would all behave in exactly the same way with respect to the strong interaction. Since quarks do not have the same mass, they do not interact in the same way (exactly like an electron placed in an electric field will accelerate more than a proton placed in the same field because of its lighter mass), and the symmetry is said to be broken.

It was noted that charge (Q) was related to the isospin projection (I_3), the baryon number (B) and flavour quantum numbers (S, C, B', T) by the Gell-Mann–Nishijima formula:[10]

$$Q = I_3 + \frac{1}{2}(B + S + C + B' + T),$$

where S, C, B', and T represent the strangeness, charm, bottomness and topness flavour quantum numbers, respectively. They are related to the number of strange, charm, bottom, and top quarks and antiquark according to the relations:

$$S = -(n_s - n_{\bar{s}}),$$
$$C = +(n_c - n_{\bar{c}}),$$
$$B' = -(n_b - n_{\bar{b}}),$$
$$T = +(n_t - n_{\bar{t}}),$$

meaning that the Gell-Mann–Nishijima formula is equivalent to the expression of charge in terms of quark content:

$$Q = \frac{2}{3}[(n_u - n_{\bar{u}}) + (n_c - n_{\bar{c}}) + (n_t - n_{\bar{t}})] - \frac{1}{3}[(n_d - n_{\bar{d}}) + (n_s - n_{\bar{s}}) + (n_b - n_{\bar{b}})].$$

Spin, orbital angular momentum, and total angular momentum

Main articles: Spin (physics), Angular momentum operator, Quantum numbers and Clebsch–Gordan coefficients

Spin (quantum number S) is a vector quantity that represents the "intrinsic" angular momentum of a particle. It comes in increments of $\frac{1}{2}$ ℏ (pronounced "h-bar"). The ℏ is often dropped because it is the "fundamental" unit of spin, and it is implied that "spin 1" means "spin 1 ℏ". In some systems of natural units, ℏ is chosen to be 1, and therefore does not appear anywhere.

Quarks are fermionic particles of spin $\frac{1}{2}$ ($S = \frac{1}{2}$). Because spin projections vary in increments of 1 (that is 1 ℏ), a single quark has a spin vector of length $\frac{1}{2}$, and has two spin projections ($S_z = +\frac{1}{2}$ and $S_z = -\frac{1}{2}$). Two quarks can have their spins aligned, in which case the two spin vectors add to make a vector of length $S = 1$ and three spin projections ($S_z = +1$, $S_z = 0$, and $S_z = -1$). If two quarks have unaligned spins, the spin vectors add up to make a vector of length $S = 0$ and has only one spin projection ($S_z = 0$), etc. Since baryons are made of three quarks, their spin vectors can add to make a vector of length $S = \frac{3}{2}$, which has four spin projections ($S_z = +\frac{3}{2}$, $S_z = +\frac{1}{2}$, $S_z = -\frac{1}{2}$, and $S_z = -\frac{3}{2}$), or a vector of length $S = \frac{1}{2}$ with two spin projections ($S_z = +\frac{1}{2}$, and $S_z = -\frac{1}{2}$).[11]

There is another quantity of angular momentum, called the orbital angular momentum, (azimuthal quantum number L), that comes in increments of 1 ℏ, which represent the angular moment due to quarks orbiting around each other. The total angular momentum (total angular momentum quantum number J) of a particle is therefore the combination of intrinsic angular momentum (spin) and orbital angular momentum. It can take any value from $J = |L - S|$ to $J = |L + S|$, in increments of 1.

Particle physicists are most interested in baryons with no orbital angular momentum ($L = 0$), as they correspond to ground states—states of minimal energy. Therefore the two groups of baryons most studied are the $S = \frac{1}{2}$; $L = 0$ and $S = \frac{3}{2}$; $L = 0$, which corresponds to $J = \frac{1}{2}^+$ and $J = \frac{3}{2}^+$, respectively, although they are not the only ones. It is also possible to obtain $J = \frac{3}{2}^+$ particles from $S = \frac{1}{2}$ and $L = 2$, as well as $S = \frac{3}{2}$ and $L = 2$. This phenomenon of having multiple particles in the same total angular momentum configuration is called *degeneracy*. How to distinguish between these degenerate baryons is an active area of research in baryon spectroscopy.[12][13]

Parity

Main article: Parity (physics)

If the universe were reflected in a mirror, most of the laws of physics would be identical—things would behave the same way regardless of what we call "left" and what we call "right". This concept of mirror reflection is called *intrinsic*

parity or *parity* (P). Gravity, the electromagnetic force, and the strong interaction all behave in the same way regardless of whether or not the universe is reflected in a mirror, and thus are said to conserve parity (P-symmetry). However, the weak interaction *does* distinguish "left" from "right", a phenomenon called parity violation (P-violation).

Based on this, one might think that, if the wavefunction for each particle (in more precise terms, the quantum field for each particle type) were simultaneously mirror-reversed, then the new set of wavefunctions would perfectly satisfy the laws of physics (apart from the weak interaction). It turns out that this is not quite true: In order for the equations to be satisfied, the wavefunctions of certain types of particles have to be multiplied by −1, in addition to being mirror-reversed. Such particle types are said to have *negative* or *odd* parity ($P = -1$, or alternatively $P = -$), while the other particles are said to have *positive* or *even* parity ($P = +1$, or alternatively $P = +$).

For baryons, the parity is related to the orbital angular momentum by the relation:[14]

$$P = (-1)^L.$$

As a consequence, baryons with no orbital angular momentum ($L = 0$) all have even parity ($P = +$).

6.4.5 Nomenclature

Baryons are classified into groups according to their isospin (I) values and quark (q) content. There are six groups of baryons—nucleon (N), Delta (Δ), Lambda (Λ), Sigma (Σ), Xi (Ξ), and Omega (Ω). The rules for classification are defined by the Particle Data Group. These rules consider the up (u), down (d) and strange (s) quarks to be *light* and the charm (c), bottom (b), and top (t) quarks to be *heavy*. The rules cover all the particles that can be made from three of each of the six quarks, even though baryons made of t quarks are not expected to exist because of the t quark's short lifetime. The rules do not cover pentaquarks.[15]

- Baryons with three u and/or d quarks are N's ($I = \frac{1}{2}$) or Δ's ($I = \frac{3}{2}$).

- Baryons with two u and/or d quarks are Λ's ($I = 0$) or Σ's ($I = 1$). If the third quark is heavy, its identity is given by a subscript.

- Baryons with one u or d quark are Ξ's ($I = \frac{1}{2}$). One or two subscripts are used if one or both of the remaining quarks are heavy.

- Baryons with no u or d quarks are Ω's ($I = 0$), and subscripts indicate any heavy quark content.

- Baryons that decay strongly have their masses as part of their names. For example, Σ^0 does not decay strongly, but $\Delta^{++}(1232)$ does.

It is also a widespread (but not universal) practice to follow some additional rules when distinguishing between some states that would otherwise have the same symbol.[10]

- Baryons in total angular momentum $J = \frac{3}{2}$ configuration that have the same symbols as their $J = \frac{1}{2}$ counterparts are denoted by an asterisk (*).

- Two baryons can be made of three different quarks in $J = \frac{1}{2}$ configuration. In this case, a prime (′) is used to distinguish between them.

  - *Exception*: When two of the three quarks are one up and one down quark, one baryon is dubbed Λ while the other is dubbed Σ.

Quarks carry charge, so knowing the charge of a particle indirectly gives the quark content. For example, the rules above say that a Λ+ c contains a c quark and some combination of two u and/or d quarks. The c quark has a charge of ($Q = +\frac{2}{3}$), therefore the other two must be a u quark ($Q = +\frac{2}{3}$), and a d quark ($Q = -\frac{1}{3}$) to have the correct total charge ($Q = +1$).

6.4.6 See also

- Eightfold way
- List of baryons
- List of particles
- Meson
- Timeline of particle discoveries

6.4.7 Notes

[1] H. Muir (2003)

[2] K. Carter (2003)

[3] W.-M. Yao *et al.* (2006): Particle listings – Θ^+

[4] C. Amsler *et al.* (2008): Pentaquarks

[5] LHCb (14 July 2015). "Observation of particles composed of five quarks, pentaquark-charmonium states, seen in $\Lambda_b^0 \to J/\psi pK^-$ decays.". *CERN website*. Retrieved 2015-07-14.

[6] R. Aaij et al. (LHCb collaboration) (2015). "Observation of J/ψp resonances consistent with pentaquark states in Λ0 b→J/ψK– p decays". *Physical Review Letters* **115** (7). Bibcode:2015PhRvL.115g2001A. doi:10.1103/PhysRevLett.115.072001.

[7] W. Heisenberg (1932)

[8] E. Wigner (1937)

[9] M. Gell-Mann (1964)

[10] S.S.M. Wong (1998a)

[11] R. Shankar (1994)

[12] H. Garcilazo *et al.* (2007)

[13] D.M. Manley (2005)

[14] S.S.M. Wong (1998b)

[15] C. Amsler *et al.* (2008): Naming scheme for hadrons

6.4.8 References

- C. Amsler *et al.* (Particle Data Group) (2008). "Review of Particle Physics". *Physics Letters B* **667** (1): 1–1340. Bibcode:2008PhLB..667....1P. doi:10.1016/j.physletb.2008.07.018.

- H. Garcilazo, J. Vijande, and A. Valcarce (2007). "Faddeev study of heavy-baryon spectroscopy". *Journal of Physics G* **34** (5): 961–976. doi:10.1088/0954-3899/34/5/014.

- K. Carter (2006). "The rise and fall of the pentaquark". Fermilab and SLAC. Retrieved 2008-05-27.

- W.-M. Yao *et al.*(Particle Data Group) (2006). "Review of Particle Physics". *Journal of Physics G* **33**: 1–1232. arXiv:astro-ph/0601168. Bibcode:2006JPhG...33....1Y. doi:10.1088/0954-3899/33/1/001.

- D.M. Manley (2005). "Status of baryon spectroscopy". *Journal of Physics: Conference Series* **5**: 230–237. Bibcode:2005JPhCS...9..230M. doi:10.1088/1742-6596/9/1/043.

- H. Muir (2003). "Pentaquark discovery confounds sceptics". New Scientist. Retrieved 2008-05-27.

- S.S.M. Wong (1998a). "Chapter 2—Nucleon Structure". *Introductory Nuclear Physics* (2nd ed.). New York (NY): John Wiley & Sons. pp. 21–56. ISBN 0-471-23973-9.

- S.S.M. Wong (1998b). "Chapter 3—The Deuteron". *Introductory Nuclear Physics* (2nd ed.). New York (NY): John Wiley & Sons. pp. 57–104. ISBN 0-471-23973-9.

- R. Shankar (1994). *Principles of Quantum Mechanics* (2nd ed.). New York (NY): Plenum Press. ISBN 0-306-44790-8.

- E. Wigner (1937). "On the Consequences of the Symmetry of the Nuclear Hamiltonian on the Spectroscopy of Nuclei". *Physical Review* **51** (2): 106–119. Bibcode:1937PhRv...51..106W. doi:10.1103/PhysRev.51.106.

- M. Gell-Mann (1964). "A Schematic of Baryons and Mesons". *Physics Letters* **8** (3): 214–215. Bibcode:1964PhL.....8..214G. doi:10.1016/S0031-9163(64)92001-3.

- W. Heisenberg (1932). "Über den Bau der Atomkerne I". *Zeitschrift für Physik* (in German) **77**: 1–11. Bibcode:1932ZPhy...77....1H. doi:10.1007/BF01342433.

- W. Heisenberg (1932). "Über den Bau der Atomkerne II". *Zeitschrift für Physik* (in German) **78** (3–4): 156–164. Bibcode:1932ZPhy...78..156H. doi:10.1007/BF01337585.

- W. Heisenberg (1932). "Über den Bau der Atomkerne III". *Zeitschrift für Physik* (in German) **80** (9–10): 587–596. Bibcode:1933ZPhy...80..587H. doi:10.1007/BF01335696.

6.4.9 External links

- Particle Data Group—Review of Particle Physics (2008).

- Georgia State University—HyperPhysics

- Baryons made thinkable, an interactive visualisation allowing physical properties to be compared

6.5 List of baryons

Baryons are composite particles made of three quarks, as opposed to mesons, which are composite particles made of one quark and one antiquark. Baryons and mesons are both hadrons, which are particles composed solely of quarks or both quarks and antiquarks. The term *baryon* is derived from the Greek *"βαρύς"* (*barys*), meaning "heavy", because, at the time of their naming, it was believed that baryons were characterized by having greater masses than other particles that were classed as matter.

Until a few years ago, it was believed that some experiments showed the existence of pentaquarks – baryons made of four quarks and one antiquark.[1][2] The particle physics community as a whole did not view their existence as likely by 2006.[3] On 13 July 2015, the LHCb collaboration at CERN reported results consistent with pentaquark states in the decay of bottom Lambda baryons (Λ^0_b).[4]

Since baryons are composed of quarks, they participate in the strong interaction. Leptons, on the other hand, are not composed of quarks and as such do not participate in the strong interaction. The most famous baryons are the protons and neutrons that make up most of the mass of the visible matter in the universe, whereas electrons, the other major component of atoms, are leptons. Each baryon has a corresponding antiparticle known as an antibaryon in which quarks are replaced by their corresponding antiquarks. For example, a proton is made of two up quarks and one down quark, while its corresponding antiparticle, the antiproton, is made of two up antiquarks and one down antiquark.

6.5.1 Lists of baryons

These lists detail all known and predicted baryons in total angular momentum $J = \frac{1}{2}$ and $J = \frac{3}{2}$ configurations with positive parity.

- Baryons composed of one type of quark (uuu, ddd, ...) can exist in $J = \frac{3}{2}$ configuration, but $J = \frac{1}{2}$ is forbidden by the Pauli exclusion principle.

- Baryons composed of two types of quarks (uud, uus, ...) can exist in both $J = \frac{1}{2}$ and $J = \frac{3}{2}$ configurations

- Baryons composed of three types of quarks (uds, udc, ...) can exist in both $J = \frac{1}{2}$ and $J = \frac{3}{2}$ configurations. Two $J = \frac{1}{2}$ configurations are possible for these baryons.

The symbols encountered in these lists are: I (isospin), J (total angular momentum), P (parity), u (up quark), d (down quark), s (strange quark), c (charm quark), b (bottom quark), Q (charge), B (baryon number), S (strangeness), C (charm), B' (bottomness), as well as a wide array of subatomic particles (hover for name). (See the *baryon* article for a detailed explanation of these symbols.)

Antibaryons are not listed in the tables; however, they simply would have all quarks changed to antiquarks, and Q, B, S, C, B', would be of opposite signs. Particles with † next to their names have been predicted by the Standard Model but not yet observed. Values in red have not been firmly established by experiments, but are predicted by the quark model and are consistent with the measurements.[5][6]

$J^P = \frac{1}{2}^+$ baryons

†^ Particle has not yet been observed.

[a] ^ The masses of the proton and neutron are known with much better precision in atomic mass units (u) than in MeV/c^2, due to the relatively poorly known value of the elementary charge. In atomic mass unit, the mass of the proton is 1.007276466812(90) u whereas that of the neutron is 1.00866491600(43) u.

[b] ^ At least 10^{35} years. See proton decay.

[c] ^ For free neutrons; in most common nuclei, neutrons are stable.

[d] ^ PDG reports the resonance width (Γ). Here the conversion τ = $^{\hbar}/\Gamma$ is given instead.

[e] ^ Some controversy exists about this data.[23]

$J^P = \frac{3}{2}^+$ baryons

†^ Particle has not yet been observed.

[h] ^ PDG reports the resonance width (Γ). Here the conversion τ = $^{\hbar}/\Gamma$ is given instead.

Baryon resonance particles

This table gives the name, quantum numbers (where known), and experimental status of baryons resonances confirmed by the PDG.[36] Baryon resonance particles are excited baryon states with short half lives and higher masses. Despite significant research, the fundamental degrees of freedom behind baryon excitation spectra are still poorly understood.[37] The spin-parity J^P (when known) is given with each particle. For the strongly decaying particles, the J^P values are considered to be part of the names, as is the mass for all resonances.

6.5.2 See also

- Eightfold way (physics)
- List of mesons
- List of particles
- Timeline of particle discoveries

6.5.3 References

[1] H. Muir (2003)

[2] K. Carter (2003)

[3] W.-M. Yao *et al.* (2006): Particle listings – Positive Theta

[4] R. Aaij et al. (LHCb collaboration) (2015). "Observation of J/ψp resonances consistent with pentaquark states in Λ0 b→J/ψK− p decays". *Physical Review Letters* **115** (7). doi:10.1103/PhysRevLett.115.072001.

[5] J. Beringer *et al.* (2012) and 2013 partial update for the 2014 edition: Particle summary tables – Baryons

[6] J.G. Körner *et al.* (1994)

[7] J. Beringer *et al.* (2012): Particle listings – p+

[8] J. Beringer *et al.* (2012): Particle listings – n0

[9] J. Beringer *et al.* (2012): Particle listings – Λ

[10] J. Beringer *et al.* (2012): Particle listings – Λ c

[11] J. Beringer *et al.* (2012): Particle listings – Λ b

[12] J. Beringer *et al.* (2012): Particle listings – Σ+

[13] J. Beringer *et al.* (2012): Particle listings – Σ0

[14] J. Beringer *et al.* (2012): Particle listings – Σ–

[15] J. Beringer *et al.* (2012): Particle listings – Σ c

[16] J. Beringer *et al.* (2012): Particle listings – Σ b

[17] J. Beringer *et al.* (2012): Particle listings – Ξ0

[18] J. Beringer *et al.* (2012): Particle listings – Ξ–

[19] J. Beringer *et al.* (2012): Particle listings – Ξ+ c

[20] J. Beringer *et al.* (2012): Particle listings – Ξ0 c

[21] J. Beringer *et al.* (2012): Particle listings – Ξ′+ c

[22] J. Beringer *et al.* (2012): Particle listings – Ξ′0 c

[23] J. Beringer *et al.* (2012): Particle listings – Ξ+ cc

[24] J. Beringer *et al.* (2012): Particle listings – Ξ b

[25] J. Beringer *et al.* (2012): Particle listings – Ω0 c

[26] J. Beringer *et al.* (2012): Particle listings – Ω– b

[27] J. Beringer *et al.* (2012): Particle listings – Δ(1232)

[28] J. Beringer *et al.* (2012): Particle listings – Σ(1385)

[29] J. Beringer *et al.* (2012): Particle listings – Σ c(2520)

[30] J. Beringer *et al.* (2012): Particle listings – Σ∗ b

[31] J. Beringer *et al.* (2012): Particle listings – Ξ(1530)

[32] J. Beringer *et al.* (2012): Particle listings – Ξ c(2645)

[33] J. Beringer *et al.* (2012): Particle listings – Ξ0 b(5945)

[34] J. Beringer *et al.* (2012): Particle listings – Ω−

[35] J. Beringer *et al.* (2012): Particle listings – Ω0 c(2770)

[36] http://pdg.lbl.gov/2014/tables/rpp2014-qtab-baryons.pdf

[37] Crede, Volker; Roberts, Winston (2013). "Progress Toward Understanding Baryon Resonances". *Rep. Prog. Phys.* **76**. doi:10.1088/0034-4885/76/7/076301. Retrieved 23 July 2015.

Bibliography

• R. Aaij et al. (LHCb collaboration) (2015). "Observation of J/ψp resonances consistent with pentaquark states in Λ0 b→J/ψK− p decays". arXiv:1507.03414 [hep-ex].

• J. Beringer *et al.* (Particle Data Group) (2012). "Review of Particle Physics". *Physical Review D* **86** (01): 010001. Bibcode:2012PhRvD...86a0001B. doi:10.1103/PhysRevD.86.010001.

• K. Nakamura *et al.* (Particle Data Group) (2010). "Review of Particle Physics". *Journal of Physics G* **37** (7A): 075021. Bibcode:2010JPhG...37g5021N. doi:10.1088/0954-3899/37/7A/075021.

• C. Amsler *et al.* (Particle Data Group) (2008). "Review of Particle Physics". *Physics Letters B* **667** (1): 1–1340. Bibcode:2008PhLB..667....1P. doi:10.1016/j.physletb.2008.07.018.

• V.M. Abazov (DØ Collaboration) (2008). "Observation of the doubly strange b baryon Ω− b" (PDF). Fermilab-Pub-08/335-E.

• K. Carter (2006). "The rise and fall of the pentaquark". *Symmetry Magazine*. Fermilab/SLAC. Retrieved 2008-05-27.

• W.-M. Yao *et al.* (Particle Data Group) (2006). "Review of Particle Physics". *Journal of Physics G* **33**: 1–1232. arXiv:astro-ph/0601168. Bibcode:2006JPhG...33....1Y. doi:10.1088/0954-3899/33/1/001.

• H. Muir (2003). "Pentaquark discovery confounds sceptics". *New Scientist*. Retrieved 2008-05-27.

• J.G. Körner, M. Krämer, and D. Pirjol (1994). "Heavy Baryons". *Progress in Particle and Nuclear Physics* **33**: 787–868. arXiv:hep-ph/9406359. Bibcode:1994PrPNP..33..787K. doi:10.1016/0146-6410(94)90053-1.

6.5.4 Further reading

• H. Garcilazo, J. Vijande, and A. Valcarce (2007). "Faddeev study of heavy-baryon spectroscopy". *Journal of Physics G* **34** (5): 961–976. doi:10.1088/0954-3899/34/5/014.

• S. Robbins (2006). "Physics Particle Overview – Baryons". *Journey Through the Galaxy*. Retrieved 2008-04-20.

• D.M. Manley (2005). "Status of baryon spectroscopy". *Journal of Physics: Conference Series* **5**: 230–237. Bibcode:2005JPhCS...9..230M. doi:10.1088/1742-6596/9/1/043.

• S.S.M. Wong (1998). *Introductory Nuclear Physics* (2nd ed.). New York (NY): John Wiley & Sons. ISBN 0-471-23973-9.

• R. Shankar (1994). *Principles of Quantum Mechanics* (2nd ed.). New York (NY): Plenum Press. ISBN 0-306-44790-8.

• E. Wigner (1937). "On the Consequences of the Symmetry of the Nuclear Hamiltonian on the Spectroscopy of Nuclei". *Physical Review* **51** (2): 106–119. Bibcode:1937PhRv...51..106W. doi:10.1103/PhysRev.51.106.

• M. Gell-Mann (1964). "A Schematic of Baryons and Mesons". *Physics Letters* **8** (3): 214–215. Bibcode:1964PhL.....8..214G. doi:10.1016/S0031-9163(64)92001-3.

• W. Heisenberg (1932). "Über den Bau der Atomkerne I". *Zeitschrift für Physik* (in German) **77**: 1–11. Bibcode:1932ZPhy...77....1H. doi:10.1007/BF01342433.

- W. Heisenberg (1932). "Über den Bau der Atomkerne II". *Zeitschrift für Physik* (in German) **78** (3–4): 156–164. Bibcode:1932ZPhy...78..156H. doi:10.1007/BF01337585.

- W. Heisenberg (1932). "Über den Bau der Atomkerne III". *Zeitschrift für Physik* (in German) **80** (9–10): 587–596. Bibcode:1933ZPhy...80..587H. doi:10.1007/BF01335696.

6.5.5 External links

- Particle Data Group – Review of Particle Physics (2008).

- Georgia State University – HyperPhysics

- Baryons made thinkable, an interactive visualisation allowing physical properties to be compared

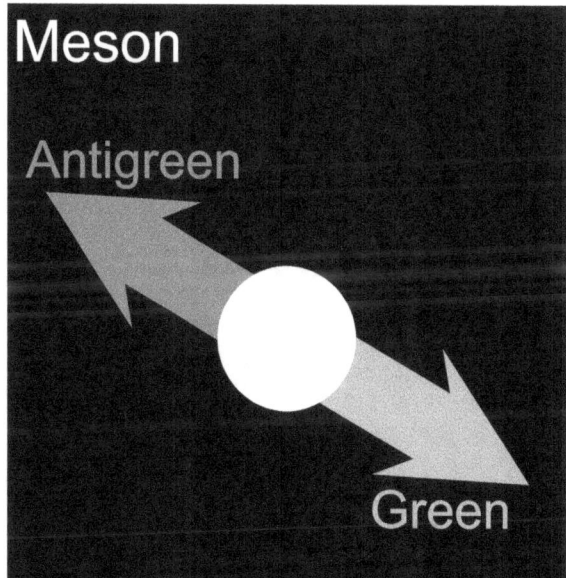

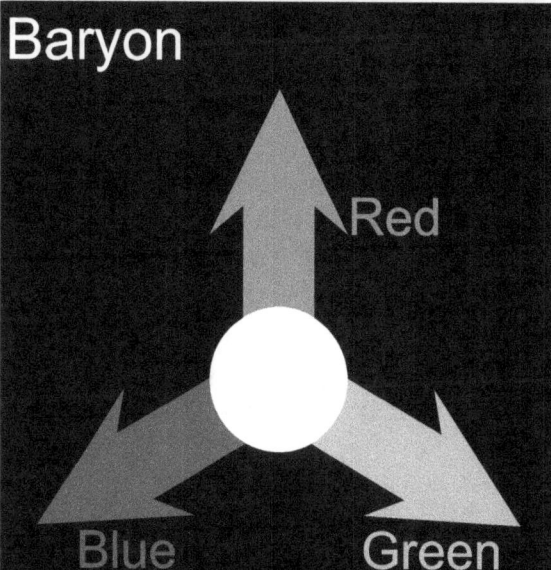

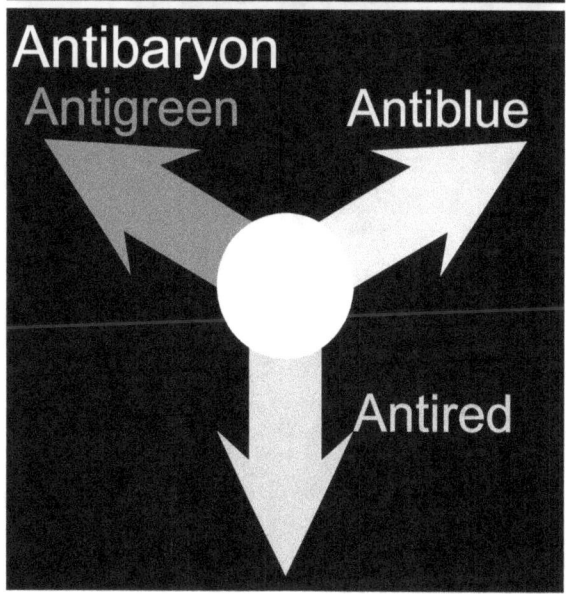

All types of hadrons have zero total color charge. (three examples shown)

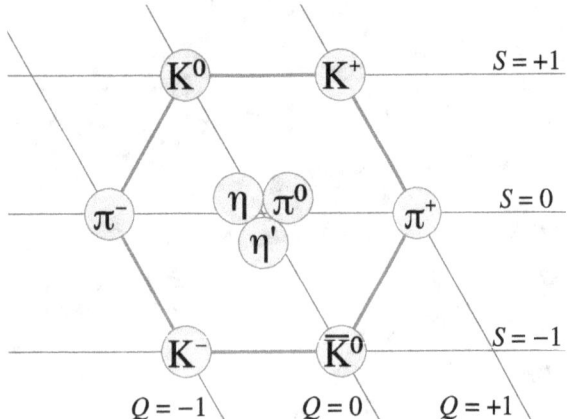

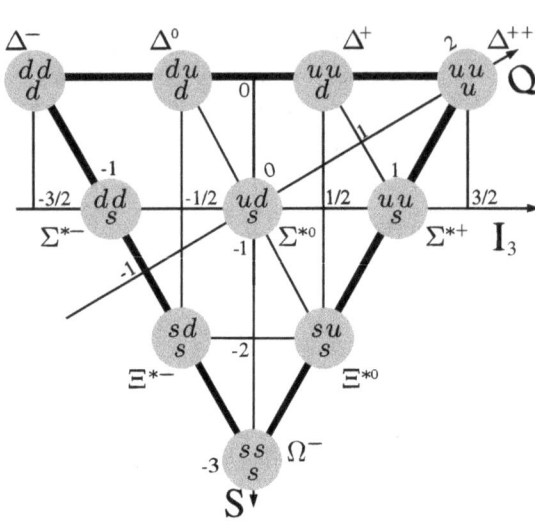

Combinations of one u, d or s quarks and one u, d, or s antiquark in $J^P = 0^-$ configuration form a nonet.

*Combinations of three **u, d** or **s** quarks forming baryons with a spin-$^3/_2$ form the* uds baryon decuplet

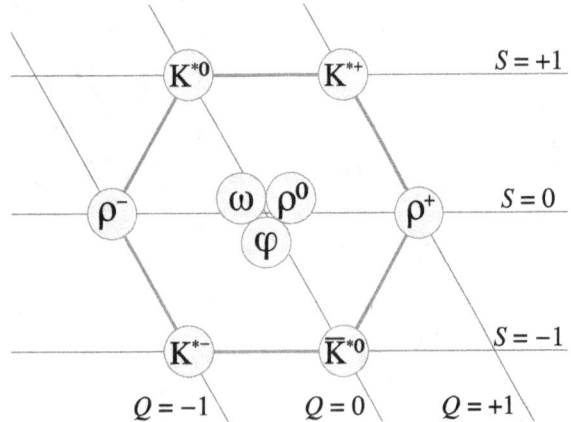

Combinations of one u, d or s quarks and one u, d, or s antiquark in $J^P = 1^-$ configuration also form a nonet.

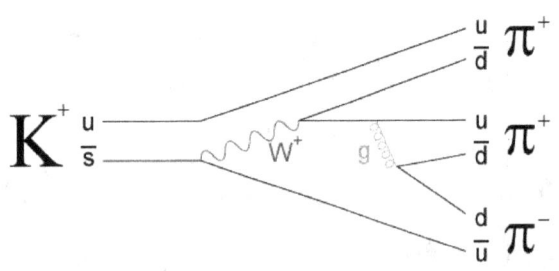

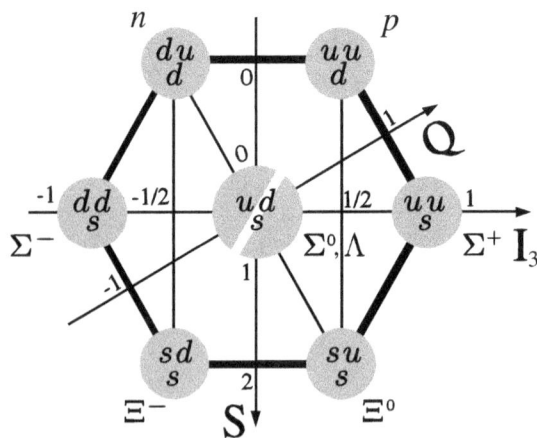

The decay of a kaon (K+) into three pions (2 π+, 1 π−) is a process that involves both weak and strong interactions.

Weak interactions: *The strange antiquark (s) of the kaon transmutes into an up antiquark (u) by the emission of a W+ boson; the W+ boson subsequently decays into a down antiquark (d) and an up quark (u).*

Strong interactions: *An up quark (u) emits a gluon (g) which decays into a down quark (d) and a down antiquark (d).*

*Combinations of three **u, d** or **s** quarks forming baryons with a spin-$^1/_2$ form the* uds baryon octet

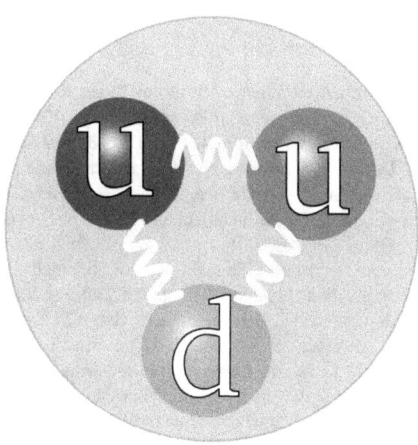

A diagram of a proton, one of the most famous baryons, containing two up quarks and one down quark

Chapter 7

Text and image sources, contributors, and licenses

7.1 Text

- **Standard Model** *Source:* https://en.wikipedia.org/wiki/Standard_Model?oldid=679087631 *Contributors:* AxelBoldt, Derek Ross, CYD, Bryan Derksen, The Anome, Ed Poor, Andre Engels, Roadrunner, David spector, Isis~enwiki, Youandme, Ram-Man, Stevertigo, Edward, Patrick, Boud, Michael Hardy, SebastianHelm, Looxix~enwiki, Julesd, Glenn, AugPi, Mxn, Raven in Orbit, Reddi, Phr, Tpbradbury, Populus, Haoherb428, Phys, Floydian, Bevo, Pierre Boreal, AnonMoos, BenRG, Jeffq, Dmytro, Drxenocide, Robbot, Nurg, Securiger, Texture, Roscoe x, Fuelbottle, Mattflaschen, Tobias Bergemann, Alan Liefting, Ancheta Wis, Giftlite, Dbenbenn, Harp, Herbee, Monedula, LeYaYa, Xerxes314, Dratman, Alison, JeffBobFrank, Dmmaus, Pharotic, Brockert, Bodhitha, Andycjp, Sonjaaa, HorsePunchKid, APH, Icairns, AmarChandra, Gscshoyru, Kate, Arivero, FT2, Rama, Vsmith, David Schaich, Xezbeth, D-Notice, Dfan, Bender235, Pt, El C, Laurascudder, Shanes, Drhex, Fogger~enwiki, Brim, Rbj, Jeodesic, Jumbuck, Alansohn, Gary, ChristopherWillis, Guy Harris, Axl, Sligocki, Kocio, Stillnotelf, Alinor, Wtmitchell, Egg, TenOfAllTrades, H2g2bob, Killing Vector, Linas, Mindmatrix, Benbest, Dodiad, Mpatel, Faethon, TPickup, Faethon34, Palica, Dysepsion, Faethon36, Qwertyca, Drbogdan, Rjwilmsi, Zbxgscqf, Macumba, Strangethingintheland, Dstudent, R.e.b., Bubba73, Drrngrvy, Agasicles, FlaBot, Naraht, Agasides, DannyWilde, Dave1g, Itinerant1, Gparker, Jrtayloriv, Goudzovski, Chobot, Bgwhite, FrankTobia, YurikBot, Bambaiah, Ohwilleke, VoxMoose, Bhny, JabberWok, Bovineone, Krbabu, SCZenz, JulesH, Davemck, Lomn, E2mb0t~enwiki, Dna-webmaster, Jrf, Dv82matt, Tetracube, Hirak 99, Arthur Rubin, Netrapt, JLaTondre, Caco de vidro, RG2, GrinBot~enwiki, That Guy, From That Show!, Hal peridol, SmackBot, YellowMonkey, Tom Lougheed, Melchoir, Bazza 7, KocjoBot~enwiki, Jagged 85, Thunderboltz, Setanta747 (locked), Skizzik, Dauto, Chris the speller, Bluebot, TimBentley, Sirex98, Silly rabbit, Complexica, Metacomet, DHN-bot~enwiki, MovGP0, QFT, Kittybrewster, Addshore, Jmnbatista, Cybercobra, Jgwacker, BullRangifer, Soarhead77, Daniel.Cardenas, Yevgeny Kats, Byelf2007, TriTertButoxy, Craig Bolon, Ajnosek, Ekjon Lok, Bjankuloski06, Tarcieri, Waggers, JarahE, Michaelbusch, Lottamiata, Newone, Twas Now, IanOfNorwich, Srain, Patrickwooldridge, J Milburn, Mosaffa, Gatortpk, Vessels42, Geremia, Van helsing, Harrigan, Phatom87, Cydebot, David edwards, Verdy p, Michael C Price, Xantharius, Crum375, JamesAM, Thijs!bot, Epbr123, Headbomb, Phy1729, Stannered, Tariqhada, Seaphoto, Orionus, Voyaging, Gnixon, Jbaranao, Jrw@pobox.com, Len Raymond, Narssarssuaq, Bakken, CattleGirl, Davidoaf, Vanished user ty12kl89jq10, Lvwarren, Taborgate, Leyo, HEL, J.delanoy, Hans Dunkelberg, Stephanwehner, Wbellido, Aoosten, Jacksonwalters, The Transliterator, DadaNeem, Student7, Joshmt, WJBscribe, Jozwolf, Hexane2000, BernardZ, Awren, Sheliak, Physicist brazuca, Schucker, Goop Goop, Fences and windows, Dextrose, Mcewan, Swamy g, TXiKiBoT, Sharikkamur, Thrawn562, Voorlandt, Escalona, Setreset, PDFbot, Pleroma, UnitedStatesian, Piyush Sriva, Kacser, Billinghurst, Francis Flinch, Moose-32, Ptrslv72, David Barnard, SieBot, ShiftFn, Robdunst, Jim E. Black, SheepNotGoats, Gerakibot, Nozzer42, Mr swordfish, Wing gundam, Bamkin, Likebox, Arthur Smart, HungarianBarbarian, Commutator, KathrynLybarger, Iomesus, C0nanPayne, Crazz bug 5, ClueBot, Superwj5, Wwheaton, Garyzx, SuperHamster, Elsweyn, Maldmac, DragonBot, Djr32, Diagramma Della Verita, Nymf, Eeekster, Brews ohare, NuclearWarfare, PhySusie, Ordovico, Mastertek, DumZiBoT, BodhisattvaBot, Guarracino, Mitch Ames, Truthnlove, Stephen Poppitt, Tayste, Addbot, Deepmath, Eric Drexler, DWHalliday, Mjamja, Leszek Jańczuk, NjardarBot, Mwoldin, Bassbonerocks, Barak Sh, AgadaUrbanit, Lightbot, Smeagol 17, Abjiklam, Ve744, Luckas-bot, Yobot, Orion11M87, AnomieBOT, JackieBot, Icalanise, Citation bot, ArthurBot, Northryde, LilHelpa, Xqbot, Sionus, Professor J Lawrence, Tomwsulcer, Edsegal, GrouchoBot, Trongphu, QMarion II, Ernsts, A. di M., Bytbox, FrescoBot, Paine Ellsworth, Aliotra, Steve Quinn, Citation bot 1, Rameshngbot, MJ94, RedBot, MastiBot, Aknochel, Sijothankam, Puzl bustr, Beta Orionis, Physics therapist, Bj norge, Innotata, Jesse V., RjwilmsiBot, Mathewsyriac, Afteread, EmausBot, Bookalign, WikitanvirBot, Wilhelm-physiker, Bdijkstra, DerNeedle, Kenmint, Dbraize, Tanner Swett, HeptishHotik, راهب شنیمن, Suslindisambiguator, Quondum, Webheh, UniversumExNihilo, Vanished user fijw983kjaslkekfhj45, Maschen, RockMagnetist, Stormymountain, Ζeta ζ, Whoop whoop pull up, Isocliff, ClueBot NG, Smtchahal, Snotbot, Tonypak, O.Koslowski, CharleyQuinton, Dsperlich, Theopolisme, ZakMarksbury, Helpful Pixie Bot, Bibcode Bot, BG19bot, Tirebiter78, AvocatoBot, Lukys~enwiki, Stapletongrey, Ownedroad9, Chip123456, ChrisGualtieri, Khazar2, Billyfesh399, Rhlozier, JYBot, Dexbot, Doom636, Rongended, Cerabot~enwiki, CuriousMind01, Cjean42, Jayanta mallick, Joeinwiki, Kowtje, JPaestpreornJeolhlna, Eyesnore, Euan Richard, Nigstomper, Particle physicist, Prokaryotes, Jernahthern, Ginsuloft, Dimension10, JNrgbKLM, Krabaey, 1codesterS, FelixRosch, Monkbot, Delbert7, BradNorton1979, Lathamboyle, Tetra quark, KasparBot, Buckbill10 and Anonymous: 357

- **List of particles** *Source:* https://en.wikipedia.org/wiki/List_of_particles?oldid=682746251 *Contributors:* AxelBoldt, Danny, Rmhermen, Stevertigo, Bdesham, Ahoerstemeier, Stan Shebs, Docu, Salsa Shark, Nikai, Evercat, Schneelocke, Charles Matthews, Jitse Niesen, CBDunkerson,

162

Bevo, Raul654, Donarreiskoffer, Robbot, Sanders muc, Merovingian, Pengo, Giftlite, Herbee, Xerxes314, Dratman, Jeremy Henty, Alensha, Bodhitha, Physicist, Hayne, Quadell, RetiredUser2, Mysidia, Icairns, Asbestos, D6, Urvabara, Discospinster, Rich Farmbrough, FT2, Qutezuce, ArnoldReinhold, Neko-chan, El C, Laurascudder, Susvolans, EmilJ, Physicistjedi, Minghong, Gbrandt, Eddideigel, Axl, Mac Davis, David Ko, Radical Mallard, RJFJR, Count Iblis, Dirac1933, TenOfAllTrades, LFaraone, Oleg Alexandrov, Linas, JarlaxleArtemis, Duncan.france, GregorB, Cedrus-Libani, Karam.Anthony.K, Palica, Rjwilmsi, Zbxgscqf, JLM~enwiki, Strait, Ems57fcva, Krash, Dan Guan, DannyWilde, Lmatt, Goudzovski, Chobot, YurikBot, Bambaiah, Vuvar1, Madkayaker, Hydrargyrum, Presscorr, Chaos, Salsb, Tavilis, SCZenz, Lexicon, TUSHANT JHA, Dna-webmaster, Tomvds, Poulpy, Cstmoore, TLSuda, NeilN, MacsBug, Tom Lougheed, McGeddon, Bazza 7, WookieInHeat, Derdeib, Yamaguchi⁇⁇, Betacommand, Bluebot, Master of Puppets, DHN-bot~enwiki, Raistuumum, Juancnuno, Kittybrewster, Acepectif, Ligulembot, TriTertButoxy, ArglebargleIV, Khazar, John, FrozenMan, JorisvS, 041744, Dr Greg, Slakr, Mets501, Scorpion0422, Cbuckley, Iridescent, TwistOfCain, Happy-melon, JRSpriggs, Flickboy, Van helsing, Lithium6, Neelix, Rotiro, Cydebot, Quibik, Christian75, Omicronpersei8, Thijs!bot, Qwyrxian, TauLibrus, Headbomb, Inner Earth, 49, Guptasuneet, Scottmsg, WinBot, Elmoosecapitan, Tyco.skinner, AubreyEllenShomo, Arch dude, Johnman239, Mwarren us, TheEditrix2, CalamusFortis, MartinBot, Sadisticsuburbanite, Bissinger, Anaxial, CommonsDelinker, Maurice Carbonaro, Zojj, OliverHarris, Joshmt, Adanadhel, Lseixas, Graphite Elbow, VolkovBot, Jmrowland, Quilbert, Anonymous Dissident, Dstary, Escalona, JPMasseo, Figureskatingfan, Inx272, Meters, Antixt, Hamish a e fowler, GoddersUK, Bluetryst, SieBot, Ishvara7, WereSpielChequers, Audrius u, VovanA, Paolo.dL, RSStockdale, Anchor Link Bot, StewartMH, Explicit, ClueBot, Unbuttered Parsnip, Nolimitownass, DragonBot, Atomic7732, TimothyRias, SkyLined, Addbot, DOI bot, Jojhutton, Favonian, LinkFA-Bot, OlEnglish, Teles, Legobot, Luckas-bot, Yobot, Dov Henis, Azcolvin429, AnomieBOT, Götz, Icalanise, Flewis, Materialscientist, OllieFury, Vuerqex, ArthurBot, Vulcan Hephaestus, Blennow, Reality006, Coretheapple, Jcimorra, RibotBOT, Ernsts, A. di M., Axelfoley12, Zosterops, FrescoBot, Paine Ellsworth, Citation bot 1, JIK1975, Tom.Reding, Diffequa, WikitanvirBot, Racerx11, 112358sam, Aegnor.erar, Hops Splurt, HESUPERMAN, Hhhippo, AvicBot, JSquish, StringTheory11, Waperkins, Bamyers99, Suslindisambiguator, L Kensington, DennisIsMe, RockMagnetist, ClueBot NG, Snotbot, Primergrey, Vio45lin, Widr, MsFionnuala, Oklahoma3477, Bibcode Bot, CityOfSilver, Cap'n G, BML0309, Dan653, Twocount, Penguinstorm300, Dexbot, LightandDark2000, Ohiggy, TwoTwoHello, Andyhowlett, Printersmoke, Orion 2013, ARUNEEK, Seino van Breugel, AspaasBekkelund, TheMagikCow, Vyom27, ParkersComments, Selva Ganapathy and Anonymous: 290

- **Quark** *Source:* https://en.wikipedia.org/wiki/Quark?oldid=681925889 *Contributors:* AxelBoldt, Derek Ross, Vicki Rosenzweig, Mav, Bryan Derksen, The Anome, Gareth Owen, Andre Engels, PierreAbbat, Peterlin~enwiki, Ben-Zin~enwiki, Zoe, Heron, Montrealais, Hfastedge, Edward, Dante Alighieri, Ixfd64, CesarB, Card~enwiki, NuclearWinner, Looix~enwiki, Ahoerstemeier, Elliot100, Docu, J-Wiki, Nanobug, Aarchiba, Julesd, Glenn, Schneelocke, Jengod, A5, Timwi, Dysprosia, DJ Clayworth, Phys, Ed g2s, Bevo, Olathe, MD87, Jni, Phil Boswell, Sjorford, Donarreiskoffer, Robbot, Sanders muc, Moncrief, Merovingian, PxT, Texture, Bkell, UtherSRG, Widsith, Ancheta Wis, Giftlite, ShaunMacPherson, Harp, Nunh-huh, Lupin, Herbee, Leflyman, Monedula, 0x6D667061, Xerxes314, Anville, Hoho~enwiki, Alison, Beardo, Moogle10000, Wronkiew, Jackol, Bobblewik, Bodhitha, Piotrus, Kaldari, Elroch, Icairns, Zfr, TonyW, Ukexpat, BrianWilloughby, Grunt, O'Dea, Jiy, Discospinster, Rich Farmbrough, Guanabot, T Long, Vsmith, Saintswithin, SocratesJedi, Mani1, Bender235, Lancer, RJHall, Mr. Billion, El C, Kwamikagami, Laurascudder, Susvolans, Triona, Axezz, Bobo192, Army1987, C S, Ziggurat, Rangelov, Matt McIrvin, Jojit fb, Nk, Pentalis, Obradovic Goran, Fwb22, Lysdexia, Benjonson, Alansohn, Gary, Gintautasm, Guy Harris, Keenan Pepper, MonkeyFoo, Lectonar, Mac Davis, Wdfarmer, Snowolf, Schapel, Knowledge Seeker, Evil Monkey, VivaEmilyDavies, CloudNine, Kusma, Kazvorpal, Kay Dekker, Crosbiesmith, Mogigoma, Linas, Mindmatrix, JarlaxleArtemis, ScottDavis, LOL, Wdyoung, Before My Ken, Tylerni7, Jwanders, Dataphiliac, AndriyK, Noetica, Wayward, Wisq, Palica, Marudubshinki, Calréfa Wéná, GSlicer, Graham87, Deltabeignet, Kbdank71, Yurik, Crzrussian, Rjwilmsi, Bremen, Marasama, SpNeo, Mike Peel, Bubba73, DoubleBlue, Matt Deres, Yamamoto Ichiro, Algebra, Dsnow75, RobertG, Nihiltres, Jeff02, RexNL, TeaDrinker, Chobot, DVdm, Jpacold, Gwernol, Elfguy, Roboto de Ajvol, YurikBot, Wavelength, Bambaiah, Sceptre, Hairy Dude, Jimp, Phantomsteve, TheDoober, Dobromila, JabberWok, CambridgeBayWeather, Chaos, Salsb, Wimt, Ugur Basak, NawlinWiki, Spike Wilbury, Bossrat, SCZenz, Randolf Richardson, Danlaycock, Tony1, DRosenbach, Robertbyrne, Dna-webmaster, WAS 4.250, Closedmouth, Pietdesomere, Heathhunnicutt, Kevin, Banus, RG2, Kamickalo, That Guy, From That Show!, Veinor, MacsBug, SmackBot, Aigarius, BBandHB, Incnis Mrsi, InverseHypercube, C.Fred, Bazza 7, Ikip, Anastrophe, Jrockley, Eskimbot, AnOddName, Jonathan Karlsson, Edgar181, Gilliam, Dauto, NickGarvey, Vvarkey, Bluebot, KaragouniS, Keegan, Dahn, Bigfun, Miquonranger03, OrangeDog, Silly rabbit, Metacomet, Tripledot, Nbarth, DHN-bot~enwiki, Sbharris, Colonies Chris, Hallenrm, Scwlong, Gsp8181, Can't sleep, clown will eat me, Mallorn, Jeff DLB, TKD, Addshore, Mqjjb30e, Cybercobra, Khukri, B jonas, Jdlambert, Lpgeffen, Nrcprm2026, Akriasas, Zadignose, Jóna Þórunn, Bdushaw, Beyazid, TriTertButoxy, SashatoBot, SciBrad, Doug Bell, Soap, Richard L. Peterson, John, Mgiganteus1, SpyMagician, Edconrad, Loadmaster, 2T, Waggers, SandyGeorgia, Ravi12346, Dbzfrk15146, Peyre, Newone, GDallimore, Happy-melon, Majora4, Chovain, Tawkerbot2, Cryptic C62, JForget, Vaughan Pratt, Hello789, ZICO, SUPRATIM DEY, Ruslik0, CuriousEric, Paulfriedman7, Logical2u, Myasuda, RoddyYoung, Typewritten, Cydebot, Abeg92, Mike Christie, Grahamec, Gogo Dodo, Jayen466, 879(CoDe), Michael C Price, Tawkerbot4, Ameliorate!, Akcarver, Gimmetrow, SallyScot, Casliber, Thijs!bot, Epbr123, NeoPhyteRep, LeBofSportif, Markus Pössel, Anupam, Sopranosmob781, Headbomb, Marek69, John254, KJBurns, MichaelMaggs, Escarbot, Eleuther, Ice Ardor, Aadal, AntiVandalBot, SmokeyTheCat, Tyco.skinner, Exteray, RobJ1981, Rsocol, Ke garne, Deflective, Husond, MER-C, CosineKitty, Andonic, East718, Pkoppenb, DanPMK, Magioladitis, WolfmanSF, Thasaidon, Bongwarrior, VoABot II, باسم, Inertiatic076, Kevinmon, Christoph Scholz~enwiki, Aka042, Giggy, Tanvirzaman, Johnbibby, Cyktsui, ArchStanton69, Ace42, Allstarecho, Shijualex, DerHexer, Elandra, Denis tarasov, MartinBot, Poeloq, Dorvaq, CommonsDelinker, HEL, J.delanoy, Nev1, Ops101ex, DrKiernan, Hgpot, Ferdyshenko, Jigesh, DJ1AM, Tarotcards, Coppertwig, TomasBat, Nikbuz, SJP, FJPB, Vainamainien, Tiggydong, Robprain, Sheliak, Cuzkatzimhut, Lights, X!, VolkovBot, CWii, ABF, John Darrow, Holme053, Nousernamesleft, Ryan032, GimmeBot, Davehi1, A4bot, Captain Courageous, Guillaume2303, Anonymous Dissident, Drestros power, Qxz, Anna Lincoln, Eldaran~enwiki, Leafyplant, Don4of4, PaulTanenbaum, Abdullais4u, Jbryancoop, Mbalelo, Gilisa, Eubulides, Chronitis, Seresin, Dustybunny, Insanity Incarnate, Upquark, Edge1212, Ollieho, AOEU Warrior, SieBot, Graham Beards, WereSpielChequers, Csmart287, Guguma5, Winchelsea, Jbmurray, Caltas, Vanished User 8a9b4725f8376, Keilana, Bentogoa, Aillema, RadicalOne, Arbor to SJ, Elcobbola, Physics one, Dhatfield, RSStockdale, Son of the right hand, Ngexpert5, Ngexpert6, Ngexpert7, Psycherevolt, Sean.hoyland, Mygerardromance, Dabomb87, Nergaal, Muhends, Romit3, SallyForth123, Atif.t2, ClueBot, The Thing That Should Not Be, Wwheaton, Xeno malleus, Harland1, Piledhigheranddeeper, Maxtitan, DragonBot, Glopso, Choonkiat.lee, Himynameisdumb, Worth my salt, Arthur Quark, Estirabot, Brews ohare, Jotterbot, PhySusie, Brianboulton, Dekisugi, ANOMALY-117, Sallicio, Yomangan, Jtle515, Katanada, DumZiBoT, TimothyRias, XLinkBot, Vayalir, Oldnoah, Saintlucifer2008, Nathanwesley3, Dragonfiremage, Devilist666, Mancune2001, Jbeans, WikiDao, SkyLined, Truthnlove, Airplaneman, Eklipse, Addbot, Eric Drexler, AVand, Some jerk on the Internet, Captain-tucker, Giants2008, Iceblock, Ronhjones, Quarksci, Mseanbrown, Looie496, LaaknorBot, Peti610botH, AgadaUrbanit, Tide rolls, Vicki breazeale, Gail, Extruder~enwiki, Abduallah mohammed, Dealer77, Luckas-bot, Yobot, Fraggle81, Cflm001, Legobot II, Amble, Mmxx, Superpenguin1984, Worm That Turned, The Vector Kid, Planlips, Fangfyre, TestEditBot, Azcolvin429, Vroo, Synchronism, Bility, Orion11M87, AnomieBOT, Xi rho, Rubinbot, Jim1138, Bookaneer, Yotcmdr, Crystal whacker, Sonic h, Materialscientist, Citation bot, Pitke, Vuerqex, Bci2, ArthurBot, LilHelpa, Xqbot, Jeffrey Mall, AbigailAbernathy, Srich32977, Alex2510, Almabot, Uscbino, Pmlineditor, RibotBOT, Shmomuffin, Gunjan verma81, Chotarocket, Ernsts, Renverse, A. di M., Weekendpartier, FrescoBot, Paine Ellsworth, DelphinidaeZeta, Steve Quinn, Citation bot 1, AstaBOTh15, Pinethicket, Jonesey95, Calmer Waters, Skyerise, Pmokeefe, Jschnur, Searsshoesales, Jrobbinz123, Lissajous, Turian, Lando Calrissian, Wotnow, Ansumang, Reaper Eternal, 564dude, Jackvancs, Bobotast, MINTOPOINT, TjBot, DexDor, Антон Глінисты, Daggersteel10, Chiechiecheist, EmausBot, John of Reading, WikitanvirBot, Duskbrood, FergalG, Slightsmile, Barak90, Wikipelli, TheLemon1234, Manofgrass, H3llBot, Stoneymufc29, GeorgeBarnick, Brandmeister, Ego White Tray, TYelliot, ClueBot NG, Gilderien, A520, Cheeseequalsyum, Timothy jordan, 123Hedgehog456, Maplelanefarm, 336, Helpful Pixie Bot, Jeffreyts11, 123456789malm, Bibcode Bot, BG19bot, Hurricanefan25, MusikAnimal, Davidiad, MosquitoBird11, Mydogpwnsall, MrBill3, Njavallil, Glacialfox, Walterpfeifer, Thebannana, CE9958, Marioedesouza, Mediran, Dexbot, Rishab021, Cjean42, Sriharsh1234, Sam boron100, Wankybanky, Wikitroll12345, RojoEsLardo, Jwratner1, NottNott, Saebre, JNrgbKLM, KheltonHeadley, AspaasBekkelund, HectorCabreraJr, Hazinho93, Quadrupedi, QuantumMatt101, Philipphilip0001, Monkbot, RiderDB, Egfraley, Tetra quark, Weed305, KasparBot and Anonymous: 705

- **Up quark** *Source:* https://en.wikipedia.org/wiki/Up_quark?oldid=666803858 *Contributors:* Bryan Derksen, Alfio, Jni, Giftlite, Xerxes314, Kjoonlee, Bookandcoffee, CharlesC, Rjwilmsi, Mike Peel, Chobot, Hairy Dude, Rt66lt, Spike Wilbury, SCZenz, Poulpy, Eog1916, Bluebot, Tamfang, T-borg, Eric Saltsman, Hetar, Lottamiata, Laplace's Demon, Merryjman, CmdrObot, Myasuda, Raoul NK, Headbomb, JAnDbot, Abyssoft, I310342~enwiki, Idioma-bot, Sheliak, Wilmot1, VolkovBot, TXiKiBoT, Anonymous Dissident, Gekritzl, AlleborgoBot, SieBot, Muhends, Bobathon71, DragonBot, DumZiBoT, TimothyRias, Addbot, Eivindbot, LaaknorBot, ChenzwBot, Naidevinci, Ehrenkater, Lightbot, Luckas-bot, Citation bot, ArthurBot, Xqbot, DSisyphBot, Ditimchanly, Almabot, A. di M., Paine Ellsworth, Citation bot 1, Trappist the monk, TjBot, Ripchip Bot, EmausBot, WikitanvirBot, StringTheory11, Quondum, Helpful Pixie Bot, Bibcode Bot, BG19bot, P76837, Oznitecki, Alexzhang2, The Great Leon, Monkbot and Anonymous: 38

- **Down quark** *Source:* https://en.wikipedia.org/wiki/Down_quark?oldid=663284022 *Contributors:* Bryan Derksen, Alfio, Timwi, Jni, Herbee, Xerxes314, Rich Farmbrough, Kjoonlee, Rjwilmsi, Mike Peel, Chobot, YurikBot, Rt66lt, Acidsaturation, Spike Wilbury, SCZenz, Poulpy, Otto ter Haar, Skizzik, Bluebot, Tamfang, Llwang, Eric Saltsman, MTSbot~enwiki, Hetar, Laplace's Demon, Myasuda, Raoul NK, Thijs!bot, Headbomb, Davidhorman, JAnDbot, Abyssoft, MartinBot, Jvineberg, I310342~enwiki, Sheliak, VolkovBot, TXiKiBoT, Anonymous Dissident, SieBot, Ngexpert6, Muhends, Bobathon71, Lawrence Cohen, Daigaku2051, Auntof6, NuclearWarfare, TimothyRias, Addbot, Lightbot, Luckas-bot, Yobot, Citation bot, ArthurBot, Xqbot, DSisyphBot, Paine Ellsworth, Citation bot 1, Tim1357, Trappist the monk, EmausBot, ZéroBot, Quondum, Rezabot, Helpful Pixie Bot, Bibcode Bot, TheMan4000, 786b6364, Monkbot and Anonymous: 22

- **Charm quark** *Source:* https://en.wikipedia.org/wiki/Charm_quark?oldid=663286006 *Contributors:* Bryan Derksen, Alfio, Bogdangiusca, Xerxes314, Bodhitha, Perey, Kjoonlee, Rjwilmsi, Mike Peel, Chobot, YurikBot, Bambaiah, Conscious, Salsb, SCZenz, Scottfisher, Poulpy, SmackBot, Delldot, Warhol13, Rezecib, Vina-iwbot~enwiki, Happy-melon, Laplace's Demon, CRGreathouse, Michael C Price, Thijs!bot, Headbomb, Nisselua, JAnDbot, Abyssoft, Uncle.wink, Bryanhiggs, HEL, I310342~enwiki, Qoou.Anonimu, Idioma-bot, Sheliak, Anonymous Dissident, Kumorifox, BeIsKr, AlleborgoBot, SieBot, Muhends, TimothyRias, Addbot, Mjamja, Lightbot, Luckas-bot, Yobot, Nallimbot, Citation bot, ArthurBot, Quebec99, Xqbot, DSisyphBot, GrouchoBot, RibotBOT, SassoBot, A. di M., Paine Ellsworth, Dogposter, D'ohBot, Citation bot 1, Citation bot 4, RedBot, MastiBot, Trappist the monk, EarthCom1000, Alph Bot, EmausBot, ZéroBot, Quondum, Anita5192, CocuBot, Rezabot, Helpful Pixie Bot, Bibcode Bot, Penguinstorm300, Hoppeduppeanut, Leowestland and Anonymous: 35

- **Strange quark** *Source:* https://en.wikipedia.org/wiki/Strange_quark?oldid=663285537 *Contributors:* Bryan Derksen, Alfio, Jni, Owain, Xerxes314, Soman, Kjoonlee, Kwamikagami, Rsholmes, Esb82, Neonumbers, Rjwilmsi, Mike Peel, Gurch, Erik4, Chobot, YurikBot, Jimp, Salsb, SCZenz, Poulpy, SmackBot, Bluebot, NCurse, Vina-iwbot~enwiki, Yevgeny Kats, Zzzzzzzzzzz, Laplace's Demon, MightyWarrior, Myasuda, Thijs!bot, Headbomb, Chillysnow, JAnDbot, Abyssoft, Bongwarrior, Albmont, McSly, I310342~enwiki, Pdcook, Sheliak, VolkovBot, SieBot, Muhends, Auntof6, Iohannes Animosus, TimothyRias, IngerAlHaosului, Addbot, ProbablyAmbiguous, Luckas-bot, Yobot, AnomieBOT, Citation bot, Sarah12sarah, Erik9bot, Thehelpfulbot, Paine Ellsworth, Rkr1991, Citation bot 1, Skyerise, Johann137, Trappist the monk, Puzl bustr, Agrasa, Wikiborg4711, EmausBot, Hhhippo, ZéroBot, Quondum, CocuBot, Helpful Pixie Bot, Bibcode Bot, Vkpd11, P76837, Matthew gib, Glaisher, RhinoMind and Anonymous: 38

- **Top quark** *Source:* https://en.wikipedia.org/wiki/Top_quark?oldid=679102575 *Contributors:* Damian Yerrick, Bryan Derksen, HPA, Haryo, Bkell, Giftlite, Xerxes314, Edcolins, Bodhitha, David Schaich, Kjoonlee, Axl, Woohookitty, Rjwilmsi, Strait, Mike Peel, Vegaswikian, Wikiliki, Goudzovski, Chobot, YurikBot, Bambaiah, JabberWok, Gaius Cornelius, Salsb, Howcheng, SCZenz, Emijrp, Physicsdavid, SmackBot, Incnis Mrsi, ZerodEgo, Mr.Z-man, Jgwacker, Pulu, Stikonas, Mets501, Peyre, RekishiEJ, Banedon, הסרפד, Headbomb, Davidhorman, Oreo Priest, AntiVandalBot, JAnDbot, Abyssoft, Maliz, HEL, Fatka, I310342~enwiki, Idioma-bot, Sheliak, Biggus Dictus, TXiKiBoT, Reibot, Kachuak, Ptrslv72, SieBot, Hatster301, Muhends, ClueBot, Niceguyedc, Noca2plus, Choonkiat.lee, Brews ohare, Kakofonous, Jtle515, TimothyRias, Prostarplayer321, SkyLined, Cockatoo, Addbot, Mr0t1633, Mjamja, ChenzwBot, Ginosbot, Zorrobot, Luckas-bot, Naudefjbot~enwiki, Dreamer08, AnomieBOT, Icalanise, Citation bot, ArthurBot, LilHelpa, DSisyphBot, Unready, GrouchoBot, RibotBOT, Soandos, Paine Ellsworth, Citation bot 1, Jonesey95, Thinking of England, Nomis2k, Higgshunter, RjwilmsiBot, Mophoplz, EmausBot, John of Reading, WikitanvirBot, Barak90, Peter.poier, Quondum, Samlever, Whoop whoop pull up, Reify-tech, Helpful Pixie Bot, Bibcode Bot, Glevum, Kephir, Mmitchell10, Quadrupedi, Monkbot, BrunoUbaldo and Anonymous: 64

- **Bottom quark** *Source:* https://en.wikipedia.org/wiki/Bottom_quark?oldid=676070719 *Contributors:* Bryan Derksen, Xerxes314, Bodhitha, Icairns, Kjoonlee, Bobo192, Pinar, WadeSimMiser, Rjwilmsi, Mike Peel, Erkcan, FlaBot, Itinerant1, Chobot, YurikBot, Bambaiah, Jimp, Conscious, Ozabluda, SpuriousQ, Salsb, SCZenz, Lexicon, Poulpy, Physicsdavid, SmackBot, Hmains, Luís Felipe Braga, Laplace's Demon, CmdrObot, Outriggr, Niubrad, הסרפד, Thijs!bot, Headbomb, JAnDbot, Abyssoft, Pkoppenb, Dr. Morbius, I310342~enwiki, Joshmt, Idiomabot, Sheliak, VolkovBot, Antixt, AlleborgoBot, BartekChom, Muhends, Auntof6, TimothyRias, Lockalbot, Addbot, Mr Sme, Luckas-bot, THEN WHO WAS PHONE?, Citation bot, ArthurBot, Xqbot, GrouchoBot, StevenVerstoep, Thehelpfulbot, Paine Ellsworth, Citation bot 1, Jonesey95, Double sharp, TjBot, EmausBot, Barak90, TuHan-Bot, ZéroBot, StringTheory11, Quondum, Chris857, ChuispastonBot, Widr, Helpful Pixie Bot, Bibcode Bot, P76837, ChrisGualtieri, Ajd268, Mfb, Monkbot, Axel Azzopardi, Kenijr and Anonymous: 33

- **Lepton** *Source:* https://en.wikipedia.org/wiki/Lepton?oldid=679886360 *Contributors:* Bryan Derksen, Andre Engels, PierreAbbat, Ben-Zin~enwiki, Heron, Xavic69, Fruge~enwiki, Fwappler, Ahoerstemeier, Julesd, Glenn, Mxn, A5, Wikiborg, Dysprosia, Radiojon, Imc, Morwen, Fibonacci, Bcorr, Phil Boswell, Donarreiskoffer, Robbot, Merovingian, Wikibot, Giftlite, Smjg, DocWatson42, Harp, Herbee, Xerxes314,

Sysin, Knutux, LiDaobing, LucasVB, ClockworkLunch, RetiredUser2, Icairns, Mike Rosoft, Chris j wood, Martinl~enwiki, Smalljim, Giraffedata, Jumbuck, RobPlatt, Neonumbers, Ahruman, Computerjoe, Simon M, Woohookitty, Mindmatrix, Rjwilmsi, Strait, Erkcan, FlaBot, Danny-Wilde, Mastorrent, Celebere, Peterl, YurikBot, Bambaiah, Jimp, Salsb, Spike Wilbury, Jaxl, SCZenz, DeadEyeArrow, Tetracube, Smoggyrob, Dmuth, Jaysbro, Sbyrnes321, That Guy, From That Show!, SmackBot, Bazza 7, KocjoBot~enwiki, Jrockley, Mom2jandk, Cool3, Hmains, Complexica, DHN-bot~enwiki, Mesons, Yevgeny Kats, TriTertButoxy, SashatoBot, Ouzo~enwiki, Happy-melon, Kurtan~enwiki, Myasuda, Cydebot, Meno25, Photocopier, Michael C Price, Casliber, Thijs!bot, Headbomb, Newton2, Mentifisto, Autotheist, Steveprutz, NeverWorker, NicoSan, MartinBot, Arjun01, HEL, J.delanoy, Numbo3, Gombang, Num1dgen, Ceoyoyo, VolkovBot, Macedonian, Mocirne, TXiKiBoT, Anonymous Dissident, Abdullais4u, Antixt, Jhb110, Thanatos666, AlleborgoBot, SieBot, ToePeu.bot, RadicalOne, Ngexpert7, Jacob.jose, Hamiltondaniel, TubularWorld, Muhends, ClueBot, ICAPTCHA, UniQue tree, Snigbrook, Fyyer, IceUnshattered, Cmj91uk, LieAfterLie, Manu-ve Pro Ski, TimothyRias, Addbot, Betterusername, AgadaUrbanit, Ehrenkater, OlEnglish, Zorrobot, Andy2308, Legobot, Luckas-bot, Ptbotgourou, Maxim Sabalyauskas, Planlips, JackieBot, Icalanise, Citation bot, ‏غامدي.م.أحمد‎24, ArthurBot, Almabot, Omnipaedista, Alexeymorgunov, ⁇, Tormine, MathFacts, Citation bot 1, MastiBot, Earthandmoon, EmausBot, John of Reading, Az29, Galaktiker, StringTheory11, Quondum, Surajt88, I hate whitespace, ClueBot NG, Scimath Genius, Braincricket, Widr, Helpful Pixie Bot, Bibcode Bot, Tyler6360534, Katagun5, Melenc, Derek-Winters, Prasanna4s, Machosquirrel, Devinhorn, KasparBot and Anonymous: 149

- **Neutrino** *Source:* https://en.wikipedia.org/wiki/Neutrino?oldid=681496713 *Contributors:* AxelBoldt, Chenyu, Bryan Derksen, Zundark, The Anome, Tarquin, Andre Engels, Xaonon, XJaM, William Avery, Roadrunner, DrBob, Heron, Cwitty, MimirZero, Spiff~enwiki, Edward, Patrick, Ken Arromdee, EddEdmondson, Ezra Wax, Gdarin, Meekohi, Bcrowell, Cyde, Arpingstone, Alfio, Looxix~enwiki, Strebe, JWSchmidt, Julesd, Glenn, Nikai, Andres, Evercat, Rob Hooft, TheSeez, Crissov, Wikiborg, Reddi, Lfh, Cos111, Tpbradbury, Fibonacci, Warofdreams, Twang, Donarreiskoffer, Drxenocide, Robbot, Findel, Zandperl, Nurg, Masao, Merovingian, Bobunf, Rursus, Meelar, Matty j, Intangir, Wikibot, Wereon, Duien, Jimduck, Bbx, David Gerard, Giftlite, Graeme Bartlett, DocWatson42, Laudaka, Mikez, Harp, Lethe, HangingCurve, Xerxes314, Anville, Dratman, Curps, Jorge Stolfi, Eequor, Mdob, Espetkov, LiDaobing, Elroch, Icairns, Doug Danner, Nickptar, Fg2, Lrenh, Deglr6328, Hmmm~enwiki, Mattman723, Helohe, Rich Farmbrough, Hydrox, Cacycle, Pjacobi, Vsmith, Dbachmann, Mani1, Pavel Vozenilek, Ralfoide, Sunborn, Neko-chan, Kharhaz, RJHall, Charm, Haxwell, RoyBoy, Smalljim, Cje~enwiki, Viriditas, Cwolfsheep, Foobaz, I9Q79oL78KiL0QTFHgyc, La goutte de pluie, Thewayforward, Thuktun, Fleurot~enwiki, Quaoar, Alansohn, Anthony Appleyard, ChristopherWillis, Calton, Axl, Mac Davis, Hdeasy, RJFJR, Dirac1933, TenOfAllTrades, Vuo, Cmprince, Pauli133, Gene Nygaard, Lyuokdea, Flying fish, Richard Arthur Norton (1958-), Woohookitty, Swamp Ig, Insaneinside, Benhocking, Nakos2208~enwiki, GregorB, SDC, Joke137, Fxer, Palica, RedBLACKandBURN, Ashmoo, Graham87, Qwertyus, Raymond Hill, Drbogdan, Rjwilmsi, Coemgenus, Strait, John187, Staecker, Jmcc150, Salix alba, Mike Peel, Vegaswikian, Oblivious, Ligulem, R.e.b., Jehochman, The wub, FlaBot, Ian Pitchford, DannyWilde, Itinerant1, RexNL, Gurch, Kolbasz, Goudzovski, Sperxios, Scythe33, Smithbrenon, Chobot, Nagytibi, DVdm, YurikBot, Bambaiah, Vuvar1, Phmer, RussBot, Ohwilleke, Witan, Xihr, Bhny, Chris Capoccia, JabberWok, Gaius Cornelius, Salsb, Grafen, Długosz, Gillis, SCZenz, Ravedave, Abb3w, CecilWard, Santaduck, Bota47, Maunus, Dna-webmaster, Ms2ger, Rhynchosaur, DrWorm, Alias Flood, Ilmari Karonen, Nimbex, Phr en, Otto ter Haar, Fragman, That Guy, From That Show!, AndrewWTaylor, Palapa, Morgan wascko, SmackBot, Trainbrain27, Reedy, Tom Lougheed, Melchoir, The Monster, Arbe, Mscuthbert, BiT, Ohnoitsjamie, Dauto, Kmarinas86, GregRM, Decowski, Yurigerhard, DHN-bot~enwiki, Sbharris, Colonies Chris, Hengsheng120, Nap~enwiki, Sergio.ballestrero, Милан Јелисавчић, Cophus, Mayrel, Wen D House, Engwar, Jdlambert, Webmaster Pete, TheMaster42, Pwjb, Akriasas, DenisRS, Rjn~enwiki, Kukini, Yevgeny Kats, Ged UK, Jjpcondor, Gobonobo, ThorAvaTahr, JorisvS, Makyen, Libera~enwiki, Aeluwas, Dicklyon, Mets501, MTSbot~enwiki, Galactor213, Fredil Yupigo, Masoninman, Newone, Richard75, Chalnoth, Mssgill, Valoem, Abeneal, Rszasz, CmdrObot, Calmargulis, Olaf Davis, Vyznev Xnebara, MrFizyx, Thubsch, Myasuda, Alton, Astralusenet, Icek~enwiki, Szdori~enwiki, Hyperdeath, Gogo Dodo, HPaul, RC Master, Q43, Michael C Price, Quibik, Christian75, DumbBOT, Joe Chick, Thijs!bot, Martin Hogbin, Naucer, Headbomb, Dtgriscom, WVhybrid, Esemono, James086, Second Quantization, Davidhorman, Weasel5i2, Jonny-mt, D.H, Greg L, Mentifisto, Luna Santin, Guy Macon, Seaphoto, Alphachimpbot, Astavats, Parande, DagosNavy, JAnDbot, Deflective, MER-C, CosineKitty, Savant13, Magioladitis, WolfmanSF, VoABot II, Nyq, Websterwebfoot, Bakken, DMcanada, Christoph Scholz~enwiki, Seleucus, Dirac66, Adrian J. Hunter, LorenzoB, NJR ZA, Khalid Mahmood, Squidonius, Pavel Jelínek, Gwern, Denis tarasov, Glrx, R'n'B, Fatka, Maurice Carbonaro, MrBell, Aqwis, Salih, Nalumc, Plasticup, Warut, Nwbeeson, Rosenknospe, Juliancolton, Mike Clough, Bonadea, Lseixas, Sheliak, Cuzkatzimhut, Jharris1993, VolkovBot, Camrn86, AlnoktaBOT, DrJohn-PCostella, TXiKiBoT, The Original Wildbear, Nxavar, MinotAuruS, Cgr1123, Awl, Michael H 34, HannesHultgren~enwiki, SuperLonghorn, BotKung, SwordSmurf, Norbu19, Richwil, Tomaxer, The assassin 47, Morangm, AlleborgoBot, Thunderbird2, Angelastic, SieBot, Fredelige, PlanetStar, Zelab, Laoris, Ergateesuk, Stratman07, Jerryobject, RadicalOne, Aaarnooo, ScAvenger lv, John fromer, Thehotelambush, ShadowPhox, Ergo4sum, AWeishaupt, Jahilia, Extensive~enwiki, Sagredo, Martarius, ClueBot, Justin W Smith, Plastikspork, Apparentslug, Der Golem, Mild Bill Hiccup, Msgarrett, Hyh1048576, Mostargue, Paulcmnt, Ajoykt, Rhmtsang~enwiki, Xmantis, JayVora, Excirial, Kain Nihil, Wacko375, TonyBermanseder, Cenarium, Alastair301, Jwfvalle, Billrob458, Askahrc, Bouhadef, Termatt56, Johnuniq, PSimeon, Ost316, Nepenthes, Jprw, PL290, Jht4060, Buchler, MystBot, SkyLined, NCDane, Stephen Poppitt, Rical, Addbot, DOI bot, Download, Tide rolls, Lightbot, Taketa, SPat, Zorrobot, GDK, Gameseeker, Olsen-Fan, Legobot, Luckas-bot, Yobot, Bunnyhop11, Zagothal, Amble, Wikipedian Penguin, Azcolvin429, LibrarianofBabel, Dickdock, Robert Treat, AnomieBOT, Fatal!ty, Wrongfilter, Jim1138, Icalanise, Dakarateka, Mahmudmasri, Citation bot, Tano-kun, Vuerqex, GB fan, Xqbot, Blennow, JimVC3, Nickkid5, Cydelin, Srich32977, GrouchoBot, Abce2, Baba476, Backpackadam, QMarion II, 78.26, Aashaa, Ernsts, A. di M., Eldudarino, FrescoBot, LucienBOT, Paine Ellsworth, Tobby72, Ajgw56, WurzelT, Citation bot 1, Merongb10, RandomDSdevel, Pinethicket, HRoestBot, Rameshngbot, Tom.Reding, Swamper777, Phil John Hawkins, PRONIZ, Rknop, Jkforde, Nieuwenh, Puzl bustr, Tawe, Higgshunter, Beladee, Lotje, Persian knight shiraz, DrSinn, Miracle Pen, EngineerFromVega, RjwilmsiBot, MalapropX14, DexDor, Ripchip Bot, Phlegat, John of Reading, MindBlender, Architeuthidae, Racerx11, RA0808, Theonhwiki, 8digits, Themorrissey, NorthernRaven, Hhhippo, ZéroBot, راهب یشنمه, StringTheory11, Waperkins, Herp Derp, H3llBot, Quondum, UniversumExNihilo, Almatinez, Aschwole, Mayur, Kranix, Maschen, Eg-T2g, LarsJanZeeuwRules, LikeLakers2, Rocketrod1960, H1tchh1ker4, Mikhail Ryazanov, ClueBot NG, PeterKirk69, 4Jmaster, Gilderien, Navasj, Law of Entropy, Confuddledone, 336, Helpful Pixie Bot, Mightyname, Electriccatfish2, Kronn8, Curb Chain, Bibcode Bot, Tirebiter78, Rm1271, Neutral current, Stehgdup, Watson system, Koska One, Cs1791, Chewkaflax, Ddanndt, Johan.lundberg, Dragonami, Ownedroad9, Quickcrazy78, Rajibganguly01, Uioplk, Acalloni, Zedshort, TF SHaDowMAn, Yaroslav Nikitenko, Neutrinoread, Jakekong, Achowat, Pritombose, Ysawires, GodsAccident, Blakee911, Pieceofchit1, BattyBot, Layth888, Friedncrispy, Dja1979, Adyyy, Dexbot, Jdjwright, Mogism, Jaxcp3, Reatlas, Provacitu74, Coladar, CensoredScribe, Juan tanesia12, Rmohapat, Rmohapatra, Christophe1946, Klingerdinger, EWPage, Monkbot, Richard Henry Eckert, Jazzwhiz101, Crystallizedcarbon, Fimatic, PerpetuaLux, Isambard Kingdom, Lxplot, TheHecster, DN-boards1, Matan Kovac, KasparBot, Alarana, Saturn comes back around, CumbleSpuzz and Anonymous: 576

- **Electron** *Source:* https://en.wikipedia.org/wiki/Electron?oldid=682603060 *Contributors:* AxelBoldt, CYD, Mav, Bryan Derksen, AstroNomer~enwiki, Ap, Ed Poor, Andre Engels, Ryrivard, William Avery, SimonP, Peterlin~enwiki, Heron, Camembert, Stevertigo, Bdesham, Patrick, D, JohnOwens, Michael Hardy, Tim Starling, Ixfd64, Fruge~enwiki, Arpingstone, PingPongBoy, Egil, NuclearWinner, Ahoerstemeier, Suisui, Jebba, JWSchmidt, Kingturtle, Aarchiba, Glenn, Scott, Kwekubo, Andres, Jordi Burguet Castell, Mxn, Agtx, Timwi, Wikiborg, Reddi, Rednblu, Markhurd, Maximus Rex, E23~enwiki, Omegatron, Secretlondon, Jusjih, BenRG, Jeffq, Donarreiskoffer, Gentgeen, Robbot, Sanders muc, Vespristiano, Merovingian, Pingveno, Blainster, Hadal, Wikibot, Wereon, Widsith, HaeB, Diberri, Dmn, Dina, Giftlite, Christopher Parham, Ferkelparade, Fastfission, Zigger, Herbee, Dissident, Xerxes314, Curps, Michael Devore, Bensaccount, Ssd, Gilgamesh~enwiki, Vadmium, Gdr, Knutux, Slowking Man, Yath, Gzuckier, Pcarbonn, Joizashmo, Karol Langner, Anythingyouwant, RetiredUser2, Bbbl67, Elroch, Icairns, JohnArmagh, JimQ, Mike Rosoft, Mindspillage, Patrick L. Goes, Discospinster, Brianhe, Rich Farmbrough, Guanabot, Hidaspal, Vsmith, Deh, Ardonik, Roybb95~enwiki, Xezbeth, Zazou, Mani1, SpookyMulder, Dmr2, ZeroOne, Kjoonlee, Goplat, Calair, Nabla, Brian0918, RJHall, Pt, Jaques O. Carvalho, El C, Huntster, Edward Z. Yang, Susvolans, Art LaPella, RoyBoy, ~K, Bobo192, Army1987, Asierra~enwiki, Flxmghvgvk, AtomicDragon, Evgeny, AllyUnion, Bert Hickman, Deryck Chan, PeterisP, Beetle B., Obradovic Goran, (aeropagitica), Pearle, Mpulier, HasharBot~enwiki, Confusedmiked, Mote, Jumbuck, Gary, ChristopherWillis, Ricky81682, Benjah-bmm27, Riana, AzaToth, DonJStevens, BernardH, Malo, David Hochron, Bart133, EagleFalconn, Schapel, Omphaloscope, RainbowOfLight, RichBlinne, H2g2bob, DV8 2XL, Gene Nygaard, Redvers, StuTheSheep, Linas, Mindmatrix, GrouchyDan, StradivariusTV, Uncle G, BillC, Kurzon, Jeff3000, HcorEric X, Eleassar777, Ozielke, Wayward, Palica, Omega21, FreplySpang, Enzo Aquarius, Rjwilmsi, Shaadow, Strait, Mike Peel, Chekaz, Bubba73, Dar-Ape, Yamamoto Ichiro, FlaBot, RobertG, Latka, DannyWilde, Nihiltres, RexNL, Kolbasz, Thecurran, Srleffler, Physchim62, Chobot, DVdm, Unclevortex, Eric B, YurikBot, Wavelength, RobotE, Bambaiah, AcidHelmNun, Jimp, Peter G Werner, Wolfmankurd, Wigie, Ventolin, JabberWok, SpuriousQ, Lucinos~enwiki, Akamad, Ori Livneh, Gaius Cornelius, Shaddack, Eleassar, Rsrikanth05, Salsb, Hawkeye7, Spike Wilbury, Jaxl, Welsh, DarthVader, Długosz, BirgitteSB, SCZenz, Retired username, Ravedave, PhilipO, Adam Rock, Mlouns, Chichui, BOT-Superzerocool, Gadget850, Bota47, Kkmurray, James Trotter~enwiki, Dna-webmaster, Ms2ger, Light current, Lycaon, Imaninjapirate, Josh3580, Kriscotta, JoanneB, Peyna, Lpm, JLaTondre, Heavy bolter, RG2, GrinBot~enwiki, Sbyrnes321, ChemGardener, Itub, SmackBot, Zazaban, Incnis Mrsi, KnowledgeOfSelf, Royalguard11, Melchoir, J.Sarfatti, KocjoBot~enwiki, Stepa, Pandion auk, Jrockley, JoeMarfice, ZerodEgo, Edgar181, Yamaguchi⬚⬚, Skizzik, Dauto, JSpudeman, Kurykh, Rajeevmass~enwiki, Persian Poet Gal, Pieter Kuiper, Jprg1966, Acrinym, Miquonranger03, MalafayaBot, Droll, Complexica, DHN-bot~enwiki, Sbharris, RAlafriz, V1adis1av, Vanished User 0001, Darthgriz98, Voyajer, Addshore, Percommode, Krich, DavidStern, Theonlyedge, Nakon, Nrcprm2026, DMacks, Daniel.Cardenas, Zeamays, Jonnyapple, Sadi Carnot, Bdushaw, Wilt, TriTertButoxy, Chymicus, UberCryxic, Bagel7, Mattfont, Heimstern, Jaganath, Ocatecir, Mr. Lefty, Ckatz, 16@r, Omnedon, Owlbuster, Waggers, SandyGeorgia, Spiel496, Funnybunny, HappyVR, Iridescent, Newone, NativeForeigner, J Di, Amakuru, Tawkerbot2, Chetvorno, Thermochap, CmdrObot, Ale jrb, Megaboz, RedRollerskate, Ruslik0, MrZap, McVities, WMSwiki, Bakanov, RobertLovesPi, Equendil, Cydebot, Acelor, Reywas92, Cantras, Bvcrist, LouisBB, Travelbird, Llort, David edwards, Tawkerbot4, Christian75, Narayanese, Ssilvers, Thijs!bot, Epbr123, Mbell, Dougsim, Nonagonal Spider, Headbomb, Yzmo, Marek69, West Brom 4ever, Tellyaddict, Cool Blue, Greg L, Sean William, VictorP, KrakatoaKatie, AntiVandalBot, WinBot, Skymt, Voyaging, Opelio, Tyco.skinner, Gef756, Chill doubt, Naturalnumber, Gdo01, Spencer, Leuko, CosineKitty, J-stan, Smith Jones, Acroterion, Magioladitis, WolfmanSF, Bennybp, Bongwarrior, VoABot II, A4, Nyq, JNW, JamesBWatson, باسم, Drondent, Slartibartfast1992, Jackal irl, Animum, Dirac66, 28421u2232nfenfcenc, Hveziris, User A1, Maliz, PoliticalJunkie, DerHexer, GregU, PEBill, MartinBot, BetBot~enwiki, Mermaid from the Baltic Sea, WizendraW, Xantolus, Thereen, CommonsDelinker, AlexiusHoratius, J.delanoy, DrKiernan, Rgoodermote, Numbo3, Acalamari, TheChrisD, Dispenser, LordAnubisBOT, JayMars, Lathrop, AntiSpamBot, TomasBat, NewEnglandYankee, Nwbeeson, SmoothK, Sunderland06, MetsFan76, Joshmt, Cometstyles, STBotD, RB972, Treisijs, D-Kuru, Dineshextreeme, Martial75, CardinalDan, Idioma-bot, Sheliak, Bondslave777, FeralDruid, X!, VolkovBot, ABF, Thisisborin9, Jacroe, Ryan032, Philip Trueman, DoorsAjar, TXiKiBoT, GimmeBot, Kriak, Hqb, GDonato, Anonymous Dissident, Crohnie, Monkey Bounce, Voorlandt, Mr. Hallman, Michael H 34, TBond, Wikiisawesome, Suriel1981, Rbdebole, Graymornings, Synthebot, Enviroboy, Rurik3, Generalguy11, !dea4u, Insanity Incarnate, Ceranthor, Yoos~enwiki, AlleborgoBot, Kalivd, EmxBot, Neparis, Swimallday, Ponyo, EJF, SieBot, Graham Beards, Scarian, CircafuciX, BotMultichill, Jauerback, Dawn Bard, Joncam, Caltas, Sergeanthuggy, Bentogoa, RadicalOne, Arbor to SJ, Prestonmag, Thadaddy3233, Oxymoron83, Antonio Lopez, KPH2293, Lightmouse, WingkeeLEE, Ealdgyth, BenoniBot~enwiki, Stustjohn, Dabomb87, PlantTrees, Dolphin51, Nergaal, Tomdobb, Muhends, WikipedianMarlith, ClueBot, Trojancowboy, GorillaWarfare, Artichoker, PipepBot, UniQue tree, The Thing That Should Not Be, Hongthay, Unbuttered Parsnip, GreenSpigot, Liekmudkipz, Mild Bill Hiccup, Correcting nonesense, NovaDog, Blanchardb, Richerman, RandomTREES, Rotational, Piledhigheranddeeper, Inala, DragonBot, Almcaeobtac, Jusdafax, MEJG, Gtstricky, Rhododendrites, Brews ohare, NuclearWarfare, Lunchscale, Jotterbot, PhySusie, Tonyfey, Lkruijsw, Kaiba, SchreiberBike, Stephing3, Thingg, Jamyricks, Aitias, Melibarr05, Kurtcobain321, Scalhotrod, Versus22, Johnuniq, MasterOfHisOwnDomain, DumZiBoT, TimothyRias, Sjodenenator, XLinkBot, Maky, Rror, Avoided, Mitch Ames, Ilikepie2221, WikHead, Mgaarafan, SkyLined, Addbot, Chizkiyahuavraham, AVand, Some jerk on the Internet, Hurleymann1, Uruk2008, DOI bot, Tcncv, Booba5, AkhtaBot, Jessepfrancis, Ronhjones, Jncraton, Moosehadley, CanadianLinuxUser, WFPM, LaaknorBot, Chamal N, CarsracBot, FiriBot, Omnipedian, LinkFA-Bot, Ehrenkater, Pnacitum, Tide rolls, Lightbot, Potekhin, UPS Truck Driver, VP-bot, Luckas-bot, Yobot, Nergality, Kan8eDie, THEN WHO WAS PHONE?, Eric-Wester, Tonyrex, AnomieBOT, Shootbamboo, DemocraticLuntz, Rubinbot, Götz, Jim1138, IRP, Piano non troppo, Icalanise, Kingpin13, Mydickishuge24, Materialscientist, The High Fin Sperm Whale, Citation bot, Neurolysis, ArthurBot, LovesMacs, Mrhellcool, Rightly, Xqbot, IrishChemistPride, IrishChemistPride2, GeometryGirl, Restu20, Srich32977, S0aasdf2sf, John5955, Alan8, ProtectionTaggingBot, Omnipaedista, RibotBOT, TonyHagale, Phillycheesesteaks, LyleHoward, A. di M., RyanOrdemann, Peter470, Thehelpfulbot, Al Wiseman, FrescoBot, Surv1v4l1st, Eadon-com, Paine Ellsworth, Tobby72, Gauravdce07, Steve Quinn, C.Bluck, Citation bot 1, MarB4, Galmicmi, Gil987, Pinethicket, HRoestBot, Voltron Hax, Raen79, Hoo man, Yos233, Allthingstoallpeople, MastiBot, Kuririmo, Noel Streatfield, Ezhuttukari, Swifterthenyou, Noisalt, Jujutacular, Euchanels, Lissajous, Dude1818, December21st2012Freak, IJBall, Jauhienij, Utility Monster, FoxBot, Sheogorath, Jdlawlis, Odatus, Bestcallumuk, Sampathsris, DARTH SIDIOUS 2, Mean as custard, RjwilmsiBot, TjBot, MinicheddarsandelephantsFTW, Benjadow, Mcmonsterbrothers, Priceracks, Csilcock, Sohaib360, Androstachys, Techhead7890, EmausBot, Optiguy54, GoingBatty, Jjasharpe, Pcorty, TuHan-Bot, Hhhippo, HiW-Bot, John Cline, Harddk, Fæ, Josve05a, StringTheory11, Wackywace, Quondum, GianniG46, Fizicist, Wayne Slam, Raynor42, Arnaugir, Jacksccsi, Brandmeister, Donner60, Negovori, RockMagnetist, Mni9791, ClueBot NG, HLachman, Hermajesty21, Jacobkh, Letoya123, Samsau ninjaguy, Ggonzalm, Moritz37, Braincricket, Helpful Pixie Bot, Geo7777, SzMithrandir, Bibcode Bot, Dfbowsmountainer, Ymblanter, Vagobot, Paolo Lipparini, Wzrd1, Lk00la1dl, JacobTrue, Socal212, Begman5, Mark Arsten, Cadiomals, Jikepaddy, Caterpillar111, Macymae, 06seagsa, BEEPTHENOOB, ItzzRevolution, Shawn Worthington Laser Plasma, Duxwing, Klilidiplomus, Uopchem251, Joe0x7F, BattyBot, Justincheng12345-bot, Cyberbot II, ChrisGualtieri, GoShow, Ankap~enwiki, Glenzo999, Barant2, BrightStarSky, Dexbot, Astromango2215, Webclient101, Mogism, 331dot, Spray787, Vanquisher.UA, Lugia2453, Kondormari, Reatlas, JellyBean4.1, Prof.Professer, The User 111, Bluemanyoung, Rohitgunturi, Ugog Nizdast, The Herald, Jwratner1, Zahid2233, MorshusApprentice, 2005-Fan, Phub Dorji, Epic Failure, Gindor, Ian98989898, Monkbot, SkateTier, Waldmannevan, Wiki1098, Wulfiedude14, Mario Castelán Castro, Fleivium, Crystallizedcarbon, Mcwikigeek, Jesus is the Light of my life, Acesoli, Soumilm, Lemmegetyou, Flying g shot, SirLagsalott, Tetra quark, Skipfortyfour, Stim 2.0, KasparBot, Fazbear7891 and Anonymous: 935

- **Electron neutrino** *Source:* https://en.wikipedia.org/wiki/Electron_neutrino?oldid=674917553 *Contributors:* Bryan Derksen, Bobrayner, Rjwilmsi, Strait, Bgwhite, Jeffhoy, Lockesdonkey, Dna-webmaster, GDallimore, Thijs!bot, Headbomb, Oreo Priest, Magioladitis, Maurice Carbonaro, Nwbeeson, Ggenellina, FourteenDays, SieBot, SkyLined, Addbot, LaaknorBot, PieterJanR, Luckas-bot, Rubinbot, Citation bot, ArthurBot, Xqbot, Carlog3, Paine Ellsworth, Citation bot 1, TjBot, EmausBot, John of Reading, Optiguy54, TuHan-Bot, JSquish, Quondum, Kasirbot, Helpful Pixie Bot, Bibcode Bot, Love's Labour Lost, Tpaine krk, Svebert, Makecat-bot, DD4235, Scipsycho and Anonymous: 10

- **Muon** *Source:* https://en.wikipedia.org/wiki/Muon?oldid=680138424 *Contributors:* AxelBoldt, The Epopt, CYD, Mav, Bryan Derksen, Zundark, Roadrunner, Bkellihan, Youandme, Tim Starling, EddEdmondson, Looxix~enwiki, Ahoerstemeier, Angela, Rob Hooft, Kbk, Donarreiskoffer, AlexPlank, Robbot, Merovingian, Rholton, Ojigiri~enwiki, Auric, Roscoe x, Bkell, Millosh, Wikibot, Ruakh, Diberri, Giftlite, Wizzy, Herbee, Xerxes314, Bodhitha, LiDaobing, Pcarbonn, DragonflySixtyseven, Deglr6328, Eb.hoop, Rich Farmbrough, Pjacobi, Vsmith, Mani1, STGM, Kjoonlee, RJHall, Army1987, Danski14, Anthony Appleyard, RobPlatt, Keenan Pepper, RJFJR, Ceyockey, Falcorian, Woohookitty, Xinghuei, Bennetto, Graham87, Vanderdecken, Rjwilmsi, Strait, Mike Peel, Bubba73, DoubleBlue, Dougluce, FlaBot, Fivemack, DannyWilde, Sp00n, Lmatt, Goudzovski, Srleffler, Chobot, YurikBot, Wavelength, Bambaiah, Limulus, JabberWok, Hellbus, Salsb, SCZenz, Ravedave, Scottfisher, Tetracube, Lt-wiki-bot, E Wing, Roberto DR, CrniBombarder!!!, Sbyrnes321, Eog1916, SmackBot, Incnis Mrsi, Melchoir, Stifle, Gilliam, Dauto, Bluebot, Tigerhawkvok, Sbharris, Colonies Chris, Can't sleep, clown will eat me, Yevgeny Kats, Dane Sorensen, JorisvS, Mets501, JoeBot, CapitalR, SchmittM, Ruslik0, Ken Gallager, Rotiro, A876, Corpx, Thijs!bot, Headbomb, D.H, Bm gub, Andrew Carlssin, Spencer, Kariteh, Deflective, Belg4mit, Swpb, Mother.earth, Nono64, HEL, Hans Dunkelberg, 5Q5, Tarotcards, Coppertwig, Bermy88, Jarry1250, Thecinimod, Sheliak, Cuzkatzimhut, VolkovBot, Larryisgood, VasilievVV, TXiKiBoT, Anonymous Dissident, Mihaip, Graymornings, SalomonCeb, SieBot, Csmart287, Gerakibot, Statue2, Mhouston, StewartMH, ClueBot, Polyamorph, DnetSvg, Esbboston, Saritepe, Stefan Ritt, BarretB, Kajabla, Addbot, Roentgenium111, Toyokuni3, Download, Ehrenkater, Lightbot, Luckas-bot, Yobot, Evaders99, Kulmalukko, AnomieBOT, Icalanise, Kingpin13, Materialscientist, Citation bot, Kotika98, ArthurBot, Xqbot, Cjxc92, Gilo1969, Srich32977, Misterigloo, Kyng, A. di M., Paine Ellsworth, Citation bot 1, Citation bot 4, Rameshngbot, Isofox, Jetstoknowhere, TobeBot, Trappist the monk, Puzl bustr, RjwilmsiBot, EmausBot, John of Reading, WikitanvirBot, Dewritech, GoingBatty, Milledit, Naviguessor, StringTheory11, Medeis, Suslindisambiguator, Quondum, Timetraveler3.14, Layona1, Aerthis, Mikhail Ryazanov, Frietjes, CaroleHenson, Kebil, Bibcode Bot, Jesusmonkey, NotWith, BattyBot, Kisokj, Liam135, MuonRay, Tony Mach, Telfordbuck, Krotera, Ajdigregorio, Seoman2snowlock, Monkbot, Jromerofontalvo, KasparBot, Corrupt Titan, QzPhysics and Anonymous: 191

- **Muon neutrino** *Source:* https://en.wikipedia.org/wiki/Muon_neutrino?oldid=680138955 *Contributors:* Bryan Derksen, The Anome, Twang, Goudzovski, Eleassar, Dna-webmaster, SmackBot, Ruslik0, Thijs!bot, Headbomb, Magioladitis, Ggenellina, FourteenDays, SieBot, Thesavagenorwegian, Muhends, Ajoykt, SkyLined, Addbot, PieterJanR, Luckas-bot, AnomieBOT, Rubinbot, Icalanise, ArthurBot, RibotBOT, Paine Ellsworth, Citation bot 1, TjBot, Mithril, EmausBot, ZéroBot, StringTheory11, Quondum, Kasirbot, Bibcode Bot, Sunitharay, Artdk, DaveW51, QzPhysics and Anonymous: 7

- **Tau (particle)** *Source:* https://en.wikipedia.org/wiki/Tau_(particle)?oldid=680139275 *Contributors:* Bryan Derksen, Iluvcapra, Ahoerstemeier, Bueller 007, Schneelocke, Dysprosia, Donarreiskoffer, Merovingian, Rorro, Davidl9999, Millosh, Harp, Herbee, Codepoet, Xerxes314, Bodhitha, CryptoDerk, Icairns, Rich Farmbrough, Pjacobi, Martpol, Sunborn, Kjoonlee, El C, Reuben, JellyWorld, RobPlatt, RJFJR, Falcorian, Dmitry Brant, Christopher Thomas, Palica, Rjwilmsi, Strait, Mike Peel, FlaBot, DannyWilde, Goudzovski, Chobot, RobotE, Bambaiah, AcidHelmNun, JabberWok, Eleassar, Salsb, SCZenz, Zwobot, Ospalh, PS2pcGAMER, Bota47, Someones life, Poulpy, Physicsdavid, Incnis Mrsi, Dauto, Pieter Kuiper, Loodog, JorisvS, MTSbot~enwiki, WISo, Q43, Thijs!bot, Headbomb, Davidhorman, Hcobb, Escarbot, RogueNinja, Yill577, Soulbot, Kostisl, STBotD, Sheliak, Joyko~enwiki, VolkovBot, Fences and windows, TXiKiBoT, Awl, Jba138, SieBot, OKBot, ImageRemovalBot, Plastikspork, Djr32, Alexbot, TimothyRias, Assosiation, BodhisattvaBot, SkyLined, J Hazard, Addbot, Eric Drexler, Ronhjones, ChenzwBot, Jklukas, Theozzfancometh, Skippy le Grand Gourou, Luckas-bot, Yobot, Grebaldar, AnomieBOT, Icalanise, Citation bot, Xqbot, Blennow, Franco3450, 吐, Paine Ellsworth, Jonesey95, Three887, Plasticspork, 3ph, Miracle Pen, RjwilmsiBot, TjBot, Ripchip Bot, EmausBot, Dcirovic, Suslindisambiguator, Quondum, Rezabot, Helpful Pixie Bot, Bibcode Bot, BG19bot, Sudsguest, YFdyh-bot, Redcliffe maven, TwoTwoHello, Akro7, KasparBot, JPPepper, QzPhysics and Anonymous: 48

- **Tau neutrino** *Source:* https://en.wikipedia.org/wiki/Tau_neutrino?oldid=644666618 *Contributors:* B.d.mills, Kfitzner, Rjwilmsi, Salsb, TriTertButoxy, Newone, Thijs!bot, Headbomb, Fences and windows, Ggenellina, FourteenDays, SieBot, SkyLined, Addbot, Ronhjones, Luckas-bot, Rubinbot, JackieBot, Icalanise, Citation bot, ArthurBot, Carlog3, LucienBOT, Paine Ellsworth, Citation bot 1, TjBot, EmausBot, K6ka, Joe Gazz84, ZéroBot, Kasirbot, Bibcode Bot and Anonymous: 5

- **Gauge boson** *Source:* https://en.wikipedia.org/wiki/Gauge_boson?oldid=662605501 *Contributors:* Bryan Derksen, Andre Engels, Michael Hardy, Ahoerstemeier, Bueller 007, LouI, Phys, Robbot, Gwrede, Rholton, Rursus, Davidl9999, Giftlite, Xerxes314, Alison, JeffBobFrank, Chinasaur, Andris, Garth 187, Beland, Setokaiba, Icairns, AmarChandra, Lumidek, Vsmith, Roybb95~enwiki, Mal~enwiki, La goutte de pluie, Nk, Kusma, Ringbang, Mpatel, Nakos2208~enwiki, Tevatron~enwiki, Kbdank71, Chobot, Roboto de Ajvol, Hairy Dude, Salsb, StuRat, ArielGold, RG2, InverseHypercube, Niels Olson, Sadi Carnot, TriTertButoxy, Ekjon Lok, Bjankuloski06en~enwiki, Phatom87, Headbomb, Tyco.skinner, Knotwork, Swpb, Maurice Carbonaro, Gombang, TXiKiBoT, Odellus, Antixt, AlleborgoBot, SieBot, Jim E. Black, Homonihilis, BOTarate, DumZiBoT, SilvonenBot, Addbot, Bertman600, NjardarBot, Numbo3-bot, Lightbot, Zorrobot, Luckas-bot, Yobot, Citation bot, ArthurBot, A. di M., Rameshngbot, RedBot, RobinK, Mary at CERN, TjBot, EmausBot, ZéroBot, StringTheory11, Mentibot, Dsperlich, CeraBot, Galactic Messiah, DerekWinters, Fisherv, KasparBot and Anonymous: 41

- **W and Z bosons** *Source:* https://en.wikipedia.org/wiki/W_and_Z_bosons?oldid=676803444 *Contributors:* AxelBoldt, Sodium, Mav, Bryan Derksen, The Anome, Ap, Andre Engels, Danny, Roadrunner, DrBob, Michael Hardy, Tim Starling, Karada, Egil, Ahoerstemeier, Ryan Cable, Julesd, Mxn, Charles Matthews, Ike9898, Saltine, Phys, Topbanana, BenRG, Finlay McWalter, Twang, Phil Boswell, Donarreiskoffer, Robbot, Pigsonthewing, Nurg, DHN, Xanzzibar, M-Falcon, Giftlite, Tremolo, Harp, Herbee, Xerxes314, Jeremy Henty, Bodhitha, LiDaobing, RetiredUser2, Icairns, Mike Rosoft, Vsmith, Gianluigi, Kjoonlee, Drhex, Obradovic Goran, Jérôme, Fkbreitl, Cameron.simpson, Gene Nygaard, Linas, LoopZilla, Graham87, Kbdank71, Rjwilmsi, Strait, Mike Peel, Lmatt, Goudzovski, Chobot, FrankTobia, Roboto de Ajvol, Ugha, Mushin, Bambaiah, Wester, Hairy Dude, Hellbus, Salsb, Seb35, Długosz, Turbolinux999, Ravedave, Scottfisher, Dna-webmaster, Modify, Argo Navis, Teply,

Sbyrnes321, SmackBot, Tom Lougheed, Jagged 85, ZerodEgo, Dauto, Bluebot, Shaggorama, Sbharris, Niels Olson, Radagast83, Acdx, John, Lottamiata, Happy-melon, Tubezone, MightyWarrior, Joelholdsworth, Tangobot, Michael C Price, Quibik, Dchristle, Realjanuary, Headbomb, Davidhorman, Nosirrom, Certain, Gökhan, JAnDbot, Tigga, Omeganian, Brimofinsanity, TheEditrix2, Trapezoidal, Magioladitis, ThoHug, Leyo, Lilac Soul, HEL, Rod57, Y2H, HiEv, Adam Zivner, Madblueplanet, Sheliak, Dextrose, Anonymous Dissident, Synthebot, Antixt, Coronellian~enwiki, SieBot, STANMAR725, Jim E. Black, Gerakibot, Martin Kealey, CutOffTies, Fratrep, ClueBot, Mild Bill Hiccup, Alexbot, Carsrac, SkyLined, Dieppu, Stephen Poppitt, Addbot, Eric Drexler, Toyokuni3, Mjamja, Ronkonkaman, Download, CarsracBot, ChenzwBot, Lightbot, M sotirov, Luckas-bot, Yobot, Jim1138, MehrdadAfshari, ArthurBot, Ernsts, A. di M., Howard McCay, FrescoBot, Paine Ellsworth, D'ohBot, Citation bot 1, Gil987, Tom.Reding, Swallerick, FoxBot, Earthandmoon, Tm1729, TjBot, Антон Гліністы, Newty23125, EmausBot, Mnkyman, StringTheory11, Quondum, MisterDub, WaterCrane, Whoop whoop pull up, ClueBot NG, Helpful Pixie Bot, Bibcode Bot, BG19bot, Bakkedal, JYBot, Mamaphyskerin, Anrnusna, MartinNicklin, Boidal-Quantized and Anonymous: 137

- **Gluon** *Source:* https://en.wikipedia.org/wiki/Gluon?oldid=681546689 *Contributors:* AxelBoldt, CYD, Bryan Derksen, Gdarin, TakuyaMurata, Card~enwiki, Looxix~enwiki, Ellywa, Ahoerstemeier, Med, Schneelocke, Phys, Phil Boswell, Donarreiskoffer, Fredrik, Merovingian, Hadal, Giftlite, Herbee, Xerxes314, Eequor, Darrien, Keith Edkins, RetiredUser2, Icairns, Mike Rosoft, AlexChurchill, HedgeHog, Kenny TM~~enwiki, David Schaich, Ioliver, Mashford, El C, Kwamikagami, Ardric47, Obradovic Goran, Alansohn, Guy Harris, Dachannien, Ricky81682, Batmanand, Velella, Kazvorpal, April Arcus, Forteblast, Mpatel, Palica, BD2412, Kbdank71, Rjwilmsi, Macumba, Strait, Mike Peel, Bubba73, Klortho, FlaBot, Srleffler, Chobot, YurikBot, Wavelength, Bambaiah, Hairy Dude, Jimp, JabberWok, Zelmerszoetrop, Salsb, SCZenz, Randolf Richardson, Ravedave, Danlaycock, Bota47, LeonardoRob0t, Anclation~enwiki, Physicsdavid, Erudy, GrinBot~enwiki, Kgf0, SmackBot, Melchoir, Cessator, Benjaminevans82, Abtal, MK8, Colonies Chris, Can't sleep, clown will eat me, Decltype, Qcdmaestro, Edconrad, Darkpoison99, FredrickS, Omsharan, Pegasusbot, Gregbard, ProfessorPaul, Thijs!bot, Headbomb, Rriegs, Oreo Priest, AntiVandalBot, Shambolic Entity, Deflective, Mujokan, Yill577, Happycool, Mother.earth, Martynas Patasius, WiiWillieWiki, HEL, Hans Dunkelberg, Gombang, Inwind, Sheliak, Jonthaler, VolkovBot, TXiKiBoT, Davehi1, Kriak, Anonymous Dissident, Imasleepviking, AlleborgoBot, EJF, SieBot, Steven Crossin, OKBot, ClueBot, Wwheaton, Qsaw, Nucularphysicist, Ottava Rima, Gordon Ecker, Rhododendrites, Brews ohare, Cacadril, RexxS, JKeck, Against the current, SkyLined, Addbot, DOI bot, Lightbot, Skippy le Grand Gourou, Luckas-bot, Planlips, AnomieBOT, Jim1138, JackieBot, Citation bot, Bci2, ArthurBot, Xqbot, Neil95, Triclops200, Omnipaedista, TorKr, ⁇⁇, Paine Ellsworth, Ivoras, Citation bot 1, Pekayer11, Rameshngbot, PNG, RjwilmsiBot, TjBot, Lilcal89012, EmausBot, Socob, JSquish, StringTheory11, Quondum, TyA, Maschen, RolteVolte, ClueBot NG, Timothy jordan, Maplelanefarm, Bibcode Bot, BG19bot, Gravitoweak, Cadiomals, Tropcho, Fraulein451, DrHjmHam, Rhlozier, D.shinkaruk, Yaara dildaara, BronzeRatio, Monkbot, Yikkayaya, KasparBot and Anonymous: 142

- **Photon** *Source:* https://en.wikipedia.org/wiki/Photon?oldid=682075782 *Contributors:* AxelBoldt, WojPob, Mav, Bryan Derksen, The Anome, Tarquin, Koyaanis Qatsi, Ap, Josh Grosse, Ben-Zin~enwiki, Heron, Youandme, Spiff~enwiki, Bdesham, Michael Hardy, Ixfd64, TakuyaMurata, NuclearWinner, Looxix~enwiki, Snarfies, Ahoerstemeier, Stevenj, Julesd, Glenn, AugPi, Mxn, Smack, Pizza Puzzle, Wikiborg, Reddi, Lfh, Jitse Niesen, Kbk, Laussy, Bevo, Shizhao, Raul654, Jusjih, Donarreiskoffer, Robbot, Hankwang, Fredrik, Eman, Sanders muc, Altenmann, Bkalafut, Merovingian, Gnomon Kelemen, Hadal, Wereon, Anthony, Wjbeaty, Giftlite, Art Carlson, Herbee, Xerxes314, Everyking, Dratman, Michael Devore, Bensaccount, Foobar, Jaan513, DÃ,ugosz, Zeimusu, LucasVB, Beland, Setokaiba, Kaldari, Vina, RetiredUser2, Icairns, Lumidek, Zondor, Randwicked, Eep², Chris Howard, Zowie, Naryathegreat, Discospinster, Rich Farmbrough, Yuval madar, Pjacobi, Vsmith, Ivan Bajlo, Dbachmann, Mani1, SpookyMulder, Kbh3rd, RJHall, Ben Webber, El C, Edwinstearns, Laurascudder, RoyBoy, Spoon!, Dalf, Drhex, Bobo192, Foobaz, I9Q79oL78KiL0QTFHgyc, La goutte de pluie, Zr40, Apostrophe, Minghong, Rport, Alansohn, Gary, Sade, Corwin8, PAR, UnHoly, Hu, Caesura, Wtmitchell, Bucephalus, Max rspct, BanyanTree, Cal 1234, Count Iblis, Egg, Dominic, Gene Nygaard, Ghirlandajo, Kazvorpal, UTSRelativity, Falcorian, Drag09, Boothy443, Richard Arthur Norton (1958-), Woohookitty, Linas, Gerd Breitenbach, StradivariusTV, Oliphaunt, Cleonis, Pol098, Ruud Koot, Mpatel, Nakos2208~enwiki, Dbl2010, Ch'marr, SDC, CharlesC, Alan Canon, Reddwarf2956, Mandarax, BD2412, Kbdank71, Zalasur, Sjakkalle, Rjwilmsi, Саша Стефановић, Strait, MarSch, Dennis Estenson II, Trlovejoy, Mike Peel, HappyCamper, Bubba73, Brighterorange, Cantorman, Egopaint, Noon, Godzatswing, FlaBot, RobertG, Arnero, Mathbot, Nihiltres, Fresheneesz, TeaDrinker, Srleffler, BradBeattie, Chobot, Jaraalbe, DVdm, Elfguy, EamonnPKeane, YurikBot, Bambaiah, Splintercellguy, Jimp, RussBot, Supasheep, JabberWok, Wavesmikey, KevinCuddeback, Stephenb, Gaius Cornelius, Salsb, Trovatore, Długosz, Tailpig, Joelr31, SCZenz, Randolf Richardson, Ravedave, Tony1, Roy Brumback, Gadget850, Dna-webmaster, Enormousdude, Lt-wiki-bot, Oysteinp, JoanneB, Ligart, John Broughton, GrinBot~enwiki, Sbyrnes321, Itub, SmackBot, Moeron, Incnis Mrsi, KnowledgeOfSelf, CelticJobber, Melchoir, Rokfaith, WilyD, Jagged 85, Jab843, Cessator, AnOddName, Skizzik, Dauto, JSpudeman, Robin Whittle, Ati3414, Persian Poet Gal, MK8, Jprg1966, Complexica, Sbharris, Colonies Chris, Ebertek, WordLife565, V1adis1av, RWincek, Aces lead, Stangbat, Cybercobra, Valenciano, EVula, A.R., Mini-Geek, AEM, DMacks, N Shar, Sadi Carnot, FlyHigh, The Fwanksta, Drunken Pirate, Yevgeny Kats, Lambiam, Harryboyles, Iron-Gargoyle, Ben Moore, A. Parrot, Mr Stephen, Fbartolom, Dicklyon, SandyGeorgia, Mets501, Ceeded, Ambuj.Saxena, Ryulong, Vincecate, Astrobayes, Newone, J Di, Lifeverywhere, Tawkerbot2, JRSpriggs, Chetvorno, Luis A. Veguilla-Berdecia, CalebNoble, Xod, Gregory9, CmdrObot, Wafulz, Van helsing, John Riemann Soong, Rwflammang, Banedon, Wquester, Outriggr, Logical2u, Myasuda, Howardsr, Cydebot, Krauss, Kanags, A876, WillowW, Bvcrist, Hyperdeath, Hkyriazi, Rracecarr, Difluoroethene, Edgerck, Michael C Price, Tawkerbot4, Christian75, Ldussan, RelHistBuff, Waxigloo, Kozuch, Thijs!bot, Epbr123, Opabinia regalis, Markus Pössel, Mglg, 24fan24, Headbomb, Newton2, John254, J.christianson, Escarbot, Stannered, AntiVandalBot, Luna Santin, Jtrain4469, Normanmargolus, Tyco.skinner, TimVickers, NSH001, Dodecahedron~enwiki, Tim Shuba, Gdo01, Sluzzelin, Abyssoft, CosineKitty, AndyBloch, Bryanv, ScottStearns, Hroðulf, Bongwarrior, VoABot II, B&W Anime Fan, SHCarter, Lgoger, I JethroBT, Dirac66, Hveziris, Maliz, Lord GaleVII, TRWBW, Shijualex, Glen, DerHexer, Patstuart, Gwern, Taborgate, MartinBot, MNAdam, Jay Litman, HEL, Ralf 58, J.delanoy, DrKiernan, Trusilver, C. Trifle, AstroHurricane001, Numbo3, Pursey, CMDadabo, Kevin aylward, UchihaFury, Pirate452, H4xx0r, Iamthewalrus35, Iamthewalrus36, Gee Eff, Chimpy07, Dirkdiggler69, Lk69, Hallamfm, Annoying editter, Yehoodig, Acalamari, Foreigner1, McSly, Samtheboy, Tarotcards, Rominandreu, ARTE, Tanaats, Potatoswatter, Y2H, Divad89, Scott Illini, Stack27, THEblindwarrior, VolkovBot, AlnoktaBOT, Hyperlinker, DoorsAjar, TXiKiBoT, Oshwah, Cosmic Latte, The Original Wildbear, Davehi1, Chiefwaterfall, Vipinhari, Hqb, Anonymous Dissident, HansMair, Predator24, BotKung, Luuva, Calvin4986, Improve~enwiki, Kmhkmh, Richwil, Antixt, Gorank4, Falcon8765, GlassFET, Cryptophile, MattiasAndersson, AlleborgoBot, Carlodn6, NHRHS2010, Relilles~enwiki, Tpb, SieBot, Timb66, Graham Beards, WereSpielChequers, ToePeu.bot, JerrySteal, Android Mouse, Likebox, RadicalOne, Paolo.dL, Lightmouse, PbBot, Spartan-James, Duae Quartunciae, Hamiltondaniel, StewartMH, Dstebbins, ClueBot, Bobathon71, The Thing That Should Not Be, Mwengler, EoGuy, Jagun, RODERICKMOLASAR, Wwheaton, Dmlcyal8er, Razimantv, Mild Bill Hiccup, Feebas factor, J8079s, Rotational, MaxwellsLight, Awickert, Excirial, PixelBot, Sun Creator, NuclearWarfare, PhySusie, El bot de la dieta, DerBorg, Shamanchill, PoofyPeter99, J1.grammar natz, Laserheinz, TimothyRias, XLinkBot, Jovianeye, Petedskier, Hess88, Addbot, Mathieu Perrin, DOI bot, DougsTech, Download, James thirteen, AndersBot, LinkFA-Bot, Barak Sh, AgadaUrbanit, Тиверополник, Dayewalker, Quantumobserver, Kein Einstein, Legobot, Luckas-bot, Yobot, Kilom691, Allowgolf~enwiki, AnomieBOT, Ratul2000, Kingpin13, Materialscientist, Citation bot, Xqbot, Ambujarind69, Mananay, Emezei, Sharhalakis, Shirik, RibotBOT, Rickproser, SongRenKai, Max derner, Merrrr, A. di M., 陈鹏, CES1596, Paine Ellsworth, Gsthae with tempo!, Nageh, TimonyCrickets, WurzelT, Steve Quinn, Spacekid99, Radeksonic, Citation bot 1, Pinethicket, I dream of horses, HRoestBot, Tanweer Morshed, Eno crux, Tom.Reding, Jschnur, RedBot, IVAN3MAN, Gamewizard71, FoxBot, TobeBot, Earthandmoon, PleaseStand, Marie Poise, RjwilmsiBot, Антон Гліністы, Ripchip Bot, Ofercomay, Chemyanda, EmausBot, Bookalign, WikitanvirBot, Roxbreak, Word2need, Gcastellanos, Tommy2010, Dcirovic, K6ka, Hhhippo, Cogiati, 1howardsr1, StringTheory11, Waperkins, Jojojlj, Access Denied, Quondum, AManWithNoPlan, Raynor42, L Kensington, Maschen, HCPotter, Haiti333, RockMagnetist, Rocketrod1960, ClueBot NG, JASMEET SINGH HAFIST, Schicagos, Snotbot, Vinícius Machado Vogt, Helpful Pixie Bot, SzMithrandir, Bibcode Bot, BG19bot, Roberticus, Paolo Lipparini, Wzrd1, Rifath119, Davidiad, Mark Arsten, Peter.sujak, Wikarchitect, Hamish59, Caypartisbot, Penguinstorm300, KSI ROX, Bhargavuk1997, Chromastone1998, TheJJJunk, Nimmo1859, EagerToddler39, Dexbot, EZas3pt14, Webclient101, Chrisanion, Vanquisher.UA, Tony Mach, PREMDASKANNAN, Meghas, Reatlas, Profb39, Zerberos, Thesuperseo, The User 111, Eyesnore, Ybidzian, Tentinator, Illusterati, Celso ad, Quenhitran, Manul, DrMattV, Anrnusna, Wyn.junior, K0RTD, Monkbot, Vieque, BethNaught, Markmizzi, Garfield Garfield, Smokey2022, Zargol Rejerfree, RAL2014, Shahriar Kabir Pavel, Sdjncskdjnfskje, Anshul1908, Professor Flornoy, Thatguytestw, Tetra quark, Harshit100, KasparBot, Chinta 01, Geek3, TheKingOfPhysics and Anonymous: 496

- **Higgs boson** *Source:* https://en.wikipedia.org/wiki/Higgs_boson?oldid=680354810 *Contributors:* AxelBoldt, CYD, ClaudeMuncey, Bryan Derksen, Manning Bartlett, Roadrunner, David spector, Heron, Ewen, Stevertigo, Edward, Boud, TeunSpaans, Dante Alighieri, Ixfd64, Gaurav, TakuyaMurata, CesarB, Anders Feder, Mgimpel~enwiki, Bueller 007, Mark Foskey, Kaihsu, Samw, Cherkash, Lee M, Mxn, Ehn, Timwi, Dcoetzee, Wikiborg, Kbk, Tpbradbury, Phys, Bevo, Topbanana, JonathanDP81, AnonMoos, Bcorr, Jerzy, BenRG, Slawojarek, Phil Boswell, Donarreiskoffer, Robbot, Josh Cherry, ChrisO~enwiki, Owain, Iwpg, Goethean, Altenmann, Nurg, Lowellian, Merovingian, Rursus, Caknuck, Hadal, Alba, Mattflaschen, David Gerard, M-Falcon, Giftlite, Graeme Bartlett, Harp, ShaneCavanaugh, Lethe, Herbee, Jrquinlisk, Xerxes314, Ds13, Fleminra, Dratman, Muzzle, Varlaam, Jason Quinn, Foobar, DÅ‚ugosz, Golbez, Bodhitha, Mmm~enwiki, Aughtandzero, Quadell, Selva, Kaldari, Fred Stober, Johnflux, RetiredUser2, Thincat, Elektron, Bbbl67, Icairns, J0m1eisler, Cructacean, Tdent, TJSwoboda, JohnArmagh, Safety Cap, ProjeX, Njh@bandsman.co.uk, Mike Rosoft, Chris Howard, Jkl, Discospinster, Rich Farmbrough, FT2, Qutezuce, Vsmith, Pie4all88, Kooo, David Schaich, Xgenei, Mal~enwiki, Dbachmann, Mani1, Bender235, ESkog, RJHall, Ylee, Pt, El C, Lycurgus, Lars~enwiki, Laurascudder, Art LaPella, Bookofjude, Brians, TheMile, Dragon76, Smalljim, C S, Reuben, La goutte de pluie, Rangelov, Sasquatch, Bawolff, Tritium6, Eritain, HasharBot~enwiki, Jumbuck, Yoweigh, Alansohn, Andrew Gray, JohnAlbertRigali, Axl, Sligocki, Kocio, Mlm42, Tom12519, Chuckupd, Atomicthumbs, Wtmitchell, KapilTagore, Endersdouble, Dirac1933, DrGaellon, Falcorian, Itinerant, DarTar, Joriki, Reinoutr, Linas, Mindmatrix, Jamsta, Sburke, Benbest, Jonburchel, Thruston, TotoBaggins, GregorB, J M Rice, CharlesC, Waldir, Christopher Thomas, Karam.Anthony.K, Tevatron~enwiki, RichardWeiss, Ashmoo, Fleisher, Kbdank71, GrundyCamellia, Drbogdan, Rjwilmsi, Nightscream, Koavf, Strait, XP1, Martaf, BlueMoonlet, MZMcBride, Mike Peel, NeonMerlin, R.e.b., Jehochman, Bubba73, Afterwriting, A Man In Black, Splarka, RobertG, Nihiltres, Norvy, Itinerant1, Gurch, Mark J, Nimur, Shawn@garbett.org, ElfQrin, DannyDaWriter, Goudzovski, Diza, Consumed Crustacean, Srleffler, Sbove, Chobot, DVdm, Bgwhite, Zentropa, Bambaiah, Wester, Hairy Dude, Huw Powell, Wikky Horse, Pip2andahalf, RussBot, Jacques Antoine, Bhny, JabberWok, Hellbus, Archelon, Eleassar, Rsrikanth05, Salsb, Big Brother 1984, NawlinWiki, Folletto, Buster79, Trovatore, Neutron, SCZenz, Daniel Mietchen, Gadget850, Bota47, Karl Andrews, Dna-webmaster, Jezzabr, Thor Waldsen, Crisco 1492, Deeday-UK, Daniel C, WAS 4.250, Paul Magnussen, Closedmouth, D'Agosta, Bondegezou, Netrapt, Egumtow, LeonardoRob0t, Ilmari Karonen, NeilN, Kgf0, Maryhit, Dragon of the Pants, SmackBot, Nahald, Moeron, Ashley thomas80, Slashme, InverseHypercube, Melchoir, Cinkcool, Baad, Jagged 85, Nickst, Frymaster, AnOddName, ZerodEgo, Giandrea, Gilliam, Ohnoitsjamie, Skizzik, Carl.bunderson, Aurimas, Dauto, JCSantos, TimBentley, RevenDS, Jprg1966, Rick7425, Cadmasteradam, Roscelese, Epastore, DHN-bot~enwiki, Sbharris, Eusebeus, Scwlong, Modest Genius, Famspear, V1adis1av, Rhodesh, Fiziker, Lantrix, Grover cleveland, Jmnbatista, Wen D House, Flyguy649, Jgwacker, Daqu, Mesons, Rezecib, Martijn Hoekstra, Pulu, BullRangifer, Andrew c, Gildir, Kendrick7, Marcus Brute, Vina-iwbot~enwiki, Yevgeny Kats, Frglee, TriTertButoxy, CIS, SashatoBot, Lambiam, Mukadderat, Hi2lok, Kuru, Khazar, Shirifan, Eikern, Tktktk, JorisvS, DMurphy, Mgiganteus1, Bjankuloski06en~enwiki, IronGargoyle, Aardvark23, Loadmaster, Smith609, Deceglie, Hvn0413, Xiphoris, Norm mit, Keith-264, Kencf0618, Britannica~enwiki, Paul venter, Newone, Twas Now, GDallimore, Benplowman, Airstrike~enwiki, Chetvorno, DKqwerty, Harold f, JForget, Laplacian, Er ouz, Jtuggle, Banedon, Ruslik0, Krioni, McVities, Keithh, Rotiro, Yaris678, Slazenger, Cydebot, Martinthoegersen, Gogo Dodo, Anonymi, Lewisxxxusa, Mat456, Jlmorgan, Hippypink, Michael C Price, Quibik, AndersFeder, Raoul NK, PKT, Thijs!bot, Keraunos, Anupam, Headbomb, Kathovo, James086, Hcobb, D.H, Logicat, Jomoal99, Northumbrian, Oreo Priest, JitendraS, -dennis-~enwiki, Widefox, Seaphoto, Orionus, QuiteUnusual, Readro, Hsstr8, Tlabshier, Tim Shuba, Yellowdesk, TuvicBot, JAnDbot, Asmeurer, Tigga, Jde123, Roman à clef, Zekemurdock, Mcorazao, Mozart998, Kborland, Bongwarrior, NeverWorker, Ronstew, Marcel Kosko, Jpod2, J mcandrews, Walter Wpg, Trugster, JMBryant, Vanished user ty12kl89jq10, CodeCat, Allstarecho, Brian Fenton, JaGa, GermanX, Alangarr, WLU, TimidGuy, Mr Shark, Pagw, Andre.holzner, Sigmundg, Ben MacDui, David Nicoson, JTiago, CommonsDelinker, Leyo, Gah4, Fconaway, Oddz, Tgeairn, J.delanoy, Fatka, Pharaoh of the Wizards, Maurice Carbonaro, Stephanwehner, Foober, Aveh8, McSly, Memory palace, NewEnglandYankee, Policron, 83d40m, Usp, Lamp90, Austinian, Izno, SoCalSuperEagle, Robprain, Cuzkatzimhut, Deor, Schucker, VolkovBot, Off-shell, ABF, Eliga~enwiki, JohnBlackburne, AlnoktaBOT, Tburket, Davidwr, Philip Trueman, Spemble, TXiKiBoT, Quatschman, The Original Wildbear, Gwib, Fatram, Kipb9, Andrius.v, Matan568, Nxavar, Nafhan, Photonh2o, Impunv, Peterbullockismyname, Cerebellum, Martin451, Praveen pillay, LoverOfArt, Abdullais4u, Justinrossetti, Cgwaldman, Bcody80, BotKung, Tennisnutt92, Dirkbb, Antixt, Francis Flinch, Moose-32, Ptrslv72, TheBendster, Masterofpsi, Jonbutterworth, Adrideba, SieBot, StAnselm, Manyugarg, PlanetStar, Jor63, Meldor, OlliffeObscurity, Jdcanfield, Yintan, Abhishikt, Flyer22, Graycrow, Infestor, Hrishirise, Cablehorn, Arthur Smart, Aperseghin, Mattmeskill, Gobbledygeek, Cthomas3, Steven Crossin, Nskillen, Sunrise, Afernand74, Jimtpat, Iknowyourider, StaticGull, Jfromcanada, MvL1234, Sphilbrick, Nergaal, Denisarona, Escape Orbit, Quinling, Martarius, PhysicsGrad2013, ClueBot, Victor Chmara, The Thing That Should Not Be, TomRed, Alyjack, Infrasonik, Mx3, Master1228, Drmies, Frmorrison, Loves martyr, Polyamorph, Nobaddude, Sjdunn9, Kitsunegami, Ktr101, Excirial, Joeyfjj, Wmlschlotterer, Pawan ctn, Lartoven, Artur80, Sun Creator, BobertWABC, PeterThe Wall, Nondisclosure, M.O.X, SchreiberBike, JasonAQuest, Another Believer, Scf1984, 1ForTheMoney, Anoopan, Wnt, Darkicebot, TimothyRias, XLinkBot, Rreagan007, Resonance cascade, JinJian, Jabberwoch, Hess88, Hybirdd, Tayste, Addbot, Proofreader77, Mortense, Jacopo Werther, DBGustavson, DOI bot, Betterusername, Ocdnctx, OttRider, Cgd8d, Leszek Jańczuk, WikiUserPedia, NjardarBot, Download, LaaknorBot, AndersBot, Favonian, AgadaUrbanit, HandThatFeeds, Tide rolls, Lightbot, ScAvenger, SPat, Zorrobot, Jarble, ScienceApe, Dnamanish, Luckas-bot, Yobot, Chreod, EchetusXe, Nsbinsnj, Evans1982, Amble, Now dance, fu.cker, dance!, Anypodetos, Nallimbot, Trinitrix, SkepticalPoet, Pulickkal, Fernandosmission, Apollo reactor, Csmallw, AnomieBOT, Novemberrain94, 1exec1, Jim1138, JackieBot, Gc9580, LlywelynII, Materialscientist, Citation bot, Brightgalrs, Onesius, ArthurBot, Northryde,

LilHelpa, Xqbot, Konor org, Noonehasthisnameithink, Engineering Guy, Yutenite, Newzebras, Universalsuffrage, DeadlyMETAL, Tomdo08, Professor J Lawrence, Br77rino, Srich32977, Arni.leibovits, StevenVerstoep, ProtectionTaggingBot, Vdkdaan, Omnipaedista, RibotBOT, Kyng, Waleswatcher, WissensDürster, Ace111, Kristjan.Jonasson, MerlLinkBot, Ernsts, Chaheel Riens, A. di M., A.amitkumar, Markdavid2000, ⁇⁇, Dave3457, FrescoBot, Weyesr1, Paine Ellsworth, Kenneth Dawson, Cdw1952, CamB424, CamB4242, Steve Quinn, N4tur4le, Jc odc-smf, Cannolis, Dolyn, Citation bot 1, Openmouth, Gil987, OriumX, Biker Biker, Gautier lebon, Pinethicket, Edderso, Boson15, Jonesey95, Three887, CarsonsDad, Calmer Waters, Jusses2, RedBot, BiObserver, Aknochel, Meier99, Trappist the monk, Puzl bustr, Proffsl, Higgshunter, Mary at CERN, Periglas, Zanhe, Lotje, Callanecc, Comet Tuttle, Jdigitalbath, Vrenator, SeoMac, ErikvanB, Tbhotch, Minimac, Coolpranjal, Mean as custard, RjwilmsiBot, TjBot, Olegrog, 123Mike456Winston789, Wease1pit, Newty23125, Techhead7890, Tesseract2, Skamecrazy123, Northern Arrow, Mukogodo, J36miles, EmausBot, John of Reading, WikitanvirBot, Stryn, Dadaist6174, Nuujinn, Montgolfière, Fotoni, GoingBatty, RA0808, Bengt Nyman, Bt8257, Gimmetoo, KHamsun, LHC Tommy, Slightsmile, NikiAnna, TeeTylerToe, Dekker451, Hhhippo, Evanh2008, JSquish, Kkm010, ZéroBot, John Cline, Liquidmetalrob, Fæ, Bollyjeff, Érico Júnior Wouters, StringTheory11, Stevengoldfarb, Sgerbic, Opkdx, Quondum, AndrewN, Tbushman, Makecat, Timetraveler3.14, Foonle77, Tolly4bolly, Wiggles007, Irenan, Nobleacuff, Brandmeister, Baseballrocks538, Chris81w, Inswoon, Maschen, Donner60, Ontyx, Angelo souti, ChuispastonBot, ChiZeroOne, Ninjalectual, Exsmokey, Herk1955, I hate whitespace, Rocketrod1960, Whoop whoop pull up, Ajuvr, Petrb, Grapple X, ClueBot NG, Perfectlight, Aaron Booth, Gareth Griffith-Jones, Siswick, MelbourneStar, Gilderien, PhysicsAboveAll, Manu.ajm, Muon, Parcly Taxel, O.Koslowski, Widr, Mohd. Toukir Hamid, Diyar se, Helpful Pixie Bot, Popcornduff, Aesir.le, Bibcode Bot, 2001:db8, Lowercase sigmabot, BG19bot, Scottaleger, Mcarmier, Jibu8, Loupatriz67, Dave4478, Frze, Ervin Goldfain, Reader505, Mark Arsten, Lovetrivedi, BarbaraMervin, Silvrous, Drcooljoe, Cadiomals, Joydeep, Altaïr, Piet De Pauw, Jeancey, Sovereign8, Visuall, Ownedroad9, Brainssturm, Jw2036, Writ Keeper, DPL bot, Nickni28, Philpill691, Lee.boston, Scientist999, Benjiboy187, Duxwing, Cengime, Skiret girdet njozet, GRighta, Downtownclaytonbrown, Diasjordan, Ghsetht, Marioedesouza, BattyBot, 1narendran, LORDCOTTINGHAM2, NO SOPA, Tchaliburton, Wijnburger, StarryGrandma, Mdann52, Dilaton, Magikal Samson, Samuelled, Dja1979, Georgegroom, BecurSansnow, EuroCarGT, MSUGRA, Rhlozier, Pscott558, Turullulla, Blueprinteditor, Misterharris~enwiki, AstroDoc, Bigbear213, Dexbot, Randomizer3, Daggerbot, DoctorLazarusLong, Caroline1981, Nitpicking polisher, SoledadKabocha, Gsmanu007, Windows.dll, Mogism, Prabal123koirala, Abitoby, Clidog, Rongended, Darryl from Mars, Cerabot~enwiki, MuonRay, TheTruth72, Capt. Mohan Kuruvilla, Gatheringstorm2, Jason7898, Mumbai999999, SkepticalKid, Cjean42, Nmrzuk, Lugia2453, Mafuee, Frosty, SFK2, Thegodparticlebook, Rijensky, Mishra866868, Rockstar999999999, Toddbeck911, Nilaykumar07, Thepalerider2012, WikiPhysTech, The Anonymouse, Ahmar Saeed, Pincrete, Apidium23, Prahas.wiki, Exenola, Pdotpwns, Epicgenius, Fireballninja, Greengreengreenred, ⁇⁇, Technogeek101, NicoPosner, Apurva Godghase, Durfyy, Soumya Mittal, American In Brazil, SaifAli13, Qwerkysteve, Spatiandas, Retroherb, Tango303, Hoppeduppeanut, Redplain, Shaelote, Quadrum, AntiguanAcademic, Simpsonojsim, Agyeyaankur, DavidLeighEllis, Ethanthevelociraptor, Qfang12, Comp.arch, Eletro1903, E8xE8, HeineBOB, Kahtar, Depthdiver, JAaron95, Mfb, Anrnusna, Stamptrader, Man of Steel 85, Cteirmn, AiraCobra, MyNamelsn'tElvis, Meganlock8, Sxxximf, Drsoumyadeepb, 22merlin, Ndidi Okonkwo Nwuneli, Monkbot, Dialga5555, Fred1810, Akro7, Pewpewpewpapapa, BradNorton1979, 21bhargav, Whistlemethis, Thinking Skeptically, Amk365, Gagnonlg, Knowledgebattle, L21234, TheNextMessiah, Naterealm224, Joey van Helsing, Adrian Lamplighter, Arnab santra, Gemadi, BATMAN1021, Isambard Kingdom, Mercedes321, DrKitts, KasparBot, JJMC89, GBjun3, TheRoamer64, Firstcause, Seventhorbitday, RobeDM and Anonymous: 962

- **Hadron** *Source:* https://en.wikipedia.org/wiki/Hadron?oldid=681027720 *Contributors:* Bryan Derksen, Manning Bartlett, Peterlin~enwiki, Edward, Erik Zachte, ESnyder2, Fruge~enwiki, TakuyaMurata, Darkwind, Glenn, Nikai, Ehn, Olya, Phys, Bevo, Topbanana, BenRG, Twang, Donarreiskoffer, Korath, Wjhonson, Merovingian, Ojigiri~enwiki, Sunray, JesseW, Xanzzibar, Giftlite, Xerxes314, Dratman, Physicist, Mikro2nd, LiDaobing, Pthompson, Icairns, Jimaginator, Mike Rosoft, Vsmith, Goochelaar, Sunborn, Livajo, El C, Kwamikagami, Shanes, Fwb22, Jumbuck, Cookiemobsta, Velella, Rebroad, Vuo, Kusma, DV8 2XL, Linas, GrouchyDan, Palica, Marudubshinki, Kbdank71, Mana Excalibur, Kinu, Strait, FlaBot, RexNL, Goudzovski, FrankTobia, YurikBot, Radishes, Bambaiah, Hydrargyrum, Salsb, NawlinWiki, Wiki alf, SCZenz, Davemck, Bota47, Scriber~enwiki, Modify, Katieh5584, Eog1916, SmackBot, McGeddon, Gilliam, Benjaminevans82, Dingar, Persian Poet Gal, Telempe, DHN-bot~enwiki, Audriusa, Acepectif, Kokot.kokotisko, JorisvS, JarahE, BranStark, SJCrew, Erraticus, Chrumps, Jtuggle, Q43, Epbr123, Wikid77, Headbomb, Escarbot, Deflective, Gcm, NE2, Trapezoidal, Naval Scene, KEKPΩΨ, NeverWorker, Wwmbes, Alexllew, Lvwarren, Jebus0, DariusU, Khalid Mahmood, Adriaan, Rustyfence, Ron2, Leyo, J.delanoy, Maurice Carbonaro, JVersteeg, Rod57, Way2Smart22, Hugh Hudson, Y2H, Ansans, Bobxii, Chris Longley, Useight, Dylan bossart, VolkovBot, TXiKiBoT, Kinkydarkbird, Anonymous Dissident, Don4of4, Wordsmith, LeaveSleaves, Antixt, Enviroboy, Insanity Incarnate, Nibios, AlleborgoBot, SieBot, Yintan, LeadSongDog, RadicalOne, Paolo.dL, OKBot, JohnSawyer, Lazarus1907, Pinkadelica, Danthewhale, Martarius, ClueBot, Amaamaddq, Authoritative Physicist, Wwheaton, Rotational, DragonBot, Sciencedude9998, Tuchomator, El planeto, Kaiba, Thingg, Koshoid, Aitias, Apparition11, Rishi.bedi, TimothyRias, InternetMeme, Jbeans, MystBot, Sgpsaros, Tayste, Addbot, Pkkphysicist, Ehrenkater, Lightbot, Luckas-bot, Yobot, Nallimbot, Dagus2000, Fangfyre, LOLx9000, Thisaccountwillbebanned, Citation bot, Xqbot, Drilnoth, Br77rino, Wikiedit33, Ajahnjohn, Omnipaedista, RibotBOT, Mashmeister, Tjbright2, My cat's breath smells like catfood, Haeinous, Citation bot 1, Javert, Gil987, I dream of horses, Jonesey95, Rameshngbot, Thinking of England, Alarichus, SkyMachine, FoxBot, Johnshnappay, Антон Глністы, Teravolt, Racerx11, Naznin farhah, Tommy2010, Rafabaez, Wikipelli, ZéroBot, StringTheory11, Hadron12, Donner60, Bobogoobo, Petrb, ClueBot NG, Gareth Griffith-Jones, Bibcode Bot, BG19bot, Dwightboone, Njavallil, Walterpfeifer, Pfeiferwalter, ChrisGualtieri, Ugog Nizdast, Lithelimbs, RoKo89, Michikohundred, KasparBot, Wwilliam726 and Anonymous: 169

- **Meson** *Source:* https://en.wikipedia.org/wiki/Meson?oldid=672777660 *Contributors:* AxelBoldt, Bryan Derksen, Josh Grosse, PierreAbbat, Ben-Zin~enwiki, Xavic69, TakuyaMurata, Fwappler, Ahoerstemeier, Ping, Phys, Bcorr, Jeffq, Donarreiskoffer, Robbot, Fredrik, Sanders muc, Merovingian, Rursus, Ojigiri~enwiki, Davidl9999, DocWatson42, Harp, Marcika, Xerxes314, Niteowlneils, Eequor, Physicist, Eroica, Icairns, Sam Hocevar, Lehi, Rich Farmbrough, Pjacobi, Tjic, Robotje, Nicke Lilltroll~enwiki, Pearle, Jumbuck, Jérôme, Bucephalus, Falcorian, Palica, Tevatron~enwiki, Mandarax, Kbdank71, Strait, Titoxd, FlaBot, Jeremygbyrne, Chobot, YurikBot, Wavelength, Bambaiah, Phmer, Jimp, Ozabluda, JabberWok, Salsb, Leutha, Długosz, SCZenz, Ravedave, Gadget850, Antiduh, Tetracube, SmackBot, Melchoir, Eskimbot, Chris the speller, DHN-bot~enwiki, Sbharris, Kevinpurcell, Mesons, DMacks, Jashank, JorisvS, Mgiganteus1, Geologyguy, Ryulong, JarahE, Myasuda, ChrisKennedy, Michael C Price, Thijs!bot, Headbomb, Escarbot, Orionus, Spartaz, Gökhan, Deflective, Magioladitis, Swpb, Khalid Mahmood, Tercer, Kostisl, Hans Dunkelberg, Tarotcards, Xiahou, JeffreyRMiles, VolkovBot, Prizrak, TXiKiBoT, Muro de Aguas, Martin451, LeaveSleaves, Antixt, SieBot, Majeston, Gerakibot, Graf Von Crayola, Humanityisthedisease, Mimihitam, Fratrep, OKBot, ClueBot, Terrorist96, Diagramma Della Verita, Brews ohare, Neville35, RMFan1, WikHead, Stephen Poppitt, Addbot, Gtakanis, Chzz, Debresser, CosmiCarl, AgadaUrbanit, Dickdock, Magog the Ogre, AnomieBOT, StratoWiki, Altruism2010, Citation bot, ArthurBot, Xqbot, Omnipaedista,

WaysToEscape, FrescoBot, Paine Ellsworth, Ironboy11, Steve Quinn, 000ojjo000, Yehoshua2, Citation bot 1, Wdcf, Thinking of England, Puzl bustr, Ale And Quail, Discovery4, Mean as custard, Dkzico007, John of Reading, WikitanvirBot, GoingBatty, Hanretty, ZéroBot, StringTheory11, Markinvancouver, ClueBot NG, Christian.kolen, Wallace Kneeland, Helpful Pixie Bot, Bibcode Bot, Glevum, DerekWinters, Mark viking, Justin567Hicks, Prokaryotes, SJ Defender, Monkbot, KasparBot and Anonymous: 87

- **List of mesons** *Source:* https://en.wikipedia.org/wiki/List_of_mesons?oldid=679943368 *Contributors:* Cherkash, Donarreiskoffer, Giftlite, Xerxes314, Michael Devore, Eequor, Rich Farmbrough, ZeroOne, Tompw, Physicistjedi, Pearle, Keenan Pepper, Zyqqh, TenOfAllTrades, Woohookitty, Ch'marr, Kbdank71, JVz, Strait, Nihiltres, Agerom, RussBot, David McCormick, SCZenz, Gadget850, Sbyrnes321, That Guy, From That Show!, SmackBot, JorisvS, Happy-melon, Charles Baynham, Chrumps, Usgnus, Cydebot, Christian75, Coccoinomane, Headbomb, Stannered, JAnDbot, Magioladitis, Mollwollfumble, Gwern, Leyo, Potatoswatter, VolkovBot, Antixt, Ocsenave, Muhends, Mikaey, SkyLined, Addbot, Yobot, Kan8eDie, 4th-otaku, Rubinbot, Citation bot, ArthurBot, Xqbot, Ulm, Carlog3, W-C, Yehoshua2, Citation bot 1, Thinking of England, John of Reading, StringTheory11, Markinvancouver, Helpful Pixie Bot, Bibcode Bot, Maysens, Srebre, YiFeiBot, Monkbot and Anonymous: 16

- **Baryon** *Source:* https://en.wikipedia.org/wiki/Baryon?oldid=681412960 *Contributors:* AxelBoldt, Tobias Hoevekamp, Bryan Derksen, Ben-Zin~enwiki, Heron, Tim Starling, Alan Peakall, Paul A, Salsa Shark, Glenn, Mxn, Charles Matthews, The Anomebot, ElusiveByte, Phys, Bevo, Traroth, Donarreiskoffer, Robbot, Korath, Kristof vt, Merovingian, Ojigiri~enwiki, Sunray, Wikibot, Giftlite, DocWatson42, Shaun-MacPherson, Herbee, Xerxes314, Dratman, DÅ‚ugosz, Kaldari, OwenBlacker, Icairns, JohnArmagh, Rich Farmbrough, Guanabot, Mani1, E2m, Tompw, El C, Bobo192, I9Q79oL78KiL0QTFHgyc, Giraffedata, Physicistjedi, Jumbuck, Gary, ABCD, Oleg Alexandrov, Woohookitty, Tevatron~enwiki, BD2412, Kbdank71, Nightscream, Ae77, MZMcBride, Chekaz, R.e.b., Erkcan, Maxim Razin, Oo64eva, Chobot, Roboto de Ajvol, YurikBot, Bambaiah, Jimp, Salsb, Ergzay, DragonHawk, SCZenz, E2mb0t~enwiki, Bota47, Simen, Sbyrnes321, Lainagier, Timotheus Canens, Bluebot, Colonies Chris, Kingdon, Shadow1, Bigmantonyd, Drphilharmonic, Kseferovic, Wierdw123, Physicsdog, Torrazzo, Verdy p, Michael C Price, Thijs!bot, Headbomb, Hcobb, Orionus, QuiteUnusual, Spartaz, Plantsurfer, Amateria1121, Diamond2, Swpb, BatteryIncluded, Hveziris, Saxophlute, Gwern, Ben MacDui, R'n'B, Ash, Tgeairn, Maurice Carbonaro, STBotD, VolkovBot, GimmeBot, NoiseEHC, Tearmeapart, BotKung, BrianADesmond, Antixt, AlleborgoBot, Lou427, SieBot, VVVBot, Gerakibot, LeadSongDog, Keilana, Paolo.dL, Doctorfluffy, TrufflesTheLamb, OKBot, Hamiltondaniel, TubularWorld, ClueBot, Artichoker, ChandlerMapBot, CalumH93, Addbot, LaaknorBot, CarsracBot, Jonhstone12, Legobot, Luckas-bot, Bugbrain 04, AnomieBOT, JackieBot, Materialscientist, Citation bot, ArthurBot, Xqbot, Omnipaedista, SassoBot, Spellage, WaysToEscape, FrescoBot, Citation bot 1, FoxBot, Noommos, EmausBot, John of Reading, JSquish, ZéroBot, StringTheory11, Stibu, Ethaniel, Markinvancouver, ClueBot NG, Koornti, Kasirbot, Rezabot, Bibcode Bot, Atomician, Zedshort, Marioedesouza, ChrisGualtieri, WorldWideJuan, CoolHandLouis, Monkbot, KasparBot and Anonymous: 106

- **List of baryons** *Source:* https://en.wikipedia.org/wiki/List_of_baryons?oldid=675785999 *Contributors:* Cherkash, GPHemsley, Donarreiskoffer, Giftlite, Rich Farmbrough, ZeroOne, Tompw, Keenan Pepper, Oleg Alexandrov, Woohookitty, Astrowob, Kbdank71, Strait, Mike Peel, R.e.b., Arctic.gnome, YurikBot, Jonrock, Jimp, Cryptic, SCZenz, Gadget850, F15mos, Banus, GrinBot~enwiki, Sbyrnes321, EvanJPW, SmackBot, Tom Lougheed, Oceanh, Jiminy pop, JorisvS, UncleDouggie, Happy-melon, JRSpriggs, Neelix, Cydebot, WillowW, Mike Christie, Abtract, Wikid77, Headbomb, Magioladitis, Mollwollfumble, Randyfurlong, Leyo, Gogobera, Beatnik Party, MichaelSchoenitzer, GimmeBot, Anonymous Dissident, Sascha.baumeister~enwiki, Antixt, PaddyLeahy, Richard Ye, Wing gundam, Thisisnotatest, Lightmouse, Dabomb87, Muhends, NuclearWarfare, SkyLined, Addbot, DOI bot, Mjamja, Debresser, Vectorboson, SamatBot, Luckas-bot, Yobot, Materialscientist, Citation bot, Carlog3, Citation bot 1, Thinking of England, EmausBot, Markinvancouver, Rmashhadi, Bibcode Bot, BG19bot, P76837, Tony Mach, YiFeiBot, Monkbot and Anonymous: 37

7.2 Images

- **File:2-photon_Higgs_decay.svg** *Source:* https://upload.wikimedia.org/wikipedia/commons/3/32/2-photon_Higgs_decay.svg *License:* CC BY-SA 3.0 *Contributors:* Own work *Original artist:* Parcly Taxel

- **File:4-lepton_Higgs_decay.svg** *Source:* https://upload.wikimedia.org/wikipedia/commons/b/b2/4-lepton_Higgs_decay.svg *License:* CC BY-SA 3.0 *Contributors:* Own work *Original artist:* Parcly Taxel

- **File:AIP-Sakurai-best.JPG** *Source:* https://upload.wikimedia.org/wikipedia/commons/2/2b/AIP-Sakurai-best.JPG *License:* Public domain *Contributors:* Own work *Original artist:* self

- **File:AirShower.svg** *Source:* https://upload.wikimedia.org/wikipedia/commons/2/2c/AirShower.svg *License:* CC BY 3.0 *Contributors:* originally from nl.wikipedia; description page is/was here. *Original artist:* Mpfiz

- **File:Ambox_important.svg** *Source:* https://upload.wikimedia.org/wikipedia/commons/b/b4/Ambox_important.svg *License:* Public domain *Contributors:* Own work, based off of Image:Ambox scales.svg *Original artist:* Dsmurat (talk · contribs)

- **File:Asymmetricwave2.png** *Source:* https://upload.wikimedia.org/wikipedia/commons/0/0d/Asymmetricwave2.png *License:* CC BY 3.0 *Contributors:* Own work *Original artist:* TimothyRias

- **File:Aurore_australe_-_Aurora_australis.jpg** *Source:* https://upload.wikimedia.org/wikipedia/commons/0/07/Aurore_australe_-_Aurora_australis.jpg *License:* CC BY-SA 3.0 *Contributors:* Own work *Original artist:*
  If you plan on using it, an email would be greatly appreciated.

- **File:Baryon-decuplet-small.svg** *Source:* https://upload.wikimedia.org/wikipedia/commons/7/78/Baryon-decuplet-small.svg *License:* Public domain *Contributors:* Own work *Original artist:* Trassiorf

- **File:Baryon-octet-small.svg** *Source:* https://upload.wikimedia.org/wikipedia/commons/b/b5/Baryon-octet-small.svg *License:* Public domain *Contributors:* Own work *Original artist:* Trassiorf

- **File:Baryon_decuplet.svg** *Source:* https://upload.wikimedia.org/wikipedia/commons/f/f6/Baryon_decuplet.svg *License:* Public domain *Contributors:* Own work (Original text: *self-made*) *Original artist:* Wierdw123 at English Wikipedia

- **File:VisibleEmrWavelengths.svg** *Source:* https://upload.wikimedia.org/wikipedia/commons/e/e2/VisibleEmrWavelengths.svg *License:* Public domain *Contributors:* created by me *Original artist:* maxhurtz

- **File:Wikinews-logo.svg** *Source:* https://upload.wikimedia.org/wikipedia/commons/2/24/Wikinews-logo.svg *License:* CC BY-SA 3.0 *Contributors:* This is a cropped version of Image:Wikinews-logo-en.png. *Original artist:* Vectorized by Simon 01:05, 2 August 2006 (UTC) Updated by Time3000 17 April 2007 to use official Wikinews colours and appear correctly on dark backgrounds. Originally uploaded by Simon.

- **File:Wikisource-logo.svg** *Source:* https://upload.wikimedia.org/wikipedia/commons/4/4c/Wikisource-logo.svg *License:* CC BY-SA 3.0 *Contributors:* Rei-artur *Original artist:* Nicholas Moreau

- **File:Wiktionary-logo-en.svg** *Source:* https://upload.wikimedia.org/wikipedia/commons/f/f8/Wiktionary-logo-en.svg *License:* Public domain *Contributors:* Vector version of Image:Wiktionary-logo-en.png. *Original artist:* Vectorized by Fvasconcellos (talk · contribs), based on original logo tossed together by Brion Vibber

- **File:Young_Diffraction.png** *Source:* https://upload.wikimedia.org/wikipedia/commons/8/8a/Young_Diffraction.png *License:* Public domain *Contributors:* ? *Original artist:* ?

7.3 Content license

- Creative Commons Attribution-Share Alike 3.0